The Earliest Europeans:
A Year in the Life

Seasonal survival strategies in the Lower Palaeolithic

Robert Hosfield

W0009772

OXBOW | books
Oxford & Philadelphia

Published in the United Kingdom in 2020 by
OXBOW BOOKS
The Old Music Hall, 106–108 Cowley Road, Oxford OX4 1JE

and in the United States by
OXBOW BOOKS
1950 Lawrence Road, Havertown, PA 19083

© Oxbow Books and the author 2020

Paperback Edition: ISBN 978-1-78570-761-2
Digital Edition: ISBN 978-1-78570-762-9 (ePub)

A CIP record for this book is available from the British Library

Library of Congress Control Number: 2020934214

Some rights reserved. No part of the print edition of the book may be reproduced or transmitted in any form or by any means, electronic or mechanical including photocopying, recording or by any information storage and retrieval system, without permission from the publisher in writing.

Materials provided by third parties remain the copyright of their owners.

Printed in the United Kingdom by Short Run Press

Typeset by Versatile PreMedia Services (P) Ltd

For a complete list of Oxbow titles, please contact:

UNITED KINGDOM	UNITED STATES OF AMERICA
Oxbow Books	Oxbow Books
Telephone (01865) 241249	Telephone (610) 853-9131, Fax (610) 853-9146
Email: oxbow@oxbowbooks.com	Email: queries@casemateacademic.com
www.oxbowbooks.com	www.casemateacademic.com/oxbow

Oxbow Books is part of the Casemate Group

Cover image (clockwise from top left): Creative Commons. Beech forest Mátra in winter (author: Susulyka; https://commons.wikimedia.org/wiki/File:Beech_forest_M%C3%A1tra_in_winter.jpg; CC BY-SA 4.0); Grib skov (author: Malene Thyssen; https://commons.wikimedia.org/wiki/File:Grib_skov.jpg; CC BY-SA 3.0); Fallow deer (author: Jiří Nedorost; https://commons.wikimedia.org/wiki/File:Fallow_Deer.jpg; CC BY-SA 3.0); Jesienny (author: Mohylek; https://commons.wikimedia.org/wiki/File:Jesienny.jpg; CC BY-SA 3.0).

An open-access on-line version of this book is available at: https://books.casematepublishers.com/The_Earliest_Europeans.pdf. The online work is licensed under the Creative Commons Attribution 3.0 Unported Licence. To view a copy of this license, visit http://creativecommons.org/licenses/by/3.0/ or send a letter to Creative Commons, 444 Castro Street, Suite 900, Mountain View, California, 94041, USA. This licence allows for copying any part of the on-line work for personal and commercial use, providing author attribution is clearly stated.

Contents

List of boxes

Preface and acknowledgements

This book was inspired by three publications in particular. The first two, Clive Gamble's *The Palaeolithic Settlement of Europe* (1986) and Mark White's *Things to do in Doggerland When You're Dead: Surviving OIS3 at the Northwestern-Most Fringe of Middle Palaeolithic Europe* (2006), emphasise, in various different ways, the ecological realities of life for Palaeolithic humans. The third, less obviously, is Ian Mortimer's *The Time Traveller's Guide to Medieval England* (2008), which offers an immersive experience of 14th century life. Together all three encouraged me to try and explore the theme of 'life on the ground' for Europe's first humans, through the lens of the changing seasons. In doing so this book is grounded in the available palaeoenvironmental and archaeological data and is therefore primarily aimed at academic researchers, final year undergraduates and taught postgraduates, but it should also be of interest to a general audience curious about Europe's first humans and how they survived.

As with all such books, there are a number of acknowledgements. First, a huge thank you, for many useful discussions and much collegiality and fun, to all my colleagues at and beyond Reading but especially to Nick Ashton, Will Attard, Martin Bell, Stuart Black, Richard Bradley, Nick Branch, Bob Chapman, Li-Chi Chiang, Wei Chu, James Cole, William Davies, Rob Davis, Dominik Fleitmann, Hannah Fluck, Jenni French, Clive Gamble, Duncan Garrow, John Gowlett, Rosa Huguet, Linda Hurcombe, Alice La Porta, Mary Lewis, Simon Lewis, Katharine MacDonald, Roger Matthews, Wendy Matthews, John McNabb, Anna Machin, Alison MacLeod, Steve Mithen, Gundula Müldner, Sam Nicholson, Matt Pope, Alex Pryor, Anne-Lyse Ravon, Danielle Schreve, the Becky Scotts (both of them!), Andy Shaw, Marcin Szymanek, Francis Wenban-Smith, Mark White, and all the participants at the *Coping with Climate: the legacy of Homo heidelbergensis* workshop (Brighton University, February 2017; funded by the AHRC: AH/N007514/1) and the *MIS 13–11: a major transformation in the European Lower Palaeolithic?* session at UISPP (Paris, June 2018). Particular thanks to Ian Gilligan, Jarod Hutson, Katharine MacDonald, Matt Pope and Gonen Sharon for their thoughtful commentaries on my paper *Walking in a Winter Wonderland? Strategies for Early and Middle Pleistocene Survival in Midlatitude Europe* (*Current Anthropology*, 2016), which challenged me to further explore a number of key issues. I would also like to thank Archaeology staff and students at the universities of Southampton, Oxford (*Palaeolithic and Quaternary Seminar Series*), and Liverpool (*Evolutionary Anthropology Seminar Series*), whose questions and feedback at various research seminars contributed to the development of the ideas in this book. Apologies to anyone who I have inadvertently forgotten in the above lists. Finally, many thanks to Duncan Garrow, John McNabb and Steve Mithen for their comments on various drafts: all remaining errors are my own.

Particular thanks to Mark Gridley (http://markgridley.carbonmade.com/) for the original reconstructions he produced for this book (Figs 3.8 [left] and 5.2). Thank you also to Jennie Chambers (Figs 5.3 and 6.15), Chris Crump (Figs 4.7, 6.12 and 6.13), Historic England/Peter Dunn (Figs 3.2 and 3.8 [right]), Jiří Hochman (www.zdenek-burian.com; Figs 3.4 and 3.5), José Antonio Peñas (Fig. 3.3), Jessica Piet (Fig. 3.6) and John Sibbick (Figs 2.1 and 6.14) for kindly permitting me to reproduce original artwork and/or their photographs. I am also very grateful for the financial support provided by the University of Reading (the School of Archaeology, Geography & Environmental Science; and the Heritage & Creativity research theme), which enabled many of these figures to be included. Thank you also to Becky Scott (TOBS) and Josie Handley for arranging permissions for the reproduction of various figures: the book would have been further delayed without you.

Thanks also to my parents and my maternal grandmother for interesting discussions about wild foods through the year, and for dragging me around the Lake District and French caves as a child – I've never forgotten the glacial landforms or the Palaeolithic art.

A big thank you to all the students in my classes at the University of Reading since 2004 (and previously at the University of Southampton), all of whom have developed my thinking about the earliest Europeans, through their questions, doubts, and occasionally plain, healthy scepticism. I would especially like to thank those students whose dissertations in various aspects of early Palaeolithic archaeology were always thought-provoking.

Thanks to the School of Archaeology, Geography and Environmental Science (SAGES, University of Reading) whose research leave policy was critical for the eventual completion of this book, and to Oxbow Books for their patience.

A particular thank you to my family, with apologies for the many, book writing-related, hours I spent at my computer, especially in the second half of 2019.

And finally, a huge thank you to John McNabb and Clive Gamble – without whom my life would have been very different. And as Clive was fond of saying to me in PhD supervisions: 'don't believe a word of it!'

For Jen and Anna, and Mum and Dad — with love,

Robert Hosfield
Reading, 2019

Figure acknowledgements

Acknowledgement is due to the following for permission to reproduce the illustrations listed below:

Fig. 1.2: Elsevier (*Earth-Science Reviews*); Fig. 1.3: Springer Nature (*Surveys in Geophysics*); Figs 1.4 & 1.5: Eric Barron, Tjeerd van Andel & David Pollard, and the McDonald Institute for Archaeological Research; Fig. 1.7: Used by permission of The Center for Sustainability and the Global Environment, Nelson Institute for Environmental Studies, University of Wisconsin-Madison.

Fig. 2.1: John Sibbick; Fig. 2.2: Google Earth (https://www.google.com/help/terms_maps/ & https://www.google.com/permissions/geoguidelines/); Fig. 2.3: Elsevier (*Quaternary Science Reviews*); Fig. 2.4: Creative Commons. Białowieża Primeval Forest, May 2007, BNP, Poland (author: Juan de Vojníkov; https://commons.wikimedia.org/wiki/File:Bialowieza_Primeval_Forest,_May_2007,_BNP,_Poland.jpg; CC BY-SA 3.0); Fig. 2.5: Archaeology Data Service (https://archaeologydataservice.ac.uk/advice/termsOfUseAndAccess.xhtml); Fig. 2.6: Elsevier (*Quaternary International*).

Chapter 3 cover image: Beech forest Mátra in winter (author: Susulyka; https://commons.wikimedia.org/wiki/File:Beech_forest_M%C3%A1tra_in_winter.jpg; CC BY-SA 4.0); Fig. 3.2: Historic England (IC015/005 – A panorama reconstruction drawing of Boxgrove Quarry by Peter Dunn; permission no: 7711); Fig. 3.3: José Antonio Peñas; Figs 3.4 & 3.5: Jiří Hochman (www.zdenekburian.com); Fig. 3.6: Jessica Piet; Fig. 3.8: Mark Gridley (left); Historic England (IC015/003 – *Palaeolithic Man* by Peter Dunn; permission no: 7566; right); Fig. 3.9: Creative Commons Attribution 4.0 International License (http://creativecommons.org/licenses/by/4.0/); Original source: Chen *et al.* (2017, Fig. 9); Fig. 3.10: Elsevier (*Quaternary International*) & Landesamt für Denkmalpflege und Archäologie Sachsen-Anhalt, Karol Schauer (State Office for Heritage Management and Archaeology Saxony-Anhalt); Fig. 3.11: Creative Commons. Clockwise from top-left: Natural shelter, Howden Moors (author: Dave Dunford; https://commons.wikimedia.org/wiki/File:Natural_shelter,_Howden_Moors_-_geograph.org.uk_-_360156.jpg; CC BY-SA 2.0); Pimmit bank erosion (author: Ivy Main; https://commons.wikimedia.org/w/index.php?curid=11842103; CC BY-SA 3.0); Pinus sylvestris group (author: Beentree; https://commons.wikimedia.org/w/index.php?curid=2228653; CC BY-SA 3.0); author's own; A big hollow tree, Omagh (author: Kenneth Allen; https://commons.wikimedia.org/wiki/File:A_big_hollow_tree,_Omagh_-_geograph.org.uk_-_1023678.jpg; CC BY-SA 2.0); Fig. 3.12: Robert L. Kelly; Fig. 3.13: University of Chicago Press (*Current Anthropology*).

Chapter 4 cover image: Creative Commons. Grib skov (author: Malene Thyssen; https://commons.wikimedia.org/wiki/File:Grib_skov.jpg; CC BY-SA 3.0); Fig. 4.2: Elsevier (*Quaternary International*); Fig. 4.3: Elsevier (*Sedimentary Geology*); Fig. 4.4: Elsevier (*Journal of Human Evolution*)

& Cambridge University Press (*Antiquity*); Fig. 4.5: Elsevier (*Journal of Archaeological Science*); Fig. 4.6: Creative Commons. Animal carcass (author: Zenith4237; https://commons.wikimedia.org/wiki/File:Animal_carcass.jpg; CC BY-SA 4.0); Fig. 4.7: Chris Crump.

Chapter 5 cover image: Creative Commons. Fallow deer (author: Jiří Nedorost; https://commons.wikimedia.org/wiki/File:Fallow_Deer.jpg; CC BY-SA 3.0); Fig. 5.1: Elsevier (*Quaternary International*); Fig. 5.2: Mark Gridley; Fig. 5.3: Jennifer Chambers.

Chapter 6 cover image: Creative Commons. Jesienny (author: Mohylek; https://commons.wikimedia.org/wiki/File:Jesienny.jpg; CC BY-SA 3.0); Fig. 6.2: Craig Williams and HarperCollins; Fig. 6.3: Elsevier Books & Cambridge University Press (*Antiquity*); Fig. 6.4: Google Earth (https://www.google.com/help/terms_maps/ & https://www.google.com/permissions/geoguidelines/); Fig. 6.5: European Environment Agency (https://www.eea.europa.eu/; Creative Commons Attribution License; CC BY 2.5 DK); Figs 6.6–6.11: Eric Barron, Tjeerd van Andel & David Pollard, and the McDonald Institute for Archaeological Research; Figs 6.12 & 6.13: Chris Crump; Fig. 6.14: John Sibbick; Fig. 6.15: Jennifer Chambers; Fig. 6.16: Creative Commons Attribution License (CC BY 4.0); original source: Solodenko *et al.* (2015, figs 4 & 7); Fig. 6.17: Elsevier (*Journal of Archaeological Science*).

Fig. 7.2: University of Chicago Press (*Current Anthropology*); Fig. 7.4: John Wiley & Sons (*Journal of Quaternary Science*).

Fig. F.1: British Geological Survey; Fig. I.1: Daniel Lieberman & John Shea; Figs N.1 & N.2: John Wiley & Sons (*Evolutionary Anthropology*); Fig. O.1: Elsevier (*Quaternary Science Reviews*).

Chapter 1

A seasonal approach

In the beginning ...

Humans, represented by members of genus *Homo*, have been living in Europe for around 1.5 million years. But who were they? How did they survive? In short, what kinds of 'humans' were these? These are the fundamental questions addressed, though the lens of the changing seasons, in the pages that follow. But why ask these questions and why should we be interested in the answers? Beyond simple curiosity I think there are two answers. The first is that the deep prehistory of Europe is a place of dramatic fluctuations and changes in climates, landscapes and environments. How Lower Palaeolithic humans adapted and responded to those many fluctuations has much to tell us about our place in the world and, sometimes, our fragility in the face of nature. As *H. sapiens* our own origins are fundamentally African and grounded in the younger period known as the Middle Stone Age. However recent genetic studies have identified evidence of interbreeding between *H. sapiens* and various archaic hominins, such as Neanderthals and Denisovans, as we dispersed across the Old World (Galway-Witham and Stringer 2018). The behaviour and adaptations of archaic Europeans in the Lower Palaeolithic period, the time of the Neanderthals' own ancestors, are thus informative both about themselves and, indirectly, us. Secondly, early humans are found across Europe, from Britain to Spain and from France to Bulgaria. Much of their archaeology, and by inference their behaviour, looks very similar, and yet, as so often, there is some devil in the details. The earliest Europeans therefore remind us of the human capacity for both local differences and broad similarities. As you will see in the pages that follow, the first Europeans were truly European.

A seasonal perspective: a Palaeolithic 'just-so' story?

This book reviews European Lower Palaeolithic life (c. 1.6–0.3 mya[1]) from the perspective of seasonal change. You might well ask why. Much of the available evidence is in the form of stone tools, and they have little to say, at least directly, about the passage of the seasons. Yet like all humans, Lower Palaeolithic hominins[2] lived within, and had to deal with, the challenges and opportunities presented by Early and Middle Pleistocene Europe.[3] At an annual scale these challenges and opportunities would include marked changes in the weather and day lengths, animals migrating to and fro, the appearance and disappearance of plant foods and a host of other cyclical patterns. For the large-brained and large-bodied hominins of early Europe, principally *H. antecessor* and *H. heidelbergensis*, these cycles would impact on all sorts of behaviours: food-getting, child-rearing, mobility around the landscape and the use (or not) of clothing, shelters and fire. While these are not behaviours that always leave clear traces in the archaeological record, they are behaviours whose likely presence or absence can be inferred, and characteristics reconstructed, based on the lived-in environments. Such an approach has been enabled by the remarkable reconstructions of Pleistocene climates and habitats which have emerged over the last few decades, and which underpin many of the arguments that follow. This book therefore adopts a heuristic approach to explore the possibilities and probabilities of seasonal life in Lower Palaeolithic Europe. It is not a book fundamentally about Lower Palaeolithic technology, or a site-by-site overview, for which many excellent sources already exist (*e.g.* Gamble 1986; Roebroeks and van Kolfschoten 1995; McNabb 2007; Pettitt and White 2012; Ashton 2017). It is however an attempt to consider the lived experiences of the earliest Europeans across the seasons, and evaluate the likely behaviours required by those lifestyles. In doing so, the book seeks to step beyond the often-uniform stereotypes of the Palaeolithic, and uncover the diversity, richness and texture of hominin lives.

When trying to reconstruct the ecological, social and material behaviours of pre-modern humans, McNabb's (2007) 'fourth option' for thinking about Palaeolithic hominins sounds a suitably cautious note:

> [they were] an animal, but one that was totally unique. Pre-modern humans were the products of ecologies and habitats for which no modern analogues now exist. Their behavioural adaptations were equally unique and we can now only project inappropriate modern human or modern animal behavioural responses onto them. (McNabb 2007, 348)

From a Lower Palaeolithic perspective the immensely long, relatively 'unchanging' nature of its iconic handaxes, Isaac's (1969) 'variable sameness',[4] might be seen as an example of these hominins' unique character. Yet if McNabb is right about the uniqueness of pre-modern humans in the Palaeolithic, is a seasonal approach useful? I offer two initial arguments in its defence. Firstly, the rich palaeoenvironmental evidence available to us suggests that Lower Palaeolithic ecologies and habitats, while unique, are to some extent knowable. A clearly seasonal climatic model (*e.g.* warmer summers, cooler winters, variations in precipitation) is evident from beetles, reptiles and other remains, while food webs and predator–prey relationships can be explored through pollen and other plant remains combined with zooarchaeological assemblages. Since no-one would dispute that hominins lived in, and were an integral part of, those worlds, the available evidence offers an environmental framework within which to try and understand them a little better. Inevitably this European stage is sometimes crisply sharp, at other times viewed more opaquely, but either way it allows us to consider how hominins, like all animals, met their fundamental needs: water, shelter and food. By connecting those needs with seasonally-changing conditions and resources on one hand, and the material traces of hominin actions on the other, it is possible to contribute new insights and understanding to a uniquely-Pleistocene behavioural 'black box'. Secondly, was Early and Middle Pleistocene life about no more than just staying warm, safe and fed, and by extension is this book no more than a suggested handbook for Lower Palaeolithic survival in Europe? This seems unlikely, given the rich and complex social lives of all animals, and sociality has been explored recently in hominin societies through explanatory frameworks such as the social brain and Theory of Mind (*e.g.* Dunbar 1998; Gamble *et al.* 2014), life history (*e.g.* Bogin and Smith 1996; Schwartz 2012), technological processes (*e.g.* Gamble 1998a; White and Foulds 2018), and care and compassion (*e.g.* Spikins *et al.* 2014; Spikins *et al.* 2019). Yet these social dynamics can also be considered with reference to seasonal variability, such as the implications of a potential clustering in conceptions and births.

What the following chapters therefore seek to demonstrate is that the unique nature of hominin sociality and cognition in the European Lower Palaeolithic can be explored and better understood by focusing on the day-to-day and seasonal fluctuations of living. The needs of survival are assessed not just in terms of material resources but also with reference to their cognitive demands, such as food-getting (planning, anticipation, cooperation and inhibition), sheltering (planning, anticipation and cooperation), and reproduction (care and cooperation).

Fundamentals of seasonality

While the seasonal specifics of Lower Palaeolithic Europe can only be understood through the analysis of Pleistocene palaeoenvironmental evidence (Chaps 2–6), the overarching drivers and trends of earth's seasonality are well known, at both orbital and regional scales.

Drivers of seasonality

The earth's seasonality concerns cyclical and largely predictable fluctuations in day length, temperature, rainfall and resource availability (Lisovski *et al.* 2017). Seasonal changes are primarily driven by the tilt of the earth's axis, around which the planet spins as it orbits the sun. Over the course of an annual orbit, this axial tilt means that the northern and southern hemispheres alternate between being closer to, and further away from, the sun, respectively resulting in summer and winter conditions (Woodward 2014; Fig. 1.1). In the higher latitudes the seasons of summer and winter are separated by spring and autumn. However, at lower latitudes near to or at the equator these seasonal effects are different, with two-season regimes (*e.g.* wet and dry) typical in those regions.

However, the earth's orbital movements have varied over time, due to the gravitational pull of the other planets, in the predictable patterns known as Croll-Milankovitch Cycles (Woodward 2014). Variations in axial tilt (obliquity) are of particular importance for seasonality. Axial tilt oscillates between 22.1° and 24.5° on a *c.* 41,000 year cycle. This is important because greater degrees of tilt result in more extreme seasons. Consequently, the specific character of seasons will have varied slightly at different times in the past, in line with these axial oscillations (Fig. 1.2). The earth's other orbital variables also impact on seasonality, although they are less significant than obliquity. The shape of the earth's orbit (eccentricity) varies between nearly circular and mildly elliptical, on *c.* 100,000 and 400,000 year cycles. When the orbit is more elliptical, the magnitude of seasonal changes increases and differences between the lengths of the seasons are more marked. Changes in eccentricity also modulate the impacts of the precession cycle. Precession refers to the wobble or circling motion of the earth's axis of rotation relative to the fixed stars, and it also varies, on a *c.* 19,000 and 23,000 year cycle. The circling motion of axial precession causes the solstices and equinoxes, *i.e.* the seasons, to shift over time, and impacts on the scale

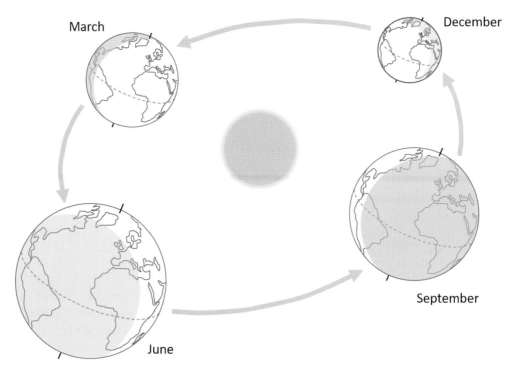

Figure. 1.1: Seasonal variations in the earth's orbit.

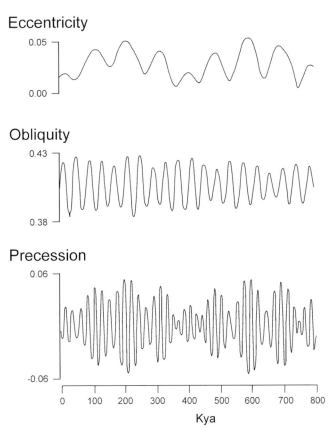

Figure 1.2: Periodicities of the eccentricity (stretch), obliquity (roll), and precession (wobble) cycles over the past 800,000 years (redrawn after Candy et al. *2014, fig. 3).*

of seasonal temperature differences. Marked differences also occur depending on the combinations of the orbital cycles: for example, when eccentricity is high then axial precession has a greater impact on seasonality. Similarly, seasonality increases when obliquity and eccentricity reach their maximum effects in tandem.

Orbital processes are not the only significant factors influencing seasonality however. While latitude correlates strongly with orbital variations in solar radiation, using latitude alone as a proxy for the full range of seasonality issues (*e.g.* precipitation and biological productivity) tends to limit our understanding of the variability (Lisovski *et al.* 2017). Therefore, other earth-based variables, such as ocean currents, sea-ice extent, wind direction, the extent of the continents and topography, also need to be considered.

For example, modern European winter climates are strongly influenced by atmospheric dynamics over the North Atlantic–European area (Fig. 1.3). These dynamics reflect the interplay between the Northern Westerlies, the Gulf Stream,

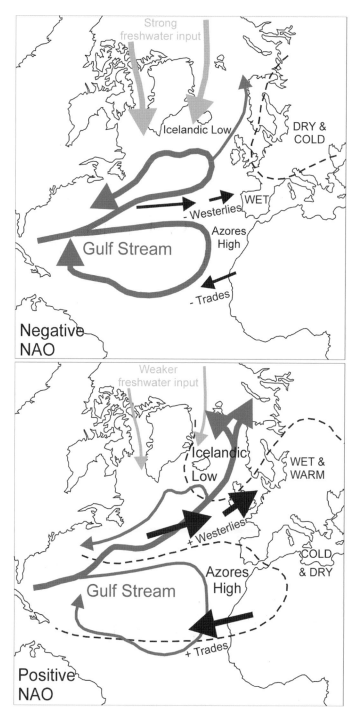

Figure 1.3: Simplified schematic of atmospheric dynamics over the North Atlantic–European area (redrawn after Wanner et al. 2001, fig. 9a & 9b).

and sea surface temperatures in the North Atlantic, expressed in the pressure gradient between the Icelandic Low and the Azores High (termed the North Atlantic Oscillation or NAO). In summary, strong freshwater input into the northern North Atlantic and resultant reduction in the strength of the North Atlantic thermohaline circulation directs warmer Gulf Stream waters into the eastern Atlantic, reducing sea surface temperatures (SSTs) in the North Atlantic. These lower SSTs in the north reduce the strength of the Icelandic Low, while the warmer waters in the eastern Atlantic reduce the strength of the Azores High and decrease the strength and trajectory of the westerlies. As a consequence of these negative NAO conditions, the Northern Westerlies are directed towards the Mediterranean, producing mild and wet winters, while the expansion of Polar Easterlies into northern Europe results in cold and dry conditions. By contrast, weaker freshwater input results in warmer Gulf Stream waters in the northern North Atlantic, strengthening the Azores High and the Icelandic Low (positive NAO conditions). The warm and humid Northern Westerlies are consequently directed further north, towards northern Europe, producing mild and wet conditions. The enhanced Azores High increases the strength of the Trade Winds, redirecting moisture away from the Mediterranean, resulting in cold and dry winter conditions in southern Europe (Wanner *et al.* 2001). Records of the 19th and 20th centuries indicate that the NAO persisted in its positive or negative state over several winters and exhibited decadal trends during those two centuries: this is a temporal pattern with implications for hominin lifespans if it also applied in the Pleistocene.

Modern trends in European seasonality

Global-scale modelling of modern data-sets highlights high levels of European seasonality if measured by variability in temperature and net primary productivity (NPP), but a much lower degree of seasonality when measured in precipitation. Modelled European seasonality is also consistently greater to the north of 44–45°N (in the Temperate Forest/Grasslands zone), with lower values in the Mediterranean zone (Lisovski *et al.* 2017, fig 2 & 3).

The 44–45°N latitude broadly captures a present-day transition from Mediterranean climates[5] to a mixture of oceanic climates[6] in western Europe, and humid continental climates[7] in eastern Europe, as defined by the Köppen climate classification system (Peel *et al.* 2007). Key climate trends in present-day Europe are (i) a broadly north–south gradient[8] in maximum (summer) temperatures (Fig. 1.4a); (ii) a northeast–southwest gradient in minimum (winter) temperatures (Fig. 1.4b); (iii) a west–east gradient in seasonal temperature ranges (Fig. 1.4c); (iv) a north–south trend in the 24 hour range in winter and summer air temperatures; (v) east–west and southeast–northwest gradients in precipitation (winter and summer respectively; Fig. 1.5); and (vi) west–east trends in the number of days with snow cover and the depth of snow cover (Barron *et al.* 2003). Measured by temperature and precipitation European seasonality is therefore especially marked in the continental interior and, to a lesser extent, the Mediterranean region,

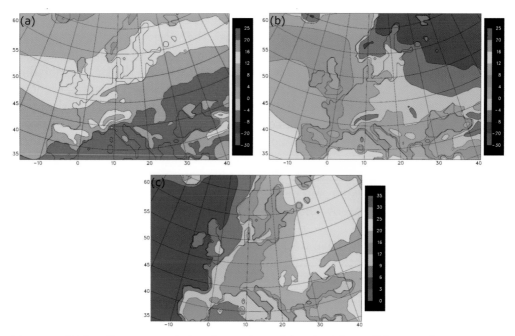

Figure 1.4: Modern European variations in maximum (a) and minimum (b) seasonal temperatures, and seasonal ranges in air temperatures (c; all temperatures in °C) (redrawn after Barron et al. 2003, appendix 5.1).

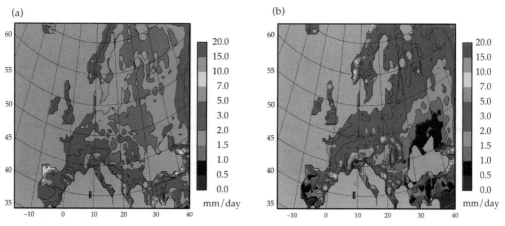

Figure 1.5: Modern European precipitation patterns for winter (a: December–February) and summer (b: June–August) (Barron et al. 2003, fig. 5.4).

with additional local and regional variations occurring in response to topography (*e.g.* mountain ranges). Modern, European-scale data-sets also reveal distinctive year-to-year variations in precipitation regimes, with summer precipitation in the west being less variable on a year-to-year basis than during the winter. In eastern Europe however the

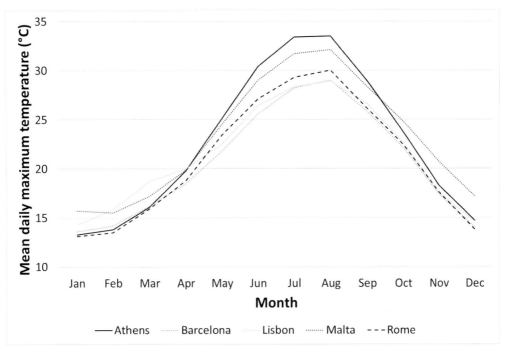

Figure 1.6: Mean daily maximum temperature variations by month in the Mediterranean climate zone. Data source: UK Meteorological Office (https://www.metoffice.gov.uk/).

reverse pattern applies, with potential implications for both drought/wildfires and flooding during the summer months (Zveryaev 2004).

Thus, the coastal lowlands encircling the Mediterranean have an essentially two-season pattern of wet winters and dry summers. Modern temperature data for a variety of Mediterranean locations (Fig. 1.6) suggests that summer could be defined here as May–October, given the clear shifts in temperature at either end of that interval, although there is also variability in the patterns and timings of the annual temperature profiles between different locations (*e.g.* contrast Athens with Barcelona and Lisbon). In the oceanic and continental climate regions to the north, the four seasons are defined following the *Societas Meteorologica Palatina* (1780): winter (December–February), spring (March–May), summer (June–August) and autumn (September–November). The season-by-season chapters of this book therefore most obviously map onto the temperate region's four season framework, but the 'spring' and 'autumn' issues can nonetheless be considered in the context of the late winter/ early summer and late summer/early winter periods in the Mediterranean region. Perhaps inevitably, the book's chapter structure also draws boundaries between the seasons in a manner which would have been meaningless to Lower Palaeolithic hominins. Where appropriate, seasons are therefore overlapped or blended (*e.g.* late spring/early summer when discussing ungulate births).

Table 1.1: Modern daylight data for selected European locations

Location	Latitude/longitude	Month	Dawn	Dusk	Daylight Hours
				(Approximate)	
London	51.4°N, 00.0°W	January	07.00–08.00	16.00–17.00	9
		July	04.00–05.00	21.00–22.00	17
Madrid	40.3°N, 03.5°W	January	08.00–09.00	18.00–19.00	10
		July	06.00–07.00	21.30–22.30	15.5
Berlin	52.3°N, 13.3°E	January	07.00–08.00	16.00–17.00	9
		July	04.00–05.00	21.00–22.00	17
Rome	41.5°N, 12.3°E	January	07.00–08.00	17.00–18.00	10
		July	05.00–06.00	20.30–21.30	15.5

data source: https://www.gaisma.com/en/

A further climatic factor concerns diurnal temperature variations. Modern data indicate notable differences in day/night temperatures, with both seasonal and geographical patterns (Barron *et al.* 2003, appendix 5.1). There are regional variations along a broadly north–south transect, with wider diurnal ranges in southern Europe, and larger variations in summer than winter: *e.g.* typical ranges of *c.* 2–4°C (winter) and 8–14°C (summer) in the Mediterranean, and *c.* 1–2°C (winter) and 4–9°C (summer) in northern Europe.

Alongside trends in temperature and precipitation, seasonality also incorporates other fluctuations. Modern daylight data indicates broad similarities across Europe, with slightly longer winter days in the south, and vice-versa in the summer (Table 1.1), although the specifics of daylight hours at any particular point in the Pleistocene past would also have been influenced by the earth's axial tilt. The length of twilight varies with the seasons, although it is longer at higher latitudes. Relative daylight levels are also further reduced beneath the canopies of closed woodland and forest habitats, which were common during the warm stage intervals of the Pleistocene.

Net primary productivity varies by both latitude and longitude, as the length of the growing season, broadly lasting from April–October/November, is controlled by mean daily air temperatures and is shorter at higher latitudes (Gamble 1986; Barron *et al.* 2003). However, summer droughts in both the Mediterranean region and the continental interior also impact on vegetation productivity, while the higher precipitation and mild winter temperatures of the oceanic west are favourable for plant growth (Fig. 1.7).

Impacts of seasonality

Annual cycles of climatic and habitat conditions therefore encompass both major variations in temperature, precipitation and daylight hours, and seasonal differences in diurnal patterns, such as cooler mornings and evenings in the otherwise warm days of late spring and early autumn. These seasonal patterns impact significantly on

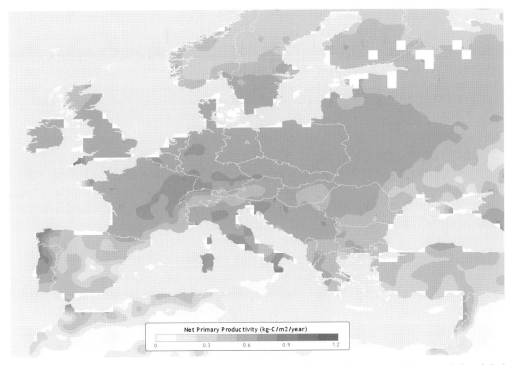

Figure 1.7: Net primary productivity in present-day Europe (Center for Sustainability and the Global Environment (SAGE), University of Wisconsin-Madison; https://nelson.wisc.edu/sage/data-and-models/atlas/maps/npp/atl_npp_eur.jpg).

organisms' adaptations, reflected in phenotypic[9] variability across the year. Seasonal variations in climatic conditions structure predictable rhythms and changes in all animal and plant species. Major changes in plant species include new or renewed spring growth, the summer and autumn harvest in fruits and nuts, and seasonal leaf loss. Amongst animals the key changes concern variations in physical condition, fluctuating aggregations and dispersals, shifting home range habitats, and the scheduling of breeding and birth. Higher latitude examples include the growing and shedding of winter and summer coats, or long-distance migrations (see also Lisovski *et al.* 2017), and the specific impacts of European seasonality on large-bodied mammals are evident in a wide range of living species (*e.g.* red deer; Fig. 1.8). While the exact timings of specific events vary between species, there are a suite of broad pan-specific trends including relatively poor winter condition, with reduced fat reserves, spring births, and enhanced summer and autumn health.

Such seasonal pressures and phenotypic adaptations should therefore also be expected in the animals of Pleistocene Europe, including hominins. From their perspective, the major 'events' and pinch points in a Lower Palaeolithic year would revolve around the relative food shortages and harsher climatic conditions of winter and early spring, the renewed plant foods available from late spring to autumn, animal new-borns in late spring, and the increasingly well-conditioned animals character-istic of summer and autumn (Fig. 1.9). Therefore, and while direct indicators of the seasonality of hominin activities are relatively rare,[10] these cyclical patterns enable the hominin year to be profitably explored through the lens of seasonally changing needs. These would have included winter survival, the rebuilding of energy stores and physical health from the late spring to early autumn, successful hominin reproduc-tion, and relocations in response to the fluctuating availability of static and mobile resources in time and space.

An emphasis on hominins as just another Pleistocene animal is explicitly stated here because, despite the mid-19th century recognition that humans have a long, 'deep time' prehistory and are the product of biological evolution by natural selection, there is sometimes still a tendency for humans to see ourselves as a step apart from the natural world. While the ongoing anthropogenic climate crisis will continue to challenge, perhaps brutally, such present-day blindness and arrogance, Pleistocene records clearly demonstrate that the earliest Europeans were part of their ecosystems in terms of their responses to dynamic, changing climates and environments. Within this context it is also important to acknowledge, from a Palaeolithic perspective, the twin dangers of anthropomorphism ('perceiving animals to be like ourselves') and anthropodenial ('a blindness to the human-like characteristics of other animals'; de Waal 1997, 51 & 52). While both concepts are often discussed in the context of non-hu-man animals rather than hominins, they are also relevant here as we are seeking to understand what kind of humans our Lower Palaeolithic ancestors were. What I have

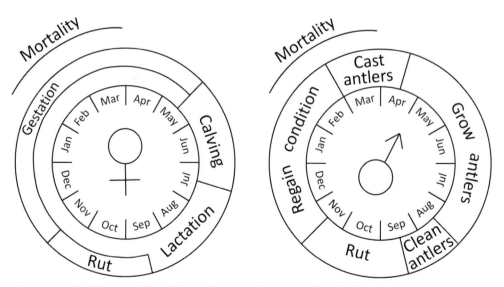

Figure 1.8: A 'red deer year' (based on modern populations on the Isle of Rum, Scotland; redrawn from http://rumdeer.biology.ed.ac.uk/deer-year).

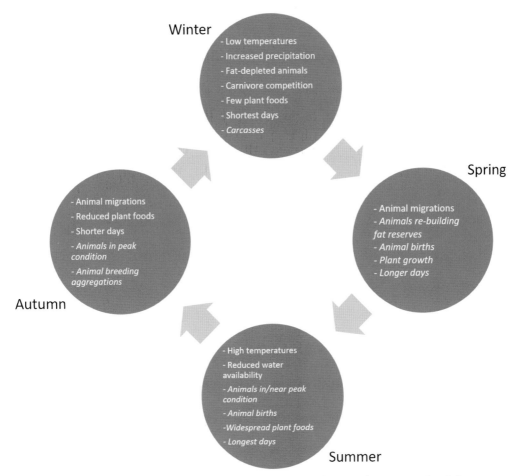

Winter
- Low temperatures
- Increased precipitation
- Fat-depleted animals
- Carnivore competition
- Few plant foods
- Shortest days
- *Carcasses*

Spring
- Animal migrations
- *Animals re-building fat reserves*
- *Animal births*
- *Plant growth*
- *Longer days*

Autumn
- Animal migrations
- Reduced plant foods
- Shorter days
- *Animals in peak condition*
- *Animal breeding aggregations*

Summer
- High temperatures
- Reduced water availability
- *Animals in/near peak condition*
- *Animal births*
- *Widespread plant foods*
- *Longest days*

Figure 1.9: Seasonal challenges (in plain text) and opportunities (in italics) in Lower Palaeolithic Europe.

sought to do, following de Waal (1997), is interpret the behaviours of *H. antecessor* and *H. heidelbergensis*[11] in the context of their habits, as reflected in the archaeological record, and natural history, as reflected in the palaeoenvironmental evidence. How the key seasonal challenges and opportunities of Early and Middle Pleistocene Europe (Fig. 1.9) were addressed by Lower Palaeolithic hominins is therefore the primary focus of this book. But I have also sought, against those same contexts, to consider the wider social complexity of the earliest Europeans, and to look beyond a life defined and dictated solely by the risks and rewards of a Pleistocene world. In doing so this book walks an interpretive tightrope familiar to Lower Palaeolithic researchers. As Dennell (2003) has argued with reference to the colonising abilities of *H. erectus*, it is important not to assume the presence of similar abilities to modern humans. The discussions of behaviours such as pyrotechnology, clothing and shelter in the chapters that follow are therefore not intended to propose or assume the existence of an essentially modern hunter/gatherer in an Early or Middle Pleistocene context. This obvious trap has previously been highlighted by McNabb (2007, chap. 13), and also by Gamble (1999, 153–72) in his re-interpretation of the much-debated Bilzingsleben 'campsite' as a hominin gathering. This book's discussions *are* intended however to focus attention onto the simple physiological and practical demands of surviving in Europe during periods of documented hominin presence, and to explore feasible strategies for doing so.

Overall, this book seeks to explore the lives of the first Europeans. In doing so, Ingold's (2013, 44) challenge to the researchers and authors of Palaeolithic archaeology feels especially pertinent. With reference to the handaxe makers of the Lower Palaeolithic he noted that 'they come across to us in the writings of modern archaeologists and anthropologists not as the powerfully built, bimanually dextrous and supremely skilled creatures that they surely were, but as clumsy hybrids stuck for over a million years in the transition from nature to culture'. The former is the view I concur with, and the perspective taken in this book, an on-the-ground exploration of the ecological, social and technological challenges of Lower Palaeolithic survival in Europe, seeks to breathe a little more life into those dynamic early northerners.

Notes

1. Million years ago.
2. The hominins are all the fossil 'human' taxa that are more closely related to modern humans than they are to any other living taxon (*e.g.* chimpanzees; Wood and Lonergan 2008).
3. 'Europe' had no specific meaning during this period and in palaeogeographical and palaeobiological terms was simply part of a wider Eurasia (Arribas and Palmqvist 1999). My use of Europe as a focus for studying survival at the mid-latitudes in the Lower Palaeolithic simply reflects Europe's rich and long history of Quaternary research, and the constraints of space.
4. Isaac (1969, 21) argued that the handaxe record suggested 'prolonged phases of relative stability with stochastic variation, and a very limited amount of gradual "progressive" change'.
5. Mediterranean climates are typically, although not exclusively, characterised by hot, dry summers and mild, wet winters.
6. Oceanic climates are characterised by mild summers for the latitude, and mild winters, with a relatively narrow annual temperature range and few extremes of temperature.
7. Humid continental climates are typified by warm–hot summers and cold winters with snow cover.
8. These gradients are ordered by increases in temperature/precipitation: *e.g.* summer temperatures increase from north to south.
9. An organism's phenotype is a set of morphological, physiological and behavioural characteristics resulting from the interaction of its genotype with the environment.
10. Nonetheless, seasonality data do exist at certain sites. At Miesenheim I, for example, indicators included specific bird species that are only summer visitors today, a red deer antler frontlet (carried from September–March/May), and ages at death for individual deer based on teeth eruption and wear stages. Collectively these indicated the period from summer to early spring (Turner 1999).
11. The very earliest European hominins may well be *H. erectus*, but the current evidence is ambiguous. There is also considerable debate as to the identity of the hominins from the later Lower Palaeolithic period after *c.* 600 kya: I have collectively referred to them here as *H. heidelbergensis sensu lato* (see Chap. 2 for details).

Chapter 2

Lower Palaeolithic Europe

Having outlined the fundamentals of mid-latitude seasonality in Chapter 1, with reference to present-day data, this chapter explores the wider context of general environmental settings and trends in the Pleistocene, specific indicators and details of Lower Palaeolithic seasonality, key hominin species and their requirements, and the fundamentals of Europe's earliest archaeological record.

The Pleistocene world

While the seasons are cyclical and predictable, an exploration of Pleistocene seasonality must also consider the context of larger-scale climate fluctuations, both cyclical and directional, that have occurred over the last two and half million years. Although often referred to as the 'ice ages', the Pleistocene environments of the earliest Europeans were marked by dramatic and regular fluctuations. These cyclical changes are often thought of in terms of the waxing and waning of ice sheets, which were driven by the earth's orbital cycles, but should also be thought of in terms of changing coastlines, river systems, plant and animal life, and climate and weather patterns. These were the macro-scale rhythms of the Pleistocene and could transform Norfolk into the 'Costa del Cromer' (Roebroeks 2005; Figure 2.1), and Spain into a cold, icy steppe. These cycles lay at the heart of the Pleistocene world, and the specific seasonal challenges faced by Lower Palaeolithic hominins for over one million years can only be fully understood when seen against this longer-term climatic framework.

Glacial and interglacial cycles

Specifically, the European Lower Palaeolithic occurred against the backdrop of the Pleistocene geological epoch, in its Early and Middle sub-divisions. The Early (*c.* 2.588–0.781 mya) and Middle Pleistocene (*c.* 0.781–0.126 mya) were characterised

Figure 2.1: Reconstruction of the Happisburgh 3 landscape, c. 850 or 950 kya (© John Sibbick & Ancient Human Occupation of Britain [AHOB] project).

globally by cycles of glacial and interglacial climates, with those cycles becoming longer and more marked in the later Middle Pleistocene, after *c.* 500 kya.[1] The impacts of these climate cycles varied across Europe, but in general terms peak interglacials[2] were associated with conditions broadly comparable to those of 'present-day' Europe (prior to anthropogenically-driven climate change), shifting in the glacials to conditions comparable to the present-day Arctic and the encircling tundra and steppe habitats of the high latitudes.

Interglacial flora

During the interglacials and warm stages (Box A) Europe was dominated by trees, although taxa and forest structure varied, particularly on a latitudinal basis, with a general trend of boreal forests in the far north, shifting through deciduous/coniferous forests to Mediterranean evergreen woodlands in the south (Van Andel and Tzedakis 1996; Woodward 2009, fig. 13.4). There were also regional contrasts alongside these latitudinal trends, reflecting the impacts of continentality, topography and precipitation. For example, Combourieu-Nebout *et al.* (2015) suggested predominantly deciduous interglacial forests in the Italian peninsula during the later Early Pleistocene (*c.* 1.8–0.78 mya) and, especially, the Middle Pleistocene, with coniferous forest in the north of the country. In northern Spain by contrast the Atapuerca sites were characterised by persistent savannah-like open woodland between *c.* 1.2–0.2 mya, with conifers, mesic,[3] and Mediterranean trees persistently present, but varying in proportions across the glacial/interglacial cycles (Rodríguez *et al.* 2011; see Fig. 2.2 for key site locations and Appendix A for site details).

A further factor is the vegetation successions which occurred during each warm stage, particularly in the north, as a consequence of species recolonising from predominantly southerly tree refugia and reflecting the climatic variability that occurred across individual warm stages. This is clearly illustrated for example in Britain, where the dominant tree species shifted over the course of MIS 11c (*c.* 424–398 kya) from birch woodland (pollen phase: Ho I) to mixed oak woodland (Ho II) and hazel/alder

Figure 2.2: Key archaeological and fossil sites in the European Lower Palaeolithic (see also Appendix A; © Google Earth 2019).

woodland (Ho III) back to pine/birch woods (Ho IV) (Ashton 2016, table 1). Further to the east the Schöningen 13-II site in north Germany highlights again both warm stage successions and local variations, with an MIS 9 (*c.* 337–300 kya) vegetation pattern of swamp forest, followed by deciduous forest, then boreal steppe forest and ending in the continental dry steppe/boreal forest associated with the famous 'spear site' (Urban and Bigga 2015). Thus, a specific location can be characterised by a changing variety of coniferous and deciduous tree types, and by shifts between more open and closed habitats, over the course of a single warm stage (Table 2.1).

Such vegetation successions highlight the presence of intra-stage variability in the Pleistocene. This is particularly evident in the ice core records that are a key archive of Pleistocene climate patterns (Box A). Put simply, 'glacials' and 'interglacials' were not uniformally cold or warm respectively, as is evident both in global and regional records and from site-specific sequences. This is the case at Hoxne for example, where Ashton *et al.* (2008a) demonstrated that the hominin occupations post-dated the peak MIS 11 interglacial (stage 11c) and the cold-climate 'Arctic Bed' interval (11b) and were instead associated with a later temperate phase of boreal woodland.[4] It is thus critical to directly associate, where possible, occupation evidence and environmental evidence when considering the lived experiences of hominins and seasonal perspectives.

Finally, there is also evidence for very short-lived environmental fluctuations. The Older Holsteinian Oscillation (OHO),[5] occurring within MIS 11 and lasting just a few hundred years, was characterised by a shift from woodland to more open, grassland conditions in Britain (*e.g.* at Marks Tey, England), while northern

Table 2.1: Examples of general vegetation successions in Middle Pleistocene Europe (Moncel et al. 2018)

	European Regions	
Climate cycle sub-stage	North[1]	Mediterranean[2]
Early warm stage	Pioneer forest: *Pinus, Betula*	Pioneer forest: *Pinus*
Interglacial maximum	Mixed oak forest: *Alnus, Corylus, Quercus, Ulmus, Carpinus* & *Ostrya*	Thermophilous forest: deciduous & evergreen *Quercus, Carpinus, Ulmus*, plus *Pinus* & Mediterranean/thermophilous taxa (e.g. *Carya* & *Pterocarya*)
Late warm stage	Coniferous forest: *Tsuga/Pinus* & *Picea*, with *Abies*[3]	Expansion of conifers (*Pinus, Abies, Picea*)
Glacial	Open vegetation: dry, herbaceous meadows (Poaceae, Asteraceae & Cyperaceae)	Open vegetation: dry meadows with steppic elements (Poaceae, Asteraceae and *Artemisia*, Chenopodiaceae)

[1]After MIS 16 there was a reduction, and then disappearance (after MIS 12), of sub-tropical taxa from the northern region (e.g. *Carya* & *Celtis*); [2]Mesothermic, relict taxa (e.g. *Carya* & *Tsuga*) persisted after MIS 12, but there was also a shift towards Mediterranean Holocene mixed forest compositions. [3]*Tsuga* and *Picea* were more typical of Poland and (with *Abies*) the Netherlands, while the UK record was characterised by *Pinus* and *Picea*, with heathland. Common English names for key plant taxa are listed in Appendix B

European continental sequences document a decline in deciduous woodland in favour of pine-dominated taiga (Candy *et al.* 2014). Shortly afterwards the Younger Holsteinian Oscillation (YHO), also within MIS 11, lasted *c.* 800 years at Ossówka lake in eastern Poland and initially resulted in the almost complete extinction of fir, followed by a slow recovery (Nitychoruk *et al.* 2018). Notably, this initiation of the YHO and the sudden disappearance of fir has been suggested to have occurred over just 50 years or so. At Ossówka the YHO has been linked to a drop in winter temperatures, late frost, or summer drought, although elsewhere different driving forces have been identified, such as a drop in summer temperatures at Dethlingen in Germany. Either way, these are all factors which would significantly impact on hominin lives at near-generational scales, presenting them with a new set of survival challenges, both at a seasonal scale and over the longer term. Even more dramatically, at Hoxne, England, the shift from Bed D to Bed C (the 'Arctic' Bed) has been associated with a reduction in mean warmest month temperatures from 15–19°C to less than 10°C, while mean coldest month temperatures declined to at least –15°C (Candy *et al.* 2014). Changes at this scale would seem likely to cause local hominin extinctions and/or significant relocations.

Moreover, such fluctuations are not limited to the north of Europe. Similar changes are evident in the high-resolution MIS 11 pollen record from Lake Ohrid in the southeast Balkans (Kousis *et al.* 2018). Significant phases of tree contraction and climatic deterioration have been documented at Lake Ohrid, including during the otherwise warmest sub-stage (MIS 11c). Lasting around 1.5 kyr, the period between 406.2–404.5 kya was characterised by a marked drop in arboreal pollen percentages and notable drops in mean annual temperature (to 3.7°C; the MIS 11c mean at Lake Ohrid is 7°C), mean coldest month temperature (-8.9°C compared to -1.5°C) and mean annual precipitation (*c.* 550 mm compared to 800 mm). To place this in context, even much smaller temperature variations (*e.g. c.* 2°C) may impact significantly on vegetation and fauna, as argued by Blain *et al.* (2009) for Gran Dolina, Spain, and there is no reason not to include hominins among the affected fauna.

Alongside temporal variability, there is also evidence for contemporary geographical variations in Early and Middle Pleistocene Europe. These patterns are more difficult to detect, because of the complications of demonstrating contemporaneity between sites of this age. However, Russo Ermolli *et al.* (2015) have demonstrated how local environmental and/or historical factors resulted in the development of distinctive woodland vegetation communities at five MIS 13 Italian sites, despite their overall warm stage similarities. The environmental factors included edaphic (soil), topographic and mesoclimatic[6] conditions, and the historical factors included the species composition of refugia and temporary changes due to disturbances. The significance of such variations has been highlighted by Margari *et al.* (2018, 155), who argued that 'populations of hominins may be unlikely to have occupied entire regions at any given time, but instead are perhaps more likely to have targeted specific habitats with appropriate local conditions'.

Box A:[1] How do we reconstruct Pleistocene climates and environments?

Our understanding of ice age (Pleistocene) climates has developed beyond all recognition over the last 30 years. This has occurred through the combination of both old and new evidence and analytical methods: pollen and plant macro-fossils, faunal assemblages (including mammals, beetles [coleoptera], molluscs, ostracods and other creatures), deep-sea marine cores, ice cores, terrestrial sediments; and landform (*e.g.* terrace) stratigraphy, multi-proxy biostratigraphy (including pollen stratigraphy), amino-acid racemization stratigraphy, magneto-stratigraphy, absolute dating (*e.g.* optically stimulated luminescence [OSL], electron spin resonance [ESR]), isotope analysis and mutual climate range and other related methods (*e.g.* Lowe and Walker 1997; Candy *et al.* 2014). Critically these methods and evidence operate at different scales: while the deep-sea marine cores highlight broad trends in Pleistocene climate (*e.g.* the repeated occurrence over the last half a million years of glacial/interglacial climatic cycles spanning 70,000–100,000 years each; Fig. A.1), the ice core records track higher resolution variations (*e.g.* demonstrating that

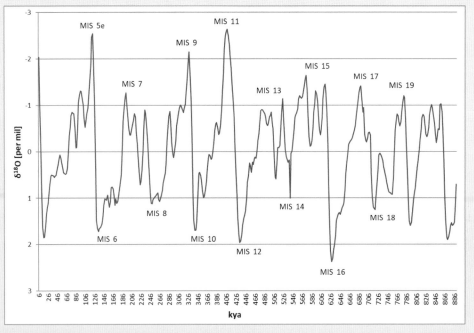

Figure A.1: Climate cycles (glacials [even-numbered] and warm stages [odd-numbered]) of the Middle and Late Pleistocene (stable oxygen isotope [deep-sea core] data from Bassinot et al. (1994, table 3); intervals between observations: 2000 years). The Y axis plots ¹⁸O isotope values and is a temperature proxy, with lower values indicating higher temperatures.

shifts in climate of up to 10°C occurred over just decadal timescales, and moreover that such dramatic shifts, both colder and warmer, occurred within the broader glacial and interglacial phases recorded in the marine cores; Fig. A.2).

An important question concerns how glacials and interglacials are defined, and by extension when they start and finish. As Candy *et al.* (2014) have highlighted, the usage of the interglacial label can itself be problematic, as its definition is not universally agreed upon. It is instead better to think of warm stages and cold stages, the start and end of which are defined by the deviation of the ^{18}O signal

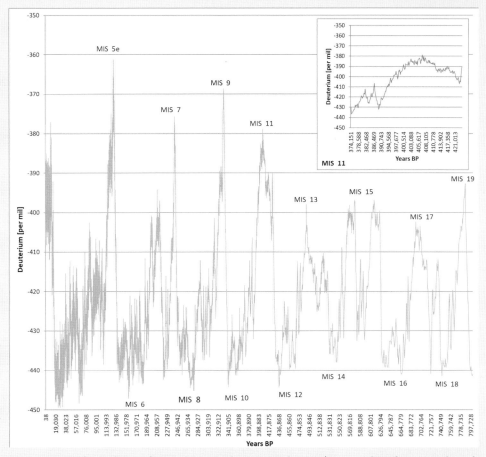

Figure A.2: High-resolution fluctuations in Pleistocene climate (ice core data from Jouzel et al. (2007); average intervals between observations: 138 years [increasing through time from 8 years [youngest pair of observations] to 1073 years [oldest pair]; inset: data for MIS 11 [424-374 kya; average intervals between observations: 241 years], highlighting high-resolution intra-stage variability). The Y axis plots deuterium (²H isotope) values and is a temperature proxy, with higher values indicating higher temperatures.

away from the mean of the Quaternary dataset (*i.e.* the '0' on the y axis on Fig. A.1):[2] periods with an ^{18}O value less than the mean are characterised by reduced global ice volumes and are described as 'warm' stages (*e.g.* MIS 11 and MIS 13); periods with ^{18}O values greater than the mean are associated with increased global ice volumes and are described as 'cold' stages (*e.g.* MIS 12 and MIS 16). The further problem is that 'warm stage' and 'interglacial' are not synonymous, although they are often used as though they were, and, moreover, interglacials have been defined in multiple ways. Candy *et al.* (2014) favoured a pollen-based definition, whereby an interglacial is defined by a period *within* a warm stage when the percentage of tree pollen is greater than the percentage of grass and shrub pollen, and when global ice volume is at its lowest. Alongside this peak interglacial, and still within the same overall warm stage, are periods of minor increases in global ice volume (*i.e.* colder conditions, known as stadials), and periods of reduced ice volume which are not as extreme as the full interglacial (these are known as interstadials). In short, each warm stage (*e.g.* MIS 5) represents an overall period of reduced global ice volume that is sub-divided into an interglacial (MIS 5e), and a series of interstadials (MIS 5c and 5a) and stadials (MIS 5d and 5b; Fig. A.1). The terms interglacial and warm stage are used in this manner throughout this book.

However, both the marine and ice core records, and available palaeoclimatic models (*e.g.* Herold *et al.* 2012; Milker *et al.* 2013; Muri *et al.* 2013; Kleinen *et al.* 2014; Rachmayani *et al.* 2016), document global and regional trends at an inevitably low spatial resolution, rather than revealing sub-regional and site-specific conditions. They are therefore of limited value for exploring Pleistocene seasonality as experienced by hominins. Moreover, as Candy and Alonso-Garcia (2018) have noted, transitions such as the Early–Middle Pleistocene Transition (EMPT) and the Mid-Brunhes Event (MBE) are spatially variable in their impacts (see also Blain *et al.* 2012). For example, regional north-eastern Atlantic records suggest that glacial/interglacial cycles from the 1–0.5 mya interval were of a similar magnitude to those after 0.5 mya, in contrast to the global marine core oxygen records (Fig. A.1).

At the smallest scale, understanding of individual sites comes instead from pollen and, critically, micro-fauna. The latter, in particular beetles, have specific environmental and climatic tolerances and evidence of a stable recent evolutionary history, thus making them ideal sources of evidence for climate reconstruction. The combined presence on Pleistocene sites of different species and/or different animal groups enables Quaternary scientists to reconstruct past conditions, based on their modern-day environmental requirements. Using the Mutual Climate Range method (MCR), the area of overlap between the various species' environmental requirements indicates the likely conditions at the site. A further benefit of micro-fauna, such as beetles and molluscs, and micro-mammals, is that they represent the genuine local habitat, whereas larger fauna such as herd animals may have been selectively accumulated through hunting or carnivore activity and therefore not be entirely representative. Larger mammals are also problematic due to their relatively wide-ranging environmental tolerances: in effect they are too resilient to reveal specific information about the local environmental conditions, especially climate.

Alongside animals, both large and small, plant pollen is another critical source of evidence for reconstructing Pleistocene environments. However, the microscopic nature of pollen further complicates the matter, as consideration must be given to how far the pollen may have been transported by wind or water and therefore whether it is representing the local habitat or the wider region. Nonetheless, the presence of different plant groups (*e.g.* the proportions of tree pollen to grass pollen) and different species (*e.g.* oak and elm as opposed to pine and birch) provide valuable information about the general climatic and landscape conditions (*e.g.* relatively cool, open grassland environments, as opposed to the closed, deciduous woodlands associated with an interglacial).

This combination of evidence, floral and faunal, enables the reconstruction of various aspects of Pleistocene sites, including seasonality indicators, such as mean annual, summer and winter temperatures, precipitation, ground cover conditions (*e.g.* the presence of leaf litter) or the nature of water bodies (*e.g.* still, stagnant or fast-flowing).

---

[1] Boxes are used throughout this book to provide background information on key issues (*e.g.* Pleistocene environments or models of hunter-gatherer mobility).

[2] The ratio of ^{18}O to ^{16}O, measured from the calcium carbonate shells of benthic (sea-bed) foraminifera within deep-sea cores, or from the water content of ice cores, provides a measure of palaeotemperatures. The ratios are also impacted by other factors, such as global ice volume and water salinity (Lisiecki and Raymo 2005).

Glacial flora

During the glacials, habitats varied from northern glaciers and polar deserts to open steppe in the Mediterranean south (Van Andel and Tzedakis 1996; Woodward 2009, fig 13.4; Combourieu-Nebout *et al.* 2015), although the south also featured localised long-term refugia in which trees were permanently present through glacials as well as warm stages (*e.g.* Tzedakis 1993; Kousis *et al.* 2018). As is demonstrated by the apparent cold-stage tree refugia at Ioannina, in contrast to the extreme glacial stage tree population contractions at the fellow Greek site of Tenaghi Philippon (Tzedakis *et al.* 2006), habitats and vegetation would also vary on more local scales, reflecting the impacts of topography: elevation, aspect, exposure and hydrology. Glacial stage reconstructions are more difficult in northern Europe, reflecting the limited biomass associated with those cold environments, and the destructive impacts of ice sheets. However, and in contrast to later Palaeolithic periods, there was relatively little cold stage occupation in northern Europe during the Lower Palaeolithic, although there are occasional examples such as at Kärlich H, Geramany, and associated with the Eartham Formation at Boxgrove, England (Haidle and Pawlik 2010; Roberts and Parfitt 1999). Thus, much of the following discussions will be focused on warm stage environments across Europe and also glacial environments in southern Europe.

Mammal fauna

Animals also varied on both geographical and chronological scales, with the combination of these factors making it difficult, and unhelpful, to refer simply to 'glacial' and 'interglacial' faunas at a European scale. However, examples of the main fauna from key warm stage sites in different parts of Europe can give some sense of the geographical similarities and variations, and of the wider animal communities to which hominins belonged (Table 2.2). In terms of chronological and potentially climate-driven variations, the long Atapuerca sequence (Sima de Elefante, Gran Dolina, Sima de los Huesos and Galería) offers a valuable perspective from southern Europe (Rodríguez *et al.* 2011). The large mammal evidence from these sites lacks species that clearly indicate harsh conditions, with the majority of species being temperate or catholic in their affinities (*e.g.* fallow deer, macaque and hippopotamus). These patterns suggest prevalent warm conditions and thus fit with the vegetation evidence outlined above and are further supported by the herpetofauna (amphibians and reptiles) and the small mammals. This broad glacial/warm stage consistency is much less apparent north of the Pyrenees however, particularly during the longer glacial/warm stage cycles of the later Middle Pleistocene (MIS 12–6) which were associated with markedly contrasting glacial (the cold-adapted *Mammuthus–Coelodonta* Faunal Complex or 'mammoth' fauna) and warm stage faunas (Kahlke *et al.* 2011). In comparison with the northern warm stage sites listed in Table 2.2 (Boxgrove, Soucy and Bilzingsleben), cold stage faunas from glacial stages (*e.g.* MIS 12 [*c.* 478–424 kya]) were characterised by species such as bison, reindeer (*Rangifer tarandus*), giant musk ox (*Praeovibos priscus*), woolly rhinoceros (*Coelodonta*) and steppe mammoth (*Mammuthus trogontherii*: Kahlke and Lacombat 2008; Kahlke 2014). However, there were also highly adaptable mammal species, for example horse, which appeared in both glacial and warm stage faunas.

Micro-fauna and seasonality indicators

In contrast to the many flexible and adaptable larger mammals, micro-fauna, in particular beetles but also mollusca, herpetofauna and small mammals, are a key source of information about local climatic conditions and, critically, seasonality (Table 2.3 & Box A). Where such assemblages can be correlated directly with hominin occupations, climate estimates indicate the various and differing seasonal challenges which were faced. At the Schöningen spear site (13 II-4; MIS 9) for example, the molluscan assemblage indicated minimum winter temperatures of -4°C and maximum summer temperatures of 16°C, combined with relatively low annual precipitation (400–450 mm). These are typical of continental conditions in central-northern Europe (Urban and

Table 2.2: Mammalian fauna from selected European Lower Palaeolithic warm stage sites

Boxgrove (UK, MIS 13)

Herbivores
Beaver (C. fiber)
Bison (Bison sp.)
Elephant (elephantidae)
Fallow deer (D. dama)
Giant deer (Megaloceros sp.)
Horse (E. ferus)
Ovicaprid (Caprinae)
Red deer (C. elaphus)
Rhinoceros (S. hundsheimensis)
Roe deer (C. capreolus)

Carnivores & omnivores
Badger (Meles sp.)
Deninger's bear (U. deningeri)[1]
Lion (P. leo)
Mink (M. lutreola)
Spotted hyena (C. crocuta)
Stoat (M. erminea)
Weasel (M. nivalis)
Wild cat (F. sylvestris)
Wolf (C. lupus)

Bilzingsleben (Germany, MIS 11)

Herbivores
Beaver (C. fiber & T. cuvieri)
Bison (B. priscus)
Clacton fallow deer (D. clactoniana)
Horse (E. mosbachensis-taubachensis)
Red deer (C. elaphus)
Rhinoceros (S. kirchbergensis & S. hemitoechus)
Roe deer (C. suessenbornensis)

Carnivores & omnivores
Cave bear (U. deningeri-spelaeus)[1]
Cave lion (P. spelaea)
Red fox (V. vulpus)
Spotted hyena (C. crocuta)
Wild boar (S. scrofa)
Wolf (C. lupus)

Soucy (France, MIS 9)

Herbivores
Aurochs (B. primigenius)
Beaver (C. fiber)
Bison (Bison sp.)
Clacton fallow deer (D. dama clactoniana)
Giant deer (Megaloceros sp.)
Horse (E. mosbachensis)
Merck's rhinoceros (D. mercki)
Red deer (C. elaphus)
Roe deer (C. capreolus)
Steppe mammoth (M. trogontherii)
Straight-tusked elephant (P. antiquus)

Carnivores & omnivores
Brown bear (U. arctos)
Wild boar (S. scrofa)
Wolf (C. lupus)

Medzhibozh 1 (Ukraine, MIS 11)

Herbivores
Beaver (C. fiber)
Fallow deer (D. dama clactoniana)
Forest rhinoceros (S. kirchbergensis)
Giant deer (Megaloceros sp.)
Red deer (C. elaphus)
Roe deer (C. suessenbornensis)

Carnivores & omnivores
Black bear (U. thibetanus)
Deninger's bear (U. deningeri)[1]
Wild boar (S. scrofa priscus)

Table 2.2: (Continued)

Atapuerca (Spain), Sima de los Huesos (MIS 11–12) & FU6		Isernia la Pineta (Italy, MIS 15)	
Herbivores	*Carnivores & omnivores*	*Herbivores*	*Carnivores & omnivores*
Ass (*E. hydruntinus*)	Badger (*M. meles*)	Bonal tahr (*H.* cf. *bonali*)	Deninger's bear (*U. deningeri*)[1]
Bonal tahr (*H. bonali*)	Deninger's bear (*U. deningeri*)[1]	Fallow deer (*D.* cf. *roberti*)	Leopard (*P. pardus*)
Fallow deer (*D. clactoniana*)	European polecat (*M. putorius*)	Hippopotamus (*H.* cf. *antiquus*)	Lion (*P. leo fossilis*)
Giant fallow deer (*M. solihacus*)	Lion (*P. leo* cf. *fossilis*)	Red deer (*C. elaphus* cf. *acoronatus*)	Wild boar (*S. scrofa*)
Horse (*E. ferus*)	Mediterranean cave lynx (*L. pardinus spelaeus*)	Roe deer (*Capreolus* sp.)	
Red deer (*C. elaphus priscus*)	Red fox (*V. vulpus*)	Rhinoceros (*S. hundsheimensis*)	
Rhinoceros (*S. hemitoechus*)	Weasel (*M. nivalis*)	Straight-tusked elephant (*P. antiquus*)	
Roe deer (*C. priscus*)	Wild cat (*F. sylvestris*)	Woodland bison (*B. schoetensacki*)	
Woodland bison (*B. schoetensacki*)			

Sources: Ballatore and Breda (2013); García and Arsuaga (2011); Lhomme (2007); Mania and Mania (2003; 2005); Parfitt (1999a); Stepanchuk and Moigne (2016). [1]Deninger's bear may have been herbivorous (García et al. 2009). Common English names for key animal taxa are also listed in Appendix C

Table 2.3: Palaeoclimatic estimates for summer (T_{max}) and winter (T_{min}) at selected European Lower Palaeolithic sites

Site	T_{min} (°C)	T_{max} (°C)	Evidence[1]	Age (MIS)	Source
		Early Pleistocene			
Barranco León (Layer D)	9.0	26.2	Amphibians & Reptiles (MER)[2]	43–49	Blain *et al.* 2016
Fuente Nueva-3	9.2	24.3	Amphibians & Reptiles (MER)[2]	43–49	Blain *et al.* 2016
Sima del Elefante (Level TE9c)	4.1	20.5	Amphibians & Reptiles (MCR)[3]	37	Blain *et al.* 2010
Happisburgh III (Bed E)	-3 – 0	+16 – +18	Coleoptera	Late 25 or late 21	Ashton & Lewis 2012; Parfitt *et al.* 2010
Gran Dolina (TD-6.2)	4.3	22.0	Herpetofauna (MCR)[3]	21	Blain *et al.* 2013
		early Middle Pleistocene			
Pakefield (Bed Cii–Ciii)	-6 – +4	+17 – +23	Coleoptera	17 or later 19	Ashton & Lewis 2012; Coope 2006
Cúllar Baza 1	+2.5 – +12.5	+21 – +27	Amphibians & Reptiles (MCR)[3]	15?	Agustí *et al.* 2009
Boxgrove (Unit 4c & Freshwater Silt Bed ≈ Units 4b & 4c)	-4 – +4	+15 – +20	Ostracods (MOTR)[4] & Herpetofauna (MCR)[3]	13	Ashton & Lewis 2012; Holman 1999; Holmes *et al.* 2010
Happisburgh I (Organic Mud)	-11 – -3	+12 – +15	Coleoptera	13?	Ashton & Lewis 2012; Coope 2006
High Lodge (Bed C1)	-4 – +1	+15 – +16	Coleoptera	13?	Coope 2006
Waverley Wood (Channel 2, Organic Mud)	-	+10 – +15	Coleoptera	13 or 15	Coope 2006; Shotton *et al.* 1993
Brooksby (Redland's Brooksby Channel)	-10 – +2	+15 – +16	Coleoptera	13 or 15	Coope 2006
		later Middle Pleistocene			
Barnham (Unit 5c; HoII)	–	+17 – +18	Herpetofauna	11c	Holman 1998
Hoxne (Stratum D; HoIIIa)	-10 – +6	+15 – +19	Coleoptera	11c	Ashton *et al.* 2008a; Coope 1993
Bilzingsleben II	-0.5 – +3	+20 – +25	Mollusca & ostracods[5]	11	Mania 1995; Mania & Mania 2003
Aridos I	+2 – +12	+20 – +28	Amphibians & Reptiles (MCR)[3]	11	Blain *et al.* 2014
Gran Dolina (TD-10 [sub-level T1])	-0.5 – +7.5	+16 – +22	Amphibians & reptiles[3]	11	Blain *et al.* 2009
Schöningen 13 II-4	-4 – -1	+16	Mollusca & ostracods	9	Urban & Bigga 2015
East Anglia (present day)[6]	-0.7 – +6.9	+14.2 – +18.0	–	–	–

Table 2.3: (Continued)

Site	T_{min} (°C)	T_{max} (°C)	Evidence[1]	Age (MIS)	Source
Bilzingsleben (present day)[7]	-3.3 – +2.0	+12.4 – +22.7	–	–	–
Madrid (present day)	+5.2	+24.0	–	–	Blain *et al.* 2014

Sources listed within the table. [1]Sensitivity tests on coleoptera-based MCR procedures suggest that winter temperature estimates are usually too warm (Pettitt and White 2012, 35); [2]Mutual Ecogeographic Range (MER) method; [3]Mutual Climate Range (MCR) method; [4]Mutual Ostracod Temperature Range (MOTR) method; [5]The specific source of the palaeo-temperature estimates is not stated, but the fauna includes molluscs and ostracods; [6]East Anglian data based on Met Office annual mean seasonal temperatures (1910–2016; http://www.metoffice.gov.uk/pub/data/weather/uk/climate/datasets/Tmean/date/East_Anglia.txt); [7]Bilzingsleben data based on Deutscher Wetterdienst (German Weather Service) monthly mean January and July temperatures (1951–2017; Erfurt-Weimar station; https://www.dwd.de/DE/leistungen/klimadatendeutschland/klarchivtagmonat.html).

Bigga 2015). By contrast, the evidence from Atapuerca TD-6.2 (MIS 21) in northern Spain indicated conditions and seasonality broadly typical of a continental Mediterranean climate, although somewhat wetter: minimum winter temperatures of 4.3°C, maximum summer temperatures of 22°C, and mean annual precipitation of 962 mm, mostly falling during spring and autumn (Blain *et al.* 2013).

Even where direct, seasonally-specific, palaeo-temperature estimates are not available, micro-fauna can offer valuable insights. The molluscan assemblages from the northern French site of La Celle provided a high-quality record of palaeoenvironmental variations during MIS 11 (Dabkowski *et al.* 2012; Limondin-Lozouet *et al.* 2015). At the beginning of the warm stage the mollusca indicated marshy, open ground. These were replaced by shade-loving species, indicating the establishment of forest, together with wet, open-ground conditions. Deciduous forest development peaked at the interglacial climatic optimum, with which the hominin occupations at La Celle were associated, after which the woodland declined and wet, marshy habitats re-appeared. The molluscan data are complemented by the geochemical data from tufa calcite, which indicated a warm and wet climatic optimum (Dabkowski *et al.* 2012). Moreover, the environmental associations of the artefacts at La Celle appear to be comparable with those at the MIS 11 sites of Beeches Pit and Saint Acheul, highlighting the ability of hominins to survive in the closed forests of the optimal interglacials (Limondin-Lozouet *et al.* 2015).

Long-term Pleistocene change

Pleistocene Europe also underwent a series of longer-term environmental and climatic changes. These were linked to the transition from shorter and more varied *c.* 41 kyr climate cycles, structured by orbital obliquity, to longer, relatively stable *c.* 100 kyr cycles – the so-called Early–Middle Pleistocene Transition (EMPT; *c.* 1.2–0.6 mya; Figure 2.3; Head and Gibbard 2005). Whereas the earlier period saw lower amplitude cycles, relatively mild conditions and a wide variety of habitats, especially in the

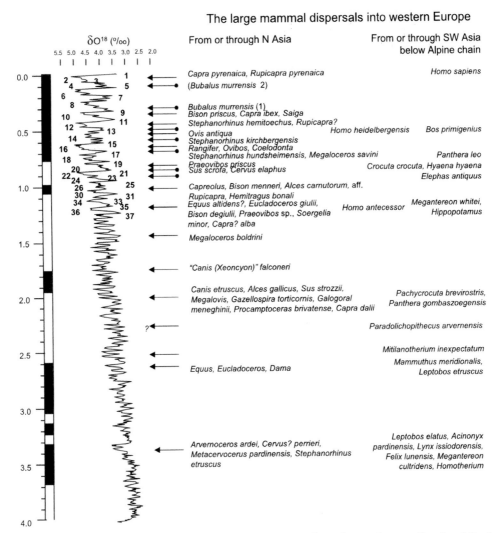

Figure 2.3: Dispersal of large mammals into Western Europe (van der Made 2011, fig. 3). While the evidence from Gombore II is strongly suggestive that H. heidelbergensis dispersed into Europe (Profico et al. 2016), the origins of H. antecessor are less certain. Bermúdez de Castro et al. (2017) suggest that H. antecessor may represent one of a succession of population waves which split away from a western Eurasian/African source and settled in Europe during the Pleistocene. Note the shift in the ¹⁸O curve from higher frequency, lower amplitude climate cycles to lower frequency, higher amplitude cycles between c. 1.2–0.6 mya (the Early-Middle Pleistocene Transition).

Mediterranean and the West (Messager *et al.* 2011; Moncel *et al.* 2018), the latter period was broadly characterised by progressive temperature decline, increasing aridity, asymmetric climate cycles (with longer and more severe glacial periods), oscillations between highly contrasting climates and landscapes (a mean annual temperature difference of c. 13°C in the northeast Atlantic sea surface temperature record), and increasingly specialised mammalian communities (Candy and Alonso-Garcia 2018; Kahlke *et al.* 2011; Sánchez Goñi *et al.* 2016). Of particular significance for this book is that the transition also resulted in greater seasonality, as is suggested by the site-specific temperature estimates from Early and Middle Pleistocene hominin sites (Table 2.3), although the strength of seasonality also fluctuated in association with orbital cycles (see Chap. 1). These changes were accentuated by the Mid-Brunhes Event (MBE; c. 450 kya), which marked the shift to increasing glacial cooling and interglacial warmth (Candy and Alonso-Garcia 2018). Such long-term trends were reflected in changing animal and plant communities. Based on the Tenaghi Philippon record, Tzedakis *et al.* (2006) noted a reduction in arboreal diversity after MIS 16 (c. 676–621 kya), with forests becoming increasingly dominated by *Quercus* (oak) and *Carpinus* (hornbeam). A similar reduction in tree taxa diversity has been noted by Combourieu-Nebout *et al.* (2015) in the Iberian Peninsula. While the specifics of those habitat changes varied from region-to-region (Table 2.4), the general trends outlined above have been commonly argued to be widespread. The impacts of reduced taxa diversity and enhanced seasonality would likely increase the challenges of European survival in the Middle Pleistocene, especially with reference to food resource variability and summer and winter climatic contrasts.

However, regional climatic patterns also differed, both from each other and from the global climate records (Candy and Alonso-Garcia 2018; Candy and Mcclymont 2013; Candy *et al.* 2015). In north-west Europe for example, climate records spanning the last 1 myr challenge the traditional model (Candy and Alonso-Garcia 2018), while Spanish evidence from Atapuerca is more in-keeping with global records (Blain *et al.* 2012). Regional climatic records in north-west Europe, specifically foraminifera from deep-sea cores in the northeastern Atlantic which document sea surface temperatures,

Table 2.4: *Prevailing and changing habitat characteristics in Europe over the last 1.2 mya (redrawn after Kahlke* et al. *2011, fig 6)*

Region	Prevailing habitats (inferred from large mammal communities)		
Apennine Peninsula	High variety of open/forested habitats	Open woodland/steppe	
Iberian Peninsula	Open savannah/lightly forested habitats	Open woodland, tree savannah/steppe	
Western Europe	High variety of open/forested habitats	Woodland/steppe	Variety of open to forested habitats
Northwest Europe	Woodland	Woodland with open patches/mixed habitats	Steppe/woodland
Central Europe	High variety of open/forested habitats	Steppe/woodland with open patches	Steppe-tundra/ woodland
Eastern Europe	High variety of forest steppe/ open steppe habitats	Open steppe/forest steppe	Steppe-tundra/ woodland
Chronology (mya)	1.2 1.0	0.8 0.6	0.4

suggest a number of differences to the patterns of the global deep-sea and ice core records (see also Box A). In particular, the magnitude of extreme glacial/warm stage cycles prior to 0.5 mya were comparable to those of the last 500,000 years (Candy *et al.* 2015, fig 1); a period of sustained warmth is indicated towards the end of the early Middle Pleistocene (MIS 15–13 [*c.* 621–478 kya], during which MIS 14 [*c.* 563–524 kya] was short-lived and minor); and strong sub-stage warming and cooling events were detected within many of the interglacials of the past 450,000 years (Candy and Alonso-Garcia 2018). The first point in particular is notable in the context of the earliest human occupations of northern Europe, which therefore appear to have occurred against a backdrop of extreme climatic cycles, thus raising key questions as to the nature of the necessary hominin adaptations. However, the second point is equally interesting in the context of the earliest widespread appearance of handaxes in western Europe (Moncel *et al.* 2015; Moncel *et al.* 2018): might the sustained warmth of MIS 15–13 have been a trigger for an extensive dispersal into Europe of handaxe-making hominins from adjacent regions with warmer climates, such as south-west Asia?

Faunal transitions

Another key component of longer-term changes were the various mammal dispersals into Europe over, and prior to, the course of the Lower Palaeolithic (van der Made 2011). Of particular note is the succession of dispersals into Western Europe[7] that occurred after *c.* 1.2 mya. While these dispersals did not feature new mammal families or orders, a variety of new species appeared. Key examples of dispersing species and timings include *S. scrofa* (wild boar), *C. elaphus* (red deer) and *C. crocuta* (spotted hyena; *c.* 900 kya), *S. hundsheimensis* (rhinoceros) and *P. leo* (lion; *c.* 700 kya), and *S. hemitoechus* (narrow-nosed rhinoceros) and *B. primigenius* (aurochs; *c.* 700 kya; see Figure 2.3 for details). Of particular note are the significant numbers of artiodactyla,[8] which were potentially valuable prey species for hominins (van der Made 2011, fig 4). The majority of these taxa originated from Asia and were adapted to open, dry environments (*e.g.* bison), reflecting the changing European conditions associated with the Early–Middle Pleistocene Transition. At the level of specific regions, faunal migrations and changes inevitably varied. For example, the first appearance of the 'mammoth' fauna in MIS 12 was initially limited to south-eastern, eastern and central Europe, with the specific conditions in western and north-eastern Europe preventing, at that time, the further spread of the *Mammuthus-Coelodonta* fauna (Kahlke *et al.* 2011).

Assessing these long-term faunal changes can be difficult, partly because of unevenness in the coverage and descriptions of the data (van der Made 2011), and also because any site-to-site comparisons, even within the same region, have to consider the possible impacts of local habitat variations. However, the long sequences at Atapuerca again offer a valuable opportunity to track evolutionary patterns in animal communities in the wider landscape across both the Early and Middle Pleistocene (Rodríguez *et al.* 2011). The sequence was characterised throughout by warm conditions and open woodland but there was, nonetheless, a significant faunal turnover at around 600 kya. While both the carnivore and the herbivore records document the replacement of a number of specific species, with a shift towards animals more adapted to arid conditions after the Mid-Brunhes Event (Blain *et al.* 2012), there was also a significant shift in the overall balance and composition of the faunal communities (Table 2.5).

Turner (1992) highlighted the significance of the Villafranchian/Galerian (*i.e. c.* 1.2 mya) faunal turnover with a focus on predatory species, particularly the balance in the earlier period of both carcass producers (the scimitar and sabre-toothed cats *H. latidens* and *M. cultridens*, alongside *A. pardinensis* [cheetah] and *P. gombaszoegensis* [jaguar]) and carcass destroyers (*e.g.* the giant short-faced hyena *P. brevirostris*). While the former produced a significant 'fleshy' carcass resource (Arribas and Palmqvist 1999), the latter are suggested to have been a significant source of competition for hominins if *Homo* was fundamentally reliant on scavenging rather than hunting in the Early Pleistocene. By contrast, the main predators after *c.* 500 kya were leopard, lion, spotted hyena and wolf, potentially opening up new niches for hominins.[9]

Landscape transformations

The cycling climates of the Early and, especially, the Middle Pleistocene would have resulted in fluctuating sea-levels and, therefore, coastline positions (*e.g.* Van Andel and Tzedakis 1996, fig. 3). These are often difficult to reconstruct but in areas lacking long-term isostatic uplift of the land it is clear that much coastal land that may have been occupied by hominins during the Pleistocene is currently submerged. Those hominin occupations of the now-submerged coastal lands were most likely to be in the south of the continent, where the glacial conditions associated with lower sea-levels would have less dramatic impacts on the habitability of the local landscapes. This is illustrated

Table 2.5: *Selected transitions in faunal species at Atapuerca (after Rodríguez et al. 2011, tables 6 & 7)*

Animal group	TE8–TD8 (c. 1.4–0.6 mya)	SH/TD10-3–TE19 (c. 430–<0.300 kya)
Carnivores	Arno River dog/Mosbach wolf (*Canis* sp. [*arnensis/mosbachensis*])	Wolf (*Canis lupus*)
	Issoire lynx (*Lynx* cf. *issiodorensis*)	Cave lynx (*Lynx pardinus spelaeus*)
	Stoat (*Mustela palerminea*)	Weasel (*Mustela nivalis*)
	European jaguar (*Panthera gombaszoegensis*)	Lion (*Panthera leo*)
	Gran Dolina bear (*Ursus dolinensis*)	Deninger's bear (*Ursus deningeri*)
	Fox (*Vulpes* cf. *V. alopecoides*; *V. praeglacialis*)	Red fox (*Vulpes vulpes*)
Herbivores	Bison (*Bison* cf. *voigtstedtensis*)	Woodland bison (*Bison schoetensacki*)
	Red deer (*Cervus elaphus* cf. *acoronatus*)	Red deer (*Cervus elaphus priscus*)
	Fallow deer (*Dama vallonnetensis*)	Clacton fallow deer (*Dama dama clactoniana*)
	Horse (*Equus altidens*)	Horse (*Equus ferus*)
	Etruscan rhinoceros (*Stephanorhinus etruscus*)	Narrow-nosed rhinoceros (*Stephanorhinus* cf. *hemitoechus*)

for example by the rich Acheulean site of Rodafnidia on the eastern Mediterranean island of Lesvos, an island that would have been connected to the nearby Anatolian coast by a glacial sea-level drop of *c.* 50 m (Galanidou *et al.* 2013; 2016). Such a drop must have occurred to enable hominin occupations of the 'sea-bed' landscapes of the local eastern Mediterranean (see also Sakellariou and Galanidou 2016) and their spread onto what is today an island.

At the other end of Europe, the long history of Pleistocene faunal finds, and occasional artefacts and hominin fossils, from the southern North Sea (Kolfschoten and Laban 1995; Mol *et al.* 2006; Bynoe *et al.* 2016; Bynoe 2018) highlights the fluctuating nature of the connection between Britain and the near continent. While the existence of a continual terrestrial connection is agreed for the Early Pleistocene and early Middle Pleistocene, the nature of the landscape and the timing of key changes after *c.* 450 kya (MIS 12; the Anglian/Elsterian glaciation) is more contested (Hijma *et al.* 2012; Gupta *et al.* 2017). The key point however is that the glacial/warm stage cycles resulted in post-Anglian periods of connection and isolation, possibly reflected in the character of handaxes and other lithic technology in the British Lower Palaeolithic record (White and Schreve 2000; Ashton and Lewis 2002; White 2015; Ashton *et al.* 2016). From a seasonal perspective, periods of high sea-level stand isolation, particularly associated with interglacial conditions, would present obvious obstacles to long-distance migrations, by both hominins and other fauna. On the longer-term timescale however, the post-Anglian cycles of connection and isolation, combined with the marked glacial/warm stage climatic variations in north-west Europe, led to both a significant cold/warm faunal turnover, and distinctive species changes in Britain (Schreve 2001; Kahlke *et al.* 2011). These were most marked between separate high sea-level warm stages, and reflected both wider evolutionary trends (*e.g.* the replacement of the giant fallow deer [*Dama dama clactoniana*] with *D. dama* between MIS 11 and 9) and the vagaries of recolonisation and environmental conditions (*e.g.* the absence of the brown bear [*Ursus arctos*] in early MIS 7 [*c.* 243–191 kya]; Schreve 2001).

The woods and the trees

A major difficulty when considering seasonality, hominin lives and survival strategies in the Lower Palaeolithic concerns the environmental context: what was their Pleistocene world actually like to live in? Margari *et al.* (2018) have emphasised the importance of understanding local conditions on the ground if we are to answer questions about hominin evolution and dispersal – and such understanding is equally, if not more, critical to explorations of day-to-day living and how it changed across the seasons. As outlined above and in Box A these environments can be reconstructed, from pollen, plant macro-fossils, sediments, macro- and micro-faunal remains, isotopes and geochemistry, and have frequently been presented in fine and sometimes spectacular detail for many of the key sites (*e.g.* Ashton *et al.* 1992; 1998; Singer *et al.*

1993; Conway *et al.* 1996; Roberts and Parfitt 1999; Mania and Mania 2005; Parfitt *et al.* 2005; 2010; Lhomme 2007; García and Arsuaga 2011; Rodríguez *et al.* 2011; Blain *et al.* 2013; Bigga *et al.* 2015; Urban and Bigga 2015). Summer and winter temperatures and rainfall can be reconstructed with some confidence (Table 2.3, see also Chaps 3 & 5), drawing in particular on micro-fauna, although there can be mis-matches in the geographical scales represented by the archaeological and palaeoclimatic/palae-oenvironmental records (Margari *et al.* 2018). The difficulties start to arise if we wish to move from static reconstructions to dynamic worlds: for example, what were the predator–prey relationships and the patterns of animal mobility? How did these vary across the seasons?

Predator–prey relationships have been modelled in a series of key recent papers (*e.g.* Rodríguez *et al.* 2012; Rodríguez-Gómez *et al.* 2016; Rodríguez and Mateos 2018), with particular emphasis on levels of competition and the prey size ranges and preferences of predators. These studies are especially valuable when considering possible changes in hominin strategies (*e.g.* from predominant scavenger to habitual hunter) across the duration of the European Lower Palaeolithic (see also Chap. 4). However, Turner's (1992, 111) warning of a 'spurious air of precision' when trying to estimate biomass is perhaps also worth recalling when reviewing such quantitative modelling approaches.

As much of the site-specific evidence associates hominin occupations with wooded landscapes, what were these European Pleistocene forests and woodlands like? Two key models have been proposed for northern Europe: the wood pasture (a relatively open structure, possibly maintained by larger herbivores, and permitting light-demanding tree species such as *Quercus* [oak]); and the high forest hypothesis (dark and dense, with low animal densities; Bradshaw *et al.* 2003). Bradshaw and Mitchell's (1999) analysis of the later Danish Eemian (MIS 5e [*c.* 123–109 kya]) suggested a limited impact of large forest herbivores, which included forest elephant, on the regional forest structure. They suggested that this might be due to large predators holding the herbivore populations at modest levels or perhaps that the diversity of grazing species had resulted in stable, low populations of individual plant species. Either way, the Danish Eemian analysis supported the argument that the forests of the optimal interglacial were closed, albeit with locally open spaces, while more open forests and woods book-ended the warm stages. A different view of the impacts of large mammals are potentially evident at Aridos 1, Spain, however, where the patchy landscape conditions suggested by the fauna (excluding the birds) seem to be in contradiction with the pollen records: Blain *et al.* (2014) interpreted the locally open habitats as the product of grazing, browsing, trampling and tree-felling by elephants and rhinoceros (see also Wenban-Smith 2013, chap. 22 & table 22.1).

This is noteworthy as the review by Bradshaw *et al.* (2003) of British Pleistocene interglacial fauna between 500–100 kya highlighted an approximately constant species richness, although each interglacial had its own distinctive characteristics: megafauna (elephants, rhinoceros and occasional hippopotami), up to six species of deer, two large bovids, big cats and hyenas, and small mammals and carnivores (Bradshaw *et al.*

2003, table 1). Of key importance is the combination of mixed feeding ungulates (*e.g.* fallow deer) and browsers, which Bradshaw *et al.* saw as indicating that even densely forested episodes included locally open spaces, and that these mosaics of open and closed habitats favoured the high mammal diversity. However, they also highlighted that forests in the past would vary on the basis of geology and specific events: for example, more open structures would be expected on sandy soils of low fertility, while seasonal flooding would probably have been characterised by different patterns to those of the present. Storms would also contribute to temporary openness, as would wildfires. Although the proportion of natural lightning is lower at higher latitudes (*c.* 78% of strikes occur between 30°N and 30°S) and fire frequencies are suggested to be lower in temperate deciduous woodlands in comparison to steppe regions, wildfires are nonetheless well documented in the modern far north (Christian *et al.* 2003; Gowlett 2016; Sorensen 2017). Since Mitchell (2005) has suggested that forest structure would most likely have dictated herbivore carrying capacity, rather than vice-versa, such phases of temporary open-ness might well have attracted greater densities of both prey and predators.

While plant species inevitably varied in other parts of Europe (*e.g.* typical Mediterranean taxa such as *Cupressaceae* [cypress] and *Olea* [olive] were predominant at Feute Neuva-3 and Barranco Leon, Spain; Blain *et al.* 2016), the themes of mosaic habitats and cyclical change are repeated. Throughout the Mediterranean the late Early Pleistocene was characterised by a general temporal pattern of warm, forested conditions, interspersed by short, cooler periods of more open vegetation. On shorter time-scales, the animal fauna at the Orce sites in south-eastern Spain indicated a mosaic blend of open, aquatic and wet woodland habitats (Blain *et al.* 2016). This habitat diversity continued into the Middle Pleistocene. At Atapuerca for example, the fauna in the lower levels (Gran Dolina TD3/4–8) included a blend of species which were suggestive of a local, or at least regional, mosaic of different habitats: for example, spotted hyena (open landscapes), European jaguar (wooded/forested areas), Etruscan rhinoceros (open woodland and/or grassland), and macaque (humid woodland; Rodríguez *et al.* 2011). Similar mosaics are evident at Atapuerca's Sima del Elefante locality (Blain *et al.* 2010) and at a range of Italian sites (Orain *et al.* 2013). In the north similar impressions of mosaic landscapes are suggested in British interglacial sites from MIS 13 and 11 (Boxgrove [Unit 4c], Swanscombe [Lower Gravel & Lower Loam] and Hoxne [AL3 West]; Parfitt 1999a; Schreve 1996; Stuart *et al.* 1993 respectively), where fallow deer and wild boar (temperate deciduous woodland) co-occurred with beaver (river floodplains) and horse and straight-tusked elephant (open grassland).

These apparent mosaics might also reflect time-averaging in the archaeological record (Stewart *et al.* 2003), or misinterpretations of the ecological preferences and tolerances of particular species. The latter is a particular problem, and recent isotopic and dental use-wear analysis (García *et al.* 2009; Julien *et al.* 2012; Pushkina *et al.* 2014; Kuitems *et al.* 2015; Rivals *et al.* 2008; 2015; Rivals and Ziegler 2018) has challenged a number of long-standing assumptions about animal diets, habitats and mobility. Of particular note in this context is the re-evaluation of a wide range of fauna from the Lower Palaeolithic sites of Mauer and Steinheim, which has highlighted both ecological flexibility and some surprising preferences in a wide variety of species (Table 2.6), and suggested their genuine co-existence in interglacial and/or interstadial periods (Pushkina *et al.* 2014).

There are also other indicators of genuine mosaic habitats. Isotopic analysis of fauna from Faunal Unit 6 at Atapuerca suggested the co-existence of trees and open landscapes, with red deer and horse associated with open forest/grassland habitats, while more forested preferences were indicated for the fallow deer in the same unit (García *et al.* 2009; García and Arsuaga 2011). Similar insights can also be drawn from non-mammalian fauna. In the case of the Sima del Elefante at Atapuerca (level TE7), the range of bird species, accumulated in the cave by raptors, highlighted the level of habitat variation at the local scale (Núñez-Lahuerta *et al.* 2016). These mixed habitats included areas near flowing water (*e.g.* common teal and grey wagtail), open, dry environments (*e.g.* crested lark and tawny pipit) and woodlands (*e.g. Corvus* [crows, rooks] and *Turdus* (thrushes) genus).

Table 2.6: *Suggested, stable isotope-based, habitat preferences for key species from Mauer (woodland & shrubland) and Steinheim (mixture of dense forest, woodland, shrubland, grassland and tundra; Pushkina* et al. *2014)*

Species	Habitat preferences
Bovids	Occupied extremely open habitats (more open than early mammoths, horses & *Coelodonta*)
C. antiquitatis (woolly rhinoceros)	More open habitats than *M. primigenius* & *M. primigenius fraasi*
C. elaphus (red deer)	Temperate, forest species, but greater variation in habitat preferences than *D. dama*; shrubland & woodland preferences similar to horses at Steinheim
D. dama (fallow deer)	Temperate, forest habitats (denser than forests associated with early mammoth & horse at Steinheim); sensitive to cold & habitat openness
E. caballus & *E. mosbachensis* (horse)	Ecologically diverse, but evidence of humid habitats with dense vegetation (perhaps alongside rhinoceros)
E. hydruntinus (ass)	Similar to *Coelodonta* (i.e. between shrubland [horse] & grassland [bovid])
M. giganteus (giant deer)	Shrubland to woodland (& wider range of forest habitats than mammoth/forest elephant); avoiding dense forest
P. antiquus (straight-tusked elephant) & *M. primigenius fraasi* (early form of *M. primigenius*)	Open shrubland (but not as open/cold habitats as those of the later, classic woolly mammoth, *M. primigenius*)
S. kirchbergensis (Merck's rhinoceros) & *S. hemitoechus* (narrow-nosed rhinoceros)	More closed habitats of woodland & shrubland than *Coelodonta*; *S. hemitoechus* slightly more flexible than *S. kirchbergensis*

Modern perspectives?

While the continuing methodological developments outlined above are starting to address some of the key ecological issues for the Lower Palaeolithic period, an alternative source of currently available data can perhaps be found in central Europe – in the shape of the Białowieża Primeval Forest (Fig. 2.4). Modern forest perspectives are valuable because they offer insights into key seasonal variations such as changing resource availability and the varying behaviours of both herbivores and carnivores (*e.g.* mobility, diet, and reproduction). Located across the Poland–Belarusian border, the Białowieża Primeval Forest (hereafter BPF), is widely argued to be Europe's best preserved temperate lowland forest (Bobiec 2002). This reflects both its distinctive history as a medieval and post-medieval royal hunting forest, and more recent management and curation: the Belarusian portion of the forest is a State National Park (since 1993) and UNESCO Man and Biosphere Reserve (since 1993); Poland's Białowieża National Park (BNP) is also a Man and Biosphere Reserve (since 1977) and a World Heritage Site (since 1979; Okarma *et al.* 1998).

The forest is 1450 km^2 in overall size, while the Białowieża National Park, in which no hunting, timber exploitation or motor transportation is permitted,[10] is 100 km^2 (Sidorovich *et al.* 1996; Musiani *et al.* 1998). The terrain is low-lying, *c.* 134–186 m

Figure 2.4: The Białowieża Primeval Forest (Source: Juan de Vojníkov [Wikipedia Commons]; details in Fig. acknowledgements).

asl. The forest's climate is essentially continental: between 1994 and 1999 the mean January and July temperatures were -2.9°C and 19.7°C (Jędrzejewski *et al.* 2001), broadly comparable to the estimates for a number of northern European Lower Palaeolithic/ Middle Pleistocene sites, although slightly higher than some of the late MIS 13 sites such as Happisburgh I, England (Table 2.3). In the same BPF study the mean annual precipitation was 611 mm, with an average of 87 days/year of snow cover, with maximum snow depths of 10–63 cm. The forest is characterised by oak–lime–hornbeam stands, with alder woods in wet locations with stagnant water, and alder and ash alongside small forest rivers and brooks (Okarma *et al.* 1995). There is an average tree stand age of 130 years, with regeneration occurring under the canopy of the old stands. In the BNP the key species are oak (20%), hornbeam (19%), spruce (16%), alder (12%), pine (11%), lime and maple (9%), birch and aspen (7%), and ash (6%; Okarma *et al.* 1995). The only open areas are marshes of sedge (*Carex* sp.) and reed (*Phragmites* sp.) which occur in narrow river valleys (0.1–1 km wide).

The forest's fauna spans ungulates (principally red deer, roe deer, wild boar, moose and bison), larger (wolf and lynx; brown bear was driven to extinction in the 19th century) and smaller carnivores (*e.g.* otter, mink, polecat, stoat and weasel), and other mammals (*e.g.* beaver; Table 2.7). The BPF's status as an old-growth forest rather than a strongly humanly-altered habitat, combined with a presence of large carnivores (>15 kg), therefore means that it can offer some potentially valuable insights into animal behaviour (Kamler *et al.* 2008) – mobility, range sizes, seasonality and dietary preferences for example. These are returned to at various points in the chapters that follow. It is critical at this point to emphasise that any quantitative data should be used only as general indicators of abundance, not absolute numbers (*e.g.* as noted by Sidorovich *et al.* (1996) in their discussion of the beaver and mustelid distributions). Moreover, the range of fauna, both herbivore and carnivore, is clearly not a perfect match for that of the European Early or Middle Pleistocene, during which animal species, densities, distributions, and predator–prey relationships would clearly have varied, both over time and with reference to the present day. For example, while the adult wild boar has no natural predator in modern Europe today (Okarma *et al.* 1995), the Siberian

Table 2.7: *Key fauna in the Białowieża Primeval Forest (Sidorovich* et al. *1996; Jędrzejewski* et al. *2001; Selva* et al. *2003).*

Carnivores	Ungulates	Rodents & Lagomorphs
Fox	European bison	Beaver
Lynx	Moose	Hare
Mink	Red deer	
Otter	Roe deer	*Corvids & Raptors*
Pine marten	Wild boar	Common Buzzard
Polecat		Raven
Stoat	*Omnivores*	White-Tailed Eagle
Weasel	Wild boar	
Wolf		

tiger (mean body weight: 180–306 kg [♂] and 100–167 kg [♀]; Mazák 1981) does prey predominantly upon it – as may its Pleistocene equivalents. Based on their modelling approaches Rodríguez *et al.* (2012) concluded that the BPF's food web is significantly less complex than various Pleistocene ecosystems, and the BPF data is therefore used cautiously. All these caveats should be kept in mind in the following chapters.

Cast of characters

One of the difficulties of discussing life in Lower Palaeolithic Europe concerns the varied cast of characters (Table 2.8). The identity of the very earliest occupants of Europe, associated with the archaeology of the Orce Basin, Spain and Pirro Nord, Italy,

Table 2.8: Key European Lower Palaeolithic fossils

Species	Key Fossil Sites	Chronology (MIS)	Remains	Cranial capacity (cc; where known)
Unknown	Barranco León, Spain	43–49	Molar tooth	NA
	Sima del Elefante, Spain	37	Mandible fragment & dentition	NA
H. antecessor	Gran Dolina TD-6, Spain	21	Cranial, dentition & post-cranial	c. 1000
H. heidelbergensis	Mauer, Germany	15	Mandible	NA
	Boxgrove, UK	13	2 incisors, tibia	NA
	Vértesszölös, Hungary	13	Cranial & dental fragments	1350
	Arago, France	12	Cranium, mandibles, dentition & post-cranial fragments	1166
	Sima de los Huesos, Spain[1]	12	All	1092–1360
	Mala Balanica, Serbia	13–11	Mandible	NA
	Aroeira, Portugal	11	Cranium	>1100
	Bilzingsleben, Germany	11	Cranial fragments & dentition	1000 (max.)
	Ceprano, Italy	11	Cranium	1050–1200
	Swanscombe, UK	11	Cranial fragments	1300
	Steinheim, Germany	11	Cranium	1140
	Petralona, Greece	6–8	Cranium	1230

(Hinton *et al.* 1938; Oakley 1952; Thoma 1972; Vlček 1978; Arsuaga *et al.* 1993; Roberts *et al.* 1994; Valoch 1995; Grün 1996; Bermúdez de Castro *et al.* 1997; Stringer and Hublin 1999; Prossinger *et al.* 2003; Falguères *et al.* 2004; Rightmire 2004; Bruner and Manzi 2005; Bridgland *et al.* 2006; Carbonell *et al.* 2008; Harvati *et al.* 2009; Manzi *et al.* 2010; Dennell *et al.* 2011; Roksandic *et al.* 2011; Wagner *et al.* 2011; Toro-Moyano *et al.* 2013; van Asperen 2013; Arsuaga *et al.* 2014; de Lumley 2015; Falguères *et al.* 2015; Daura *et al.* 2017; Demuro *et al.* 2019). [1]Sima de los Huesos cranial capacity data excludes specimens described as immature by Arsuaga *et al.* (2014, table S1). See Figure 2.2 for locations of key sites.

is uncertain, but given the Orce chronology (c. 1.4 mya), *H. erectus* or a *H. erectus*-like hominin seems the most likely candidate (Toro-Moyano *et al.* 2013; Agustí *et al.* 2015). The slightly younger fossils from the Sima del Elefante are currently assigned only to *Homo* sp. (Carbonell *et al.* 2008). Just after 1 mya *H. antecessor* appears on the scene, but at the moment is only known from Gran Dolina (TD-6; Bermúdez de Castro *et al.* 1997; Carbonell *et al.* 2005). It is therefore uncertain whether other early sites, such as Happisburgh III (Parfitt *et al.* 2010), are also associated with this species, although the dimensions of the Happisburgh III footprints do not exclude *H. antecessor* (Ashton *et al.* 2014).

A further change occurred at around 600 kya, with the appearance of *H. heidelbergensis* in Europe. The definition, origins and distribution of this species is much debated, with possible specimens in Europe, Africa and perhaps also Asia, a problematic type-specimen (the Mauer individual is only represented by a mandible), and uncertain evolutionary relationships (Rightmire 1998; Rosas and Bermúdez de Castro 1998; Mounier *et al.* 2009; Moncel 2010; Dennell *et al.* 2011; Stringer 2012; Mosquera *et al.* 2013; Buck and Stringer 2014b; Roksandic *et al.* 2018). The question of whether *H. heidelbergensis* was purely ancestral to Neanderthals or was the last common ancestor of both Neanderthals and ourselves remains unresolved. While the European fossils post-dating c. 600–700 kya are sometimes collectively referred to as *H. heidelbergensis*, a wide range of other species names have also been proposed and used (Roksandic *et al.* 2018, table 1), including *H. cepranensis* (Ceprano, Italy; Bruner and Manzi 2005), *H. erectus* (*e.g.* Vértesszőlős, Hungary; Thoma 1972), *H. erectus heidelbergensis* (Mauer, Germany; Mounier *et al.* 2009), *Homo erectus tautavelensis* (Arago, France; de Lumley 2015), *H. sapiens steinheimensis* (Steinheim, Germany; Prossinger *et al.* 2003), and various early Neanderthal labels, including pre- and proto- (*e.g.* Swanscombe, England and Montmaurin, France; Stringer and Hublin 1999; Vialet *et al.* 2018). Resolution is difficult, in part because only two large European samples exist (Sima de los Huesos and Arago), and because *H. heidelbergensis* was originally defined on the Mauer mandible (Stringer 2012). Dental variations between the Sima de los Huesos and Arago samples raise the interesting possibility that there were broadly contemporary lineages of *H. heidelbergensis* and early Neanderthals in the European later Middle Pleistocene (Martinón-Torres *et al.* 2012). However, since this book is primarily concerned with Lower Palaeolithic behaviour from a seasonal perspective, I have chosen to simplify matters by collectively referring to later Lower Palaeolithic European hominins as *H. heidelbergensis sensu lato*, with an emphasis on their shared characteristics as large-brained and large-bodied Middle Pleistocene hominins. In that sense I have followed the approach of Dennell *et al.* (2011):

> The term "*H. heidelbergensis*" is thus a convenient abbreviation for a longer statement along the lines that whilst most European Middle Pleistocene hominin specimens share some features with *H. erectus*, *H. neanderthalensis* and even some specimens regarded as "archaic *H. sapiens*" – leaving aside for the moment how each is or can be defined – they seem nonetheless to be sufficiently distinct to be placed in a separate category that was ancestral in Europe to Neanderthals. (Dennell *et al.* 2011, 1513)

What we do know is that *H. heidelbergensis* was large-brained, within the lower range of modern human variation (Table 2.8): Robson and Wood (2008) suggested an average cranial capacity of 1204 cc (1130–1278 cc; compared to a modern human value of 1478 cc), while Arsuaga *et al.* (2014) reported a mean of 1232 cc for the Sima de los Huesos sample. It was large bodied, with a tall, wide body plan, and with a blend of more primitive (*e.g.* the large, robust chin-less jaw) and derived features (*e.g.* distinctive, triangular shovel shaped, incisors: Martinón-Torres *et al.* 2012; Buck and Stringer 2014b; Table 2.9). Based on the exceptional sample from the Sima de los Huesos ('Pit of the Bones'¹¹), pooled sex average height has been estimated as 163.6 cm, with an average weight of 69.1 kg. However, there was clearly significant variation: the height and weight of the largest Sima de los Huesos male has been estimated at 168.9–171.2 cm and 90.3–92.5 kg, while the weight of the individual represented by the Boxgrove tibia was calculated to be well over 80 kg (Bonmatí *et al.* 2010; Arsuaga *et al.* 2015; Buck and Stringer 2015). For reference, this makes *H. heidelbergensis* slightly taller, but slightly lighter, than the average Neanderthal (whose pooled sex values are 160.0 cm and 72.1 kg), although Robson and Wood (2008) suggested slightly heavier estimates for *H. heidelbergensis*, averaging 71 kg for the species. Female and male mean stature and weight estimates from the Sima de los Huesos sample are 157.7 cm (♀) and 169.5 cm (♂), and 57.6 kg (♀) and 76.8 kg (♂) (Arsuaga *et al.* 2015, tables S3 & S4). Female and

Table 2.9: Key attributes of H. antecessor and H. heidelbergensis (Stringer 2012; Arsuaga et al. 2014; Bermúdez de Castro et al. 2017).

Species	Cranio-dental traits	Post-cranial traits
H. antecessor	• Endocranial volume at upper end of *H. erectus* range • Derived *H. sapiens* facial features (e.g. projecting nose & presence of canine fossa) • Mix of primitive (e.g. premolar crown & root morphology) and derived (e.g. permanent canines) dental characteristics • Relatively derived mandibles (size & morphology)	• Hand & foot bone morphology closer to *H. sapiens* than Neanderthals • Long bone morphology (e.g. clavicle and humerus) shares features with Neanderthals
H. heidelbergensis	• Large endocranial volume, overlapping both *H. erectus* (lower range) & *H. sapiens/H. neanderthalensis* (higher range) • Derived Neanderthal facial & dental features (e.g. mid-facial projection; "shovel-shaped" incisors) • Primitive cranial vault (e.g. low position of maximum cranial breadth; strongly angled occipital bone)	• Wide *Homo* bauplan • Thick bones and significant musculature • Larger costal skeleton relative to stature (compared with *H. sapiens*) • Broad shoulders • Large, robust pelvis • Powerful precision grip & fine precision grasping capabilities

male differences to their Neanderthal equivalents are comparable to the pooled sex differences with regards to stature (*i.e. c.* +3 cm). However, while male weights are comparable (Neanderthal male: 76.3 kg), *H. heidelbergensis* females are slightly lighter (Neanderthal females: 61.6 kg). Whichever estimates are used however, *H. heidelbergensis'* levels of sexual dimorphism are not unusual compared to modern humans (a ratio of 1.08, comparable to modern levels), in contrast to earlier hominins such as the australopithecines (Robson and Wood 2008; Arsuaga *et al.* 2015).

Life spans are notoriously difficult to estimate, but anterior tooth wear rates suggest that the functionality of the teeth of the Sima de los Huesos populations would stop during an individual's 5th decade, potentially limiting their life expectancies (Bermúdez de Castro *et al.* 2003a). The impacts of these dental trends would be especially marked in the absence, and possibly in spite, of any significant social care in the form of pre-processed food provision. Focusing on the traits of the Sima de los Huesos material is potentially problematic, since the fossils show clear evidence of selected Neanderthal traits which are not universally shared in contemporary fossils from the late Lower Palaeolithic (Stringer 2012). Moreover, there may be a broader west/east division in the general distribution and character of late Lower Palaeolithic fossils, with Neanderthal-like traits typical in western Europe (Roksandic *et al.* 2018). However, in light of the sample's highly valuable insights into life history and post-cranial attributes the Sima data is utilised here, albeit cautiously.

The fossil evidence for *H. antecessor* is more limited, but stature estimates are nonetheless possible (Carretero *et al.* 1999; Gómez-Olivencia *et al.* 2010). Carretero *et al.* (1999) suggested heights between 170.9 and 174.5 cm, depending on the bones used, with a pooled sex mean of 172.5 cm, and upper limb proportions that are more similar to *H. ergaster*/modern humans than to Neanderthals. A cranial capacity of around 1000 cc has been suggested, albeit based on a single sample (Bermúdez de Castro *et al.* 1997; Robson and Wood 2008). The species is a fascinating mix of traits, both primitive and derived (Table 2.9), with the latter features revealing similarities with both *H. sapiens* (*e.g.* the morphology of the mid-face; Arsuaga *et al.* 2001, fig 2) and Neanderthals (*e.g.* markedly shovel-shaped upper incisors). *H. antecessor's* evolutionary position remains much debated. Bermúdez de Castro *et al.* (2017) suggested that it may be a western European side branch of an African/Western Eurasian *Homo* clade that produced the last common ancestor of Neanderthals and *H. sapiens*. *H. antecessor* may therefore represent just one of a wave of populations that arose as part of that clade and dispersed into Europe.

Dietary needs

Discussions of stature and brain size lead into a consideration of the dietary requirements of Lower Palaeolithic *Homo* in Europe, with reference to both diet quality and quantity, as explored in the expensive tissue hypothesis (Aiello and Wheeler 1995). Although many of the discussions of Palaeolithic diets in the late 20th and early 21st centuries have focused on animal foods (*e.g.* isotope studies and hunting/scavenging debates; Lee and DeVore 1968; Blumenschine 1991; Richards *et al.* 2000; Domínguez-Rodrigo 2002),[12] more recent research has shifted away from protein-dominated perspectives (*e.g.* Butterworth *et al.* 2016). These views have arisen from both studies of living/recent hunter-gatherer populations and from Palaeolithic research, in particular the evidence from dental calculus (*e.g.* K. Hardy *et al.* 2016; 2017; 2018), dental wear (high rates have been linked to highly abrasive diets; Bermúdez de Castro *et al.* 2003a), and nutritional modelling (with a key focus on the dangers of excess protein; *e.g.* Speth and Spielmann 1983), although direct evidence for plant food consumption in the Lower Palaeolithic still remains scarce. Exploration of broad food category contributions to hunter-gatherer diets (Cordain *et al.* 2000) provides a useful modern perspective on variations between different environments: percentage proportions of gathered plant: hunted animal: fished animal foods were 16–25%: 26–35%: 46–55% (for a northern coniferous forest) and 36–45%: 16–25%: 36–45% (for a temperate forest). Of particular significance to Europe, Cordain *et al.* (2000) also stressed the marked decrease in plant food consumption amongst modern hunter-gatherers above and below 40° north (this latitudinal line divides the Iberian Peninsula roughly in half) and 40° south (see also Lee 1968). However, in those examples plant foods were fundamentally replaced by fished foods, for which there is minimal Lower Palaeolithic evidence. A different strategy would therefore be required, and a key and ongoing debate in Lower and Middle Palaeolithic studies has concerned whether hunted animal or plant foods filled the 'gap'. Aiello (2007) has also stressed the value of a mixed diet, highlighting that animal meat satisfies nutritional requirements, as it includes essential amino and fatty acids, fat-soluble vitamins and minerals, but is relatively low in bulk. This enables complementary use of high-carbohydrate, but lower overall nutritional quality, plants (*e.g.* tubers or underground storage organs [USOs][13]) – providing the energy for the larger bodies of the later Pleistocene. It seems unlikely that these principals would not apply in Lower Palaeolithic Europe, when hominin bodies were broadly comparable in size to those of Neanderthals, and Bruce Hardy (2010) has explored the availability of USOs in Europe during the Middle Palaeolithic. With general reference to carbohydrates, Karen Hardy *et al.* (2015) have also stressed various nutritional reasons why they may have been critical in human evolution (see also Box B).

Data on Eurasian wild foods with reference to hunter-gatherers remains relatively limited,[14] and there are likely considerable differences between modern domesticated food plants and their ancient wild ancestors (Copeland 2016). Nonetheless Eaton *et al.* (1997) provided a useful comparison of the average nutrient contributions of plant and animal hunter-gatherer foods in general (Table 2.10; with additional data from Hockett and Haws 2003). While the ranges of values associated with individual foodstuffs are inevitably wide these data nonetheless highlight some key differences (note for example the relative values for ascorbate, calcium and sodium), but also broadly comparable energy contributions from plants and

Box B: Mixed diets in the Palaeolithic?

In seeking to understand the specific dietary strategies adopted within Lower Palaeolithic Europe, broad overviews of hunter-gatherer and Palaeolithic diets can still be useful (Eaton *et al.* 1997; Eaton and Konner 1985; Cordain *et al.* 2000). Eaton *et al.* (1997) suggested a generalised Palaeolithic diet with a 37:41:22% ratio of protein, carbohydrate and fat, emphasising the differences between Palaeolithic and modern dietary carbohydrates: fruits and vegetables dominate the former, compared to the widespread 'empty calorie' sugars in the latter. They also noted the likely low levels of saturated fatty acids (reflecting the properties of game meat), terrestrial availability of key polyunsaturated fatty acids (PUFAs), and high levels of protein and fibre in Palaeolithic diets, compared to those of the modern western world.

The paucity of direct plant food evidence, and perhaps also the 'apex predator' characterisation of Neanderthals that emerged in the 1990s and 2000s (*e.g.* Richards *et al.* 2000; Bocherens *et al.* 2005), has resulted in sustained debates and disagreements over the nature of diet in the earlier Palaeolithic, and especially the relative contributions of plant and animal foods (*e.g.* Butterworth *et al.* 2016; Speth 2017; Guil-Guerrero 2018; Hardy 2018). Yet although Arctic peoples are often argued to be examples of successful high protein diets, the detailed data contradicts this view. Their diet consists of *c.* 50% fat, 30–35% protein (and this figure is lower for pregnant women) and 15–20% carbohydrates (mostly as glycogen from meat, if it is frozen soon after slaughter). Moreover, vitamins, minerals and carbohydrates can be acquired from the stomach contents of terrestrial prey (and tundra plants and kelp) – *i.e.* they eat more carbohydrates than generally thought (see also Buck and Stringer 2014a). Finally, fats, especially from marine sources, provide energy for non-glucose-dependent tissues – thus sparing glucose for where it its needed, such as the large human brain (Hardy *et al.* 2015). While the importance of a meat-dominated diet is often highlighted with reference to those essential fatty acids that are key for brain growth and function (*e.g.* docosahexaenoic acid [DHA]), these can also be sourced directly from other dietary elements and/or synthesized from other fatty acids, for example α-linolenic acid [ALA] (Mann 2018). This is found in ocean fish, eggs, seed oils, and various leafy plant foods (Hardy *et al.* 2015). Hardy *et al.* (2015) also emphasised that the energy benefits of meat may, at least on occasions, be offset by the energy demands of pursuing and catching the animal. This is relevant to the segregated, patchy resources of the higher latitudes, particularly in light of the rich ethnographic evidence for the unpredictable returns associated with large animal hunting (*e.g.* Bird 1999; Bliege Bird and Bird 2008). Finally, it is clear from the available environmental evidence (Table 2.3) that the majority of Lower Palaeolithic occupations were characterised by temperate woods rather than Arctic-type conditions, with enhanced plant food availability and potential for a more balanced plant: animal food diet.

Hardy *et al.* (2015) have stressed various specific nutritional reasons why carbohydrates may have been critical in human evolution, with a particular focus on digested glycaemic carbohydrates as our main source of glucose (Copeland 2016). As well as providing energy glucose is key in fetal growth, supports us during periods of hardship, and is the only energy source for sustaining running speeds above 70% of maximal oxygen consumption, which may be significant if endurance played a significant role in pursuing, if not tracking, prey (see also Chap. 6). Finally, 35–40% is the upper limit to the energy that can be gained from proteins, as above this protein toxicity occurs. Thus, a carbohydrate-less diet is unlikely, and it seems likely that the encephalisation of the Middle Pleistocene would have required an increased supply of pre-formed glucose.

The role of dietary fats has been strongly emphasised by Ben-Dor *et al.* (2011), who argued that fats effectively filled the gap left by the 35–40% 'protein ceiling' and a suggested 'carbohydrate ceiling'.[1] Obligatory animal fat requirements of 44–62% were suggested by Ben-Dor *et al.* (2011, table 2) for *H. erectus*,[2] and elephant fat was specifically highlighted as a key Levantine resource, prior to their disappearance *c.* 400 kya (see also Reshef and Barkai 2015; Agam and Barkai 2016). Speth (1991a) similarly highlighted the importance of fats as an efficiently metabolised and concentrated energy source, and source of essential fatty acids, while Cordain *et al.* (2000) also stressed the hunting of larger animals with greater fat reserves as one means of circumventing the 'protein ceiling' (although the size threshold of these larger animals was not clearly defined). The complication from a European Lower Palaeolithic perspective is that evidence for the exploitation of the largest animals (*e.g.* elephant and rhinoceros) is sporadic, although by no means invisible.

One final potential food stuff is worth briefly considering. While there is little clear evidence for the exploitation of aquatic foods in the Lower Palaeolithic world there are still occasional examples, such as the exploitation of aquatic mammals, reptiles and fish at Koobi Fora FwJj20 (Braun *et al.* 2010). Such foods would be potentially rich sources of LCPUFAs (long chain polyunsaturated fatty acids), including the omega 3 fatty acid DHA, which are critical to brain growth. While Koobi Fora is a long way from Europe, it highlights that Lower Palaeolithic hominins *sensu lato* were perfectly capable of recognising, and utilising, aquatic food sources, and is potentially significant given the number of lakeside and riverbank activity sites in the European record. The specific potential of European coastal settings has previously been highlighted by Cohen *et al.* (2012), who noted the value of foods such as shellfish and seaweed (and perhaps also beached marine mammals; Speth 1991a) in a mid-latitude winter.

---

[1] The 'carbohydrate ceiling' was proposed on the basis of various issues including collection/processing times, foraging returns, the lack of specialised dentition in *H. erectus* (*contra* earlier hominins), the apparently late expansion in the salivary amylase gene, and the small hominin gut implied by the expensive tissue hypothesis (Aiello and Wheeler 1995; Aiello 2007; Ben-Dor *et al.* 2011).

[2] These estimates were based on animal fat calories divided by total calories obtained from animal sources only. The figures are reduced to 27–44% when all food sources are included.

Table 2.10: Nutrient values for plant and animal foods. ¹Hockett and Haws (2003): Animal food data based on average muscle values of horse, bison, red deer, rabbit, wild boar and reindeer (comparable data for organs, based on average values for beef liver, brains and kidneys, as follows: protein: 20 g; fat: 10 g; carbohydrates: 4 g; energy: 195 kcal); plant food data based on over 200 edible plant foods found in the Mediterranean region; ²Eaton et al. (1997): total sample sizes as follows: plant foods (n=236), animal foods (n=85).

	Plant foods				Animal foods			Source
	n	*Mean*	*Min.*	*Max.*	*n*	*Mean*	*s.d.*	
Protein (g/100g)	–	8.3	–	–	–	22.0	–	1
Fat (g/100g)	–	2.5	–	–	–	2.8	–	1
Carbohydrates (g/100g)	–	14.8	–	–	–	0	–	1
Energy (kcal/100g)	–	132	–	–	–	119	–	1
Vitamins (mg/100g)								2
Riboflavin	89	0.168	0.001	1.14	26	0.399	0.246	2
Folate	11	0.018	0.0028	0.0618	3	0.00567	0.00170	2
Thiamin	101	0.015	0	0.94	28	0.215	0.197	2
Ascorbate	123	33	0	414	18	4.79	5.43	2
Carotene	51	0.328	0	6.55	–	–	–	2
Vitamin A	59	1.08	0	8.41	6	0.461	0.368	2
Vitamin E	24	1.93	0.007	9.08	–	–	–	2
Minerals (mg/100g)								2
Iron	167	2.90	0.1	31	22	4.15	2.77	2
Zinc	91	1.12	0.1	9.5	11	2.67	0.860	2
Calcium	181	103	1	650	28	22.7	30.9	2
Sodium	139	13.5	0	352	16	59	23.6	2
Potassium	112	448	5.1	1665	16	317	43.3	2
Fiber (g/100g)	132	6.15	0	44.9	–	–	–	2
Energy (kcal/100g)	184	109	4	563	44	126	46.8	2

animals. More specifically, Cordain *et al.* (2000) noted the high energy densities in nuts and seeds (*c.* 12 kJ/g, or *c.* 287 kcal/100 g) relative to other plant foods (albeit based on an Australian data-set), while Butterworth *et al.* (2016) have highlighted the broad variations in energy and nutrient contributions between different plant food categories: *e.g.* foliage (rich in amino-acids), fruits (sugars) and USOs, nuts and seeds (carbohydrates).

However, the European latitudes would probably shift the dietary balance towards a higher percentage of non-plant food input (Cordain *et al.* 2000). The animal food percentage might also have been periodically raised during the colonisations and re-colonisations of 'new' environments (*e.g.* range expansions into northern Europe after a glacial interval). This is because animal foods may well have had a greater importance than usual in such scenarios, as they could enable hominins to focus on a common dietary niche in different conditions (Leonard *et al.* 2010).

In short, it is extremely difficult, albeit on the evidence of modern human physiology and current/recent hunter-gatherers, to envisage a Lower Palaeolithic diet without a *c.* 60% or above contribution of fats and carbohydrates, as a key means of ensuring that protein limits are not exceeded (Hockett and Haws 2003; see also Box E), combined with other sources of essential vitamins and minerals. Thus, even if animal foods did dominate European Lower Palaeolithic diets (Cordain *et al.* 2000), what appears to be critical is what parts of animals were eaten (*e.g.* fats, marrow, organs) – and thus primary carcass access would appear to be a critical part of a European dietary strategy. This appears especially significant with regards to the high levels of carnivore competition inferred for the Early Pleistocene, including the presence of bone-breaking scavengers such as *P. brevirostris* (Turner 1992; Rodriguez-Gomez *et al.* 2017). However, these dietary needs, in particular for carbohydrates, could be partly met by a wide range of other foods with low technology/processing requirements for which we don't, or only rarely, have evidence (Bliege Bird and Bird 2008). These could include birds (and their eggs), small mammals, reptiles, amphibians, invertebrates, fruits, seeds and nuts, fungi, stems and shoots, roots, honey/nectar (another solution of Cordain *et al.* (2000) to the 'protein ceiling'), and tree sap.

There is a further critical behavioural component to the carbohydrate discussion. Hardy *et al.* (2015) argued, in line with Wrangham (2009), that cooking greatly increases the glucose-releasing potential of starchy plant materials once they are in the human gut. Moreover, they suggested that cooking-driven increases in the availability of digestible starches may be linked to genetic changes in salivary amylases (which begin the hydrolysis of starch in the mouth), although these may post-date the separation of modern humans from the Neanderthal/Denisovan lineage and thus not be relevant to Lower Palaeolithic Europeans. Alongside the benefits of increased energy from starchy foods, cooking would also have reduced chewing time, improved the palatability and digestibility of polyphenol-rich plant foods (which spans items as diverse as hazelnuts and wild blueberries), and enhanced reproductive function and infant survival (Hardy *et al.* 2015). Some form of processing, *e.g.* pounding or cooking, can also address the problems associated with toxins and/or tough plant tissues (Butterworth *et al.* 2016). It is not difficult to envisage the resulting benefits for an early European hominin, although Henry *et al.* (2018) offered a more cautious interpretation of the value of cooking, based on an assessment of the energetic benefits of cooked over raw foods (plant and animal) relative to the costs of fuel gathering. The major problem, to be discussed in Chapter 3, is where are those fires and why are there not more burnt bones? Moreover, the evolutionary importance of cooking can be challenged by the potential dietary significance of rotting animal foods (Speth 2017), gastrophagy (Buck *et al.* 2016; Buck and Stringer 2014a), and perhaps also mechanical processing of raw foods (Planer 2018; Box C).

In light of the very partial representation of different types of foods in the archaeological record, the complexities involved in estimating hominin calorific requirements (Leonard and Robertson 1997), the substantial inter-specific variations

Box C: Rotted meats, gastrophagy and pounded foods – alternatives to cooking in the Lower Palaeolithic?

The value of rotting meat and fish has been highlighted by Speth (2017) with reference to its nutritional benefits and, in this context, its provision of pre-digested protein and fat which reduces chewing and digestion costs and which, unlike cooking, can be provided passively (*i.e.* meat can be left to rot in the ground or in water while hominins engage in other activities). It is also an effective method of preservation in humid environments, where the prevention of spoiling by drying can be difficult and fuel-demanding (see also Chap. 3). It is therefore an intriguing option in light of the evidence for significant animal food yields at kill-butchery sites such as Boxgrove, Schöningen and Gran Dolina (Chap. 6). The potential importance of rotting meat, if not necessarily fish, in Lower Palaeolithic diets is perhaps also suggested by its widespread ethnohistoric use at both high and low latitudes: Speth convincingly demonstrates that such foods were not a health hazard, and were not fall-back or marginal foods, but rather highly desirable. While ethnohistoric examples often involve pits, which are very scarce and difficult to detect in the Lower Palaeolithic record, Fisher's pond and peat bog experiments in Michigan (Fisher 1995; see also Speth 2017) highlight the usability of natural storage features that provide a 'shelf-life' of several months (in an environment with hot summers and short, cold winters in those particular experiments).

Gastrophagy, focusing on semi-digested contents from ungulate stomachs, intestines and/or chyme[1] reduces the costs to hominins of processing and digesting – a cost reduction which Wrangham (2009) has primarily linked to the very early adoption of cooking. Mechanical processing, or 'mashing-up', of raw foods would make them easier to chew and digest (as all new parents will know), again mimicking some of the benefits of cooking (Planer 2018) and reducing plant food processing and consumption times. The required processing technology (*e.g.* hammerstones, cobbles) is widespread throughout the period (*e.g.* Barsky *et al.* 2015; Mosquera *et al.* 2016), even if the direct archaeological signature is likely to be near-invisible. All of these methods would bring the added potential benefit of lowering fuel needs, by reducing a reliance on cooking, although fire could be used to meet a number of other needs (Chap. 3). In potential support of a non or partial-cooking hypothesis is the key observation that the leaves and stems, flowers, seeds, fruits (hard and soft mast) and rhizomes of many European plant species can all be consumed raw, and such practices are well documented in recent times (*e.g.* Tardío *et al.* 2005). In short, cooking may not necessarily have been a core requirement for meeting Lower Palaeolithic dietary needs.

If seasonal shortages in plant food availability at higher latitudes are accepted (but see Hardy 2018), the potential nutritional challenges could partly have been met through gastrophagy – reindeer chyme for example provides vitamins C and E, magnesium, calcium and iron, and is a carbohydrate source (Buck *et al.* 2016). Guil-Guerrero (2018) has also highlighted the carbohydrate-richness of juvenile mammal stomachs, due to their curdled milk contents, and various benefits of other specific animal parts are also highlighted in Chapter 3 (Fig. 3.13). In potentially expanding access to a wide range of macro- and micronutrients such food sources might have been especially key for meeting the dietary demands of pregnant and lactating females and weanlings (Box E & Chap. 4), and may have been an important component of a hominin niche in Lower Palaeolithic Europe. It is noteworthy that amongst the Hadza animal guts are neither rare or fall-back foods (Buck *et al.* 2016). As a source it would have been available throughout the year, though it would have required primary rather than secondary access to the carcass. The consumption of rotting meat may also have been a significant dietary component, as lactic acids bacteria (LAB) creates vitamin B12, riboflavin and folate, and may also preserve the lipid-rich brains of mammals, which provide the key long chain polyunsaturated fatty acids DHA and arachidonic acid (AA; Speth 2017). As an added bonus, LAB also inhibits the invasion of the meat by unwanted pathogens (*e.g. C. botulinum*) and prevents fats from becoming rancid by inhibiting the auto-oxidation of lipids. Finally, by avoiding the need for cooking, rotted and fermented meat preserves vitamin C, which is present in various internal parts of mammals. This can be difficult to access at specific times of year in mid-latitude seasonal environments, although the Schöningen evidence indicates a range of potential plant food sources of vitamin C: pine and birch bark, berries of common bearberry, European elder, raspberry, and leaves of *Ranunculus* and *Chenopodium* (Bigga *et al.* 2015).

---

[1] The pulpy acidic fluid which passes from the stomach to the small intestine, consisting of gastric juices and partly digested food.

in the energetic and nutritional benefits of food stuffs (*e.g.* Cordain *et al.* 2000), and the impacts of individuals' age- and sex-based differences (Dennell 1979), this book does not seek to explicitly model diet and nutrition in detail. Rather it considers the range of potential food-stuffs that were both available, as indicated by palaeoenvironmental records, and definitely exploited – drawing on an increasingly wide range of archaeological indicators.

Nonetheless it is possible to explore the calorific requirements of Lower Palaeolithic hominins. Any suggested figures are estimates, not least because there are a number of different approaches to estimating daily requirements (*e.g.* DEE [daily energy expenditure]; Froehle and Churchill 2009; Leonard and Robertson 1997), and a number of difficult to quantify variables, including body weight and physical activity levels (PAL).[15] However both non-climatic (Kleiber's and Schofield's equations) and 'temperate' climate estimates (Froehle's equation) all suggest broadly similar DEE values: c. 2,240–2,490 kcal/day and c. 3,420–3,570 kcal/day for female and male *H. heidelbergensis* respectively[16] (Table 2.11). These values are broadly comparable to those previously published by Froehle and Churchill (2009) for Neanderthals in temperate conditions (♀: 2,297–2,547 & ♂: 3,227–3,527, with the larger span of Neanderthal values reflecting the wider range of available body size estimates). However, the cold condition Neanderthal estimates were noticeably larger (♀: 3,180–3,190 & ♂: 4,469–4,877), a difference mainly driven by the use of higher PAL values by Froehle and Churchill (2009). When these PAL values were applied to *H. heidelbergensis*, based on the Schöningen (II-4) mean annual temperature estimate of 6°C (defined as 'cold' by Froehle and Churchill 2009, table 3), the DEE values increase to 2,990 (♀) and 4,576 (♂). What is perhaps of most interest here are the modest differences between DEE predictions (using Froehle's equation)

Table 2.11: Daily energy expenditure (DEE) estimates for H. heidelbergensis *(after Froehle and Churchill 2009).*

BMR Equation	Sex	Mass (kg)[1]	Location	T_{mean} (°C)	BMR (kcal/day)	PAL[2]	DEE (kcal/day)
Kleiber[3]	F	57.6	n/a	n/a	1464	1.70	2488
	M	76.8			1816	1.93	3505
Schofield[4]	F	57.6			1339	1.70	2276
	M	76.8			1852	1.93	3574
Froehle[5]	F	57.6	Barranco León D	16.8[6]	1318	1.70	2241
	M	76.8			1770	1.93	3416
	F	57.6	Gran Dolina TD-6.2	12.3[7]	1335	1.70	2270
	M	76.8			1795	1.93	3465
	F	57.6	Lake Ohrid	7.0[8]	1355	1.70	2304
	M	76.8			1825	1.93	3522
	F	57.6	Schöningen (level II-4)	6.0[9]	1359	2.20	2990
	M	76.8			1830	2.50	4576

[1]Body mass estimates: Arsuaga *et al.* (2015); [2]Physical activity levels (PAL) based on values for tropical/temperate hominins, with the exception of Schöningen, which uses the 'cold' climate values (Froehle and Churchill 2009, 103); [3]Kleiber BMR equation (M = mass in all the following equations): BMR = $70*(M^{0.75})$; [4]Schofield BMR equations: BMR (Female) = $14.8*M) + 486.4$; BMR (Male) = $(15.1*M) + 691.9$; [5]Froehle equations: BMR (Female) = $(9.2*M) - (3.8*T_{mean}) + 852$; BMR (Male) = $(14.7*M) - (5.6*T_{mean}) + 735$; [6]Mean annual temperature (MAT) estimate for Barranco León D (Guadix-Baza Basin, Spain, c. 1.2–1.5mya) from Blain *et al.* (2016); [7]MAT estimate for Gran Dolina TD-6.2 (Atapuerca, Spain, MIS 21) from Blain *et al.* (2013); [8]MAT estimate for Lake Ohrid (Balkans, MIS 11c) from Kousis *et al.* (2018); MAT estimate for Schöningen level II-4 (Germany, MIS 9) from Urban and Bigga (2015; a higher annual temperature estimate of 8.1°C is also suggested in this paper).

for varying temperate regions of Europe when PAL is assumed to be consistent. While mean annual temperature (MAT) varies by *c.* 10°C between Barranco León D (Early Pleistocene, southern Spain) and Lake Ohrid (late Middle Pleistocene, Balkans), the impacts on DEE remain very modest. It is only when PAL estimates are increased, as in the Schöningen example, that DEE increases markedly. While the model assumptions, and in particular the division between 'temperate' and 'cold', are clearly subject to error, the exercise does raise interesting questions as to what extent Lower Palaeolithic hominins inhabiting cooler (interstadial?) environments and living more active lifestyles may have expended significantly greater energy.

The period of the European Lower Palaeolithic therefore encompasses a number of key changes in hominin anatomy, in particular significant encephalisation (brain size increase) at some point after 1 mya. It is therefore likely that hominin cognition, behaviours and strategies also changed over the course of the Lower Palaeolithic (Box D), as is tentatively suggested by changes in lithic technology, such as the widespread appearance of handaxes and other Acheulean traits in Western Europe after *c.* 600 kya (Ashton 2015; Moncel *et al.* 2015), and changes in the geographical distribution of hominins and their archaeology (Carbonell *et al.* 1996; Dennell and Roebroeks 1996; Hosfield and Cole 2018; Roebroeks and van Kolfschoten 1994). Yet alongside changes in material behaviours, encephalisation also had significant implications for the social lives of hominins.

Life history

Modern human life history stages (childhood, juvenile, adolescence, adulthood) are associated with the unusual collection of traits which characterise humans: prolonged gestation, growth and maturation; extremely short inter-birth intervals; helpless newborns; a short period of breastfeeding/early weaning; extended offspring dependency; an adolescence growth-spurt; delayed reproduction; and the menopause (Bogin and Smith 1996; Schwartz 2012). This collection of traits is markedly different to the great apes, highlighting the questions of when they arose, and whether they emerged piecemeal, or together. Of particular importance to the Lower Palaeolithic occupation of Europe, and especially in light of the region's marked seasonality, may be the emphasis in the human model on early weaning. An early weaning strategy places infants at risk, as they are unskilled at finding appropriate foods of sufficient high quality to fuel their brain growth but which are also suitable for their small, deciduous teeth. They are also essentially defenceless and can be competing with other adults for high quality foods (Aiello and Key 2002; Kennedy 2003). Early weaning therefore has notable dietary strategy implications (Box E), but it also has significant implications for infant care. In short, because early weaning is associated with shorter inter-birth intervals, other forms of childcare are required for the 'weanlings', in order to avoid excessive energy demands on females (Aiello and Key 2002). Potential solutions include alloparenting from grandmothers (*e.g.* Hawkes *et al.* 1998) or older siblings, or in the form of increased male provisioning: the implications of both of these strategies are considered in Chapter 4.

Box D: Instinctive actions, or detached thinking?

While some human behaviours appear instinctive, others seem to involve detached or abstract thought. Compare, for example, fleeing a large, aggressive animal with planning out a day's activities in advance. But when and how in our evolution did the latter ability emerge? Brain sizes in *H. heidelbergensis* (Table 2.8) approach that of *H. sapiens* (*c.* 1350 cc), although figures for *H. antecessor* are somewhat smaller. But brains are about much more than their volume, whether scaled to body size or not: the organisation of the brain is key. While the evidence for hominin brain evolution is frequently controversial, Falk (2012) has suggested that the early stages of the re-organisation of the pre-frontal cortex, linked in humans to various activities including recollection and anticipating the future, were occurring in the australopithecine ancestors of early *Homo*. This would imply the potential for at least a degree of planning by European Lower Palaeolithic hominins. From a material culture perspective, multi-stage activities such as tool-making have often been held up as examples of planning (*e.g.* Haidle 2009), although a predominantly hard-wired genetic basis has recently been proposed for handaxe making (Corbey *et al.* 2016; see also Chap. 5: Box N). The resolution of debates around the mode and tempo of brain and cognitive evolution is not the primary focus of this book (for recent discussions of these issues see, *e.g.*, Neubauer 2014; Coolidge and Wynn 2016; Bruner 2018), but the cognitive abilities of anticipation and planning are certainly important to the seasonal approach adopted here. The changing characteristics and challenges of the seasons would have demanded shifting behaviours (*e.g.* migrations or relocations, contracting and expanding territories, group fission and fusion, dietary changes and food storage). Were all of these behaviours purely hard-wired, the outcomes of natural selection operating on a pool of varying behaviours in different hominin groups?

Some behaviours may indeed have been essentially instinctive, arising from long-term selection pressures, such as seasonally-timed territorial relocations in response to deteriorating or improving conditions, or the building up of internal stores through 'gorging' on foods when they were abundant. However, it is proposed here that European Lower Palaeolithic hominins were also capable of anticipating, and planning for, future seasonal needs, for example cold winter conditions and general food shortages. In short, food storage or the preparation of insulating animal hides were not purely instinctive or innate behaviours. In the following chapters this case is made both on the basis of what was required to meet seasonal challenges, and from the evidence in the Lower Palaeolithic archaeological record that suggests both an ability to undertake multi-stage activities and, albeit rarely, seasonal variations in behaviour.

Box E: Dietary demands of reproduction

The need for a balanced diet becomes especially evident when pregnancy and child rearing is considered. While focused on Neanderthals, Hockett's (2012) analysis modelled calorific requirements for a pregnant female as varying between 3357–5500 calories/day, depending on activity level assumptions, with additional requirements during lactation. To test the impacts of a terrestrial mammal-dominated diet, bison, deer and hare were used as representative species, with fatty cuts and organs and other marrow and fat sources included. The model highlighted fundamental problems in a terrestrial mammal-dominated diet, with potentially toxic levels of protein (55–60%, compared to an RDA [recommended daily allowance] of 10–35%); probably toxic levels of vitamin A, niacin, iron, zinc and selenium; severe under-consumption of carbohydrates and vitamin C; and probable shortage of calcium (which is further exacerbated because very high doses of protein block calcium absorption). In short, diets dominated by terrestrial mammal foods are insufficiently diverse, and too much of their muscle and internal organs in the diet results in the over- and under-consumption of essential nutrients – other types of food are necessary.

At the same time, animal protein is also valuable as a weaning food, as infants need dietary protein sources that consist of essential amino acids for 37% of their weight, in contrast to 15% by weight in the sources for adults (Aiello 2007). Other key dietary needs include folates, especially for pregnant females, and calcium for bone health (Kuhnlein *et al.* 2008). Graves-Brown (1996) has similarly stressed the importance of continuous and reliable nutrition to female reproductive success: to increase the probability of conception and healthy pregnancy, to fuel lactation, and to support the growth and development of children weaned at an early age. With broader regard to individual health, wild greens and fruits also have a greater concentration of secondary compounds, with potential medicinal benefits. Their properties include antimicrobial, anti-inflammatory, antioxidant, anti-atherosclerotic and astringent characteristics, amongst many others (Leonti *et al.* 2006).

Finally, it is increasingly clear that maternal health and early life adversity has significant impacts both on an infant's adult life and multi-generational trends, as explored through the Developmental Origins of Health and Disease (DOHaD) hypothesis (Barker 2012; Gowland 2015). While the scale and resolution of the Pleistocene fossil record make this difficult to explicitly assess in the Lower Palaeolithic, it highlights the sustained impact of short-term (seasonal?) fluctuations in food provision or brief periods of disease.

The change to a fully modern life history therefore brings costs, but the addition of an adolescence stage can also enable valuable socialisation – learning the 'rules' of social life (*e.g.* hierarchies of food access, recognising the moods and emotions of individuals), observing and participating in the sexual and infant-care practices that are critical to success in adulthood (Kennedy 2003), and learning and practising the knowledge and skills required by foraging and other tasks (*e.g.* MacDonald 2007; Milks *et al.* 2019). Thus, a modern human model of life history, while costly in terms of the 'immediate' demands on the labour of others, would likely have produced sub-adults and adults with the skills and knowledge to meet the social, ecological and technological challenges of seasonal Europe, and thus be ultimately beneficial when considered holistically.

It is possible to assess life history and its evolution amongst extinct hominins because of the evidence for strong correlations between a range of life history traits: brain size, body size, age at sexual maturity, age at first birth, gestational length, lifespan and dental development (Kennedy 2003). The reconstruction of life histories is not straightforward however. Robson and Wood (2008) made the important distinction between life history variables (LHVs: gestation length, age at weaning, age at first reproduction, inter-birth interval, mean life span and maximum life span), which cannot, with the possible exception of weaning age, be detected for fossil hominins, and life history-related variables (LHRVs: body mass, brain mass, dental crown and root formation times and dental eruption times), which can. This is significant because Robson and Wood argued that the life history-related variables do not correspond perfectly with life history and that hominin reconstructions, therefore, should be made with some care: for example, using tooth crown and root formation times to estimate age at weaning or age at first birth. Nonetheless, based on dental data Bermúdez de Castro *et al.* (1999; 2003b) argued that *H. antecessor* and *H. heidelbergensis* would have experienced an essentially human life cycle, including an adolescent stage, although childhood and adolescence may have been slightly shorter than amongst modern humans (Dean *et al.* 2001; Hublin *et al.* 2015). In contrast, Robson and Wood (2008) concluded that while the life history pattern of *H. heidelbergensis* might be that of modern humans, *H. antecessor* offers less evidence for a modern human-type pattern, with a brain mass akin to later *H. erectus* and varied dental trends. Overall they suggested that life history changes towards the modern human model are piecemeal, with shifts in body mass appearing earlier in the hominin linage, *e.g.* just after 2 mya, and with dental developments emerging later. The implications of these different life history interpretations are explored further below, particularly with reference to the opportunities for sub-adult learning (Chap. 5).

Hominin communities

A key benefit of encephalisation data concerns its use in the prediction of hominin community sizes: the social brain hypothesis (Dunbar 1998; 2003; 2007; 2009; Dunbar and Shultz 2007; Gamble *et al.* 2014). Estimates for *H. heidelbergensis* community sizes are *c.* 120–130, compared to *c.* 90–110 for *H. antecessor* and late *H. erectus* (figures vary slightly depending on which specimens and cranial capacities are selected; Dunbar 2003, fig. 2; Gamble *et al.* 2014, fig. 3.5). However, these numbers refer to a social unit, the *community*, with all of whom each individual has a personal relationship, not necessarily to day-to-day living groups (Dunbar 2003). Indeed Gamble *et al.* (2014) suggested much smaller estimates for such groups: around 15 for daily foraging groups, and 50 for overnight camping bands (based on the modern human community number of 150). These smaller estimates for day-to-day groups are also suggested by the size of many Palaeolithic sites, particularly from the Middle and Upper Palaeolithic, and also by comparisons with historical and ethnographic hunter-gatherer societies (Gamble and Boismier 1991; Kelly 1995; Gamble *et al.* 2014). Queries have also been raised regarding the community size predictions of the social brain hypothesis, for example because there are certain species that deviate markedly from the neocortex/total brain size scaling trend (Steele 1996). However Steele's method, using female adult body size and total adult brain volume, did produce similar numbers for early *H. erectus* (although not for later *Homo* species).

   Key components of the social brain hypothesis are the notions of theory of mind (ToM) and intentionality. ToM refers to the ability to comprehend one's own mental state and that of others, and to recognise that the mental state(s) of others may differ from your own. Intentionality measures the complexity of that recognition: the ability to project your own theory of mind onto others within your group, to comprehend and/or predict the belief states of others, and to link individuals together into a cognitive chain (McNabb 2007). Most modern humans can operate at fifth-order intentionality (although much day-to-day social interaction probably operates at third-order or below):

> I believe (1st order) that you think [2nd order] that I wonder [3rd order] whether you suppose [4th order] that I believe [5th order] that something is the case. (Dunbar 2007, 100)

Third-order (Dunbar 2003, fig. 4) or 4th-order intentionality (Gamble *et al.* 2014, table 5.2) has been predicted for *H. heidelbergensis* (with 3rd-order intentionality predicted for *H. antecessor*'s cranial capacity of *c.* 1000 cc), and suggests the potential for complex social interactions (*e.g.* 'I *believe* that you *think* that I *wonder* whether you will share your foraged food with me'). McNabb (2007) specifically suggested that *H. heidelbergensis* could have conducted visual displays whose learned social significance could have been interpreted by the remainder of the group. However Cole (2015) explored ToM and intentionality through notions of identity, material culture, visual display and social communication (*i.e.* using artefacts to broadcast your identity and have others 'buy into' that identity), and concluded on the basis of handaxe symmetry levels that those artefacts were not embedded within social communication systems (although other studies have suggested higher levels of symmetry; White and Foulds 2018). Unsurprisingly, McNabb (2007) and Cole (2015) both argued that the mainte-nance of, respectively, encoded social messages (*i.e.* cultural traditions) and/or abstract contexts (*e.g.* ideologies) would have required fifth-order intentionality, and thus been beyond *H. heidelbergensis*.

Language?

A further potential implication, and benefit, of the social brain hypothesis concerns the emergence of language[17] and its role in social bonding. A reliance on primate-style one-to-one grooming as the sole method of social bonding becomes problematic as group size increases: it leaves too little time for other activities such as foraging. Dunbar (2003, fig. 3) therefore suggested, based on the inferred time demands of grooming in larger groups, that language likely appeared, in some form, by around *c.* 0.5 mya. This is firmly within the European Lower Palaeolithic, raising key questions as to whether the changing hominin distributions and/or strategies evident towards the end of the Lower Palaeolithic, such as the widespread dispersal into northern Europe (Dennell and Roebroeks 1996), were supported by, or even dependent upon, language. While direct evidence for language is inevitably limited at this time, it is notable that the hyoid bones[18] from the Sima de los Huesos are human-like in size and morphology (Martínez *et al.* 2008), while ear bones from the same site suggest a hearing bandwidth which covers human speech frequencies (Martínez *et al.* 2004; 2013a). Cervical vertebrae evidence from the Sima also suggest vocal tract proportions comparable to the La Ferrassie 1 (France) Neanderthal (Martínez *et al.* 2013a). Finally, it has also been demonstrated that the Sima de los Huesos population was as right-handed as modern populations – of significance because of the possible associations between handedness, brain lateralisation and specialisation and language (Lozano *et al.* 2009).[19]

Potential similarities and differences between the various Lower Palaeolithic hominin species of Europe, in behavioural, social and biological terms, are therefore discussed in the following chapters (particularly Chap. 7).

Nature of the Lower Palaeolithic record

The archaeology of this period is dominated by lithic artefacts, with occasional glimpses of what was presumably a far more widespread organic component (*e.g.* wooden spears and bone knapping tools; Mania and Mania 2003; Rosell *et al.* 2011; Schoch *et al.* 2015; van Kolfschoten *et al.* 2015b; Zutovski and Barkai 2016). Butchered animal remains have been found sufficiently frequently to support reconstructions of dietary strategies (*e.g.* Parfitt and Roberts 1999; Yravedra *et al.* 2010; Saladié *et al.* 2011; Huguet *et al.* 2013; van Kolfschoten *et al.* 2015a; Lebreton *et al.* 2017), and cut-marks on bone and other traces are being increasingly complemented by use-wear and residue analysis (B. Hardy *et al.* 2018). What is also important, and in many ways highly challenging, to an understanding of seasonal behaviour is the relative dominance of re-worked assemblages, usually although not exclusively in fluvial settings, over primary context archaeology (Roebroeks and van Kolfschoten 1995).

Much re-worked archaeology, in particular, was recovered from sand and gravel contexts during 19th and 20th century commercial quarrying (*e.g.* Cooper and Symonds 2014; Harris *et al.* 2019), and as a consequence of both Pleistocene taphonomy and collecting practices, formal tools often dominate the record, particularly handaxes and other large cutting tools (LCTs[20]). Moreover, such re-worked artefacts have been removed, to a greater or lesser degree, from their original settings and contexts by Pleistocene agents such as flowing water. Nonetheless we do have occasional moments of spectacular, high resolution archaeology, as represented by the open-air site excavations of preserved landsurfaces at Boxgrove (Roberts and Parfitt 1999; Fig. 2.5) and Schöningen (Conard *et al.* 2015; Serangeli *et al.* 2015a). Such sites give insights into hunting/scavenging and butchery tasks (Parfitt and Roberts 1999;

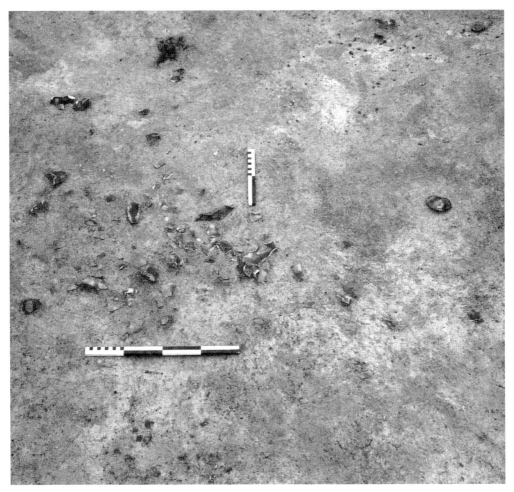

Figure 2.5: Detail of refitting scatter from Q2/A, Unit 4c (palaeosol horizon), Boxgrove (Roberts and Parfitt 1999, fig. 258).

Box F: Where do we find our archaeology?

A longstanding problem in Palaeolithic archaeology, especially in its earlier periods, concerns the very partial preservation and archaeological investigation of its landscapes (Pope *et al.* 2016). This actually reflects two key factors: unevenness in the distribution, and critically the long-term preservation, in so-called 'sediment traps', of Pleistocene deposits across the landscape; and unevenness in the distribution of those sediments which have been exposed, whether through commercial developments (typically aggregates extraction), archaeological excavation, or natural erosion (or a combination of two or even all three of these processes). Consequently we tend to know most about hominin behaviour in river valley environments, where extensive 'stone and bone' archaeology has been found in the commercially-valuable gravels and sands left behind by Pleistocene rivers (*e.g.* Fig. F.1), and a little about hominin behaviour around lakes and in coastal and estuarine environments (in the UK this is primarily thanks to dramatic coastal erosion on the East Anglian coast, and the distinctive raised beach sequences on the West Sussex Coastal Plain), and in caves (Arago and Gran Dolina provide spectacular, albeit rare, insights).

However, we know very little about what hominins were doing away from the inland rivers, for example on the plateaux between river valleys or in the woodlands above the floodplains (Blundell 2020; although excellent recent work is changing this situation in northern France – *e.g.* Hérisson *et al.* 2016). This reflects both the vulnerability of sediments and any archaeology in such locations to subsequent erosion, and the difficulties of predicting where the archaeology might be.

As a consequence, there is a widespread recognition amongst archaeologists that we are only seeing a part of the full range of landscapes which were exploited by hominins, and therefore only a part of their full range of behaviours. A classic example of this partial view concerns the very limited evidence for hominin open-air campsites in the Lower Palaeolithic: where are they? Did they exist at all (at least in a form that we might recognise)? It should not come as a surprise that we rarely find clear traces of them in river valley sediments, such as those of the Pleistocene River Thames. Such sediments were lain down, and subsequently disturbed and modified, by rivers, in conditions which were hardly conducive to the preservation of identifiable campsite traces such as 'domestic' activity areas and perhaps also hearths. Moreover, river banks and floodplains might have also been unattractive settings for campsites, given the risks presented by floodwaters, the terrain itself, and other predators. Similar arguments can be made for the absence of apparent campsite traces across the Boxgrove landscape, where the accumulation of butchered animal remains presumably attracted other dangerous predators. Instead we have rare glimpses of possible campsites in those less

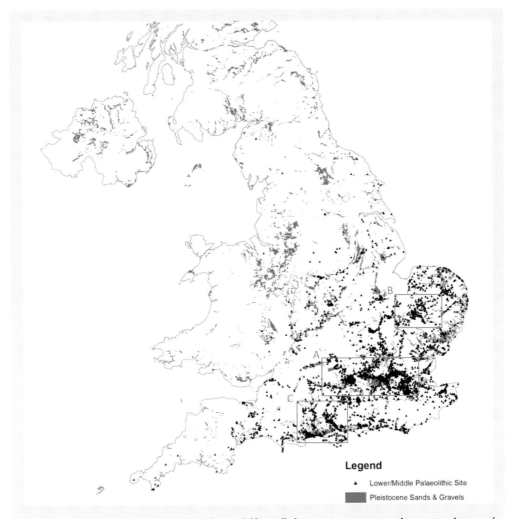

Figure F.1: *Major Pleistocene sand and gravel (fluvial) deposits in Britain and associated Lower/ Middle Palaeolithic sites and findspots. Note site/findspot concentrations along the valleys of the post-MIS 12 River Thames (A), pre- and post-MIS 12 East Anglian rivers (B) and the Solent River (C). Based upon BGS Geology 50k (DiGMapGB-50), with the permission of the British Geological Survey.*

visible landscape locations discussed above: for example, in the closed woodland and spring-source environments at Beeches Pit (Preece *et al.* 2006), or possibly on a lakeshore at Bilzingsleben (Mania and Mania 2005).

Finally, if you're wondering how we know what an early Palaeolithic camp-site should look like ... I am not sure that we do! We'll return to this point in the chapters that follow ...

Huguet *et al.* 2013; van Kolfschoten *et al.* 2015a; Rodriguez-Hidalgo *et al.* 2017), raw material acquisition (del Cueto *et al.* 2016), tool production, use and mobility (Hallos 2005; Pope and Roberts 2005), and, very occasionally, possible domestic spaces (Mania and Mania 2005). Cave sites are rare, although not unknown (*e.g.* Arago and Gran Dolina TD-6.2; de Lumley *et al.* 2004; Saladié *et al.* 2014), although some artefact assemblages found within caves have been re-worked from occupations and/ or activities that were originally located outside of the cave mouths (*e.g.* Sima del Elefante, level TE9c; Huguet *et al.* 2017).

However, the record is nonetheless biased in terms of the types of landscape settings which are represented (Box F), and while this is slowly being addressed (Pope *et al.* 2016) it remains an ongoing problem. Moreover, we have an increasing sense of how the dynamic and complex behaviour of the Lower Palaeolithic can leave a very modest lithic signature behind. At Schöningen 13 II-4 for example, the lithic assemblage amounts to just *c.* 1500 artefacts, of which nearly 90% consist of small flakes, chips and retouching debris (Serangeli and Conard 2015). Yet this is also a locality with the remains of over three dozen horses, most probably accumulating across a series of separate hominin hunting events (Voormolen 2008; Starkovich and Conard 2015), with evidence for butchery, hide-working, wood-working and tool resharpening (Rots *et al.* 2015; Serangeli and Conard 2015). At Schöningen it has been possible to reconstruct the richness of these activities through the quality of the preservation, but the site should also serve to remind us of the fragile and often archaeologically invisible nature of early human behaviour.

Nonetheless, the complementary combination of a small number of primary context sites and an abundance of re-worked artefact assemblages offers a valuable perspective on life in the Lower Palaeolithic: both of the day-to-day behavioural strategies that were used by hominins to adapt to the challenges of Europe, and the long-term evolutionary consequences that are reflected in the geographical and chronological distributions of artefacts (Gamble 1996a).

The earliest occupations of Europe

The earliest occupations of Europe are found in the south, with the oldest sites dating from *c.* 1.3–1.6 mya: Barranco León and Fuente Nueva-3 in Spain's Orce Basin, Pirro Nord in Italy, and possibly Lézignan Le-Cèbe in southern France (Gibert *et al.* 1998; Arzarello *et al.* 2007; Arzarello *et al.* 2012; Agustí *et al.* 2015; Bourguignon *et al.* 2016). Prior to *c.* 700 kya, the majority of early sites remain to the south of the Alps and the Pyrenees (*e.g.* Sima del Elefante, the earlier layers at Gran Dolina, Vallparadís, and Barranc de la Boella, Spain; Monte Poggiolo, Italy: Carbonell *et al.* 1999; 2008; Martínez *et al.* 2010; Arzarello and Peretto 2010; Vallverdú *et al.* 2014), with only occasional earlier sites further north (*e.g.* Lunery-Rosières, Pont-de-Lavaud, and Le Vallonnet, France; Happisburgh III, UK; and Kärlich A & B, Germany: Despriée *et al.* 2010; Haidle and Pawlik 2010; Parfitt *et al.* 2010; Despriée *et al.* 2011; Ashton *et al.* 2014). After *c.* 700 kya the occupation evidence becomes richer across western Europe in both the south and the north, although with fewer sites to the east of the Rhine. There is a mixture of iconic Palaeolithic sites (*e.g.* Boxgrove, Hoxne, Swanscombe, Mauer, Schöningen, Bilzingsleben, St Acheul, Cagny, Abbeville, Arago, Sima de los Huesos, Torralba, Ambrona, Isernia la Pineta, Notarchirico: Roebroeks and van Kolfschoten 1995; Gamble 1999, fig. 4.1), some but not all in primary context, and a rich array of 'lesser' artefact assemblages, typically associated with rivers across western Europe (*e.g.* the Thames, Bytham, Solent, Somme, Rhine, Guadalquivir, Guadiana, Tagus, Duero, Tiber and Aniene: Santonja and Villa 1990; 2006; Tuffreau and Antoine 1995; Wymer 1999; Bahain *et al.* 2007; Haidle and Pawlik 2010; Moncel *et al.* 2015). The record in eastern Europe is more modest in both the Early and Middle Pleistocene. Although a small number of significant sites are known (*e.g.* Rusko, Trzebnica, Račiněves, Stránská skála I, Vértesszőlős, Korolevo, Kozarnika, Dealul Guran, Marathousa, Rodafnidia, and Petralona: Gamble 1999, fig. 4.1; Burdukiewicz 2003; Fridrich and Sýkorová 2003; Koulakovska *et al.* 2010; Sirakov *et al.* 2010; Iovita *et al.* 2012; Valoch 2013; Tourloukis and Harvati 2018), other claimed sites have been disputed (*e.g.* Doronichev 2008). These chronological and geographical patterns in site and assemblage distribution have been dubbed the modified short chronology (Dennell and Roebroeks 1996) or the punctuated long chronology (Hosfield and Cole 2018), and a long-standing question in Palaeolithic studies has concerned the explanation(s) for this changing European pattern. Factors such as increased encephalisation, environmental shifts, changing life history, and/or innovative behaviours in pyrotechnology, cultural insulation, subsistence strategies and lithic/organic technology may all play a role (Roebroeks 2001; 2006; Gowlett 2006; Ashton 2015; Hosfield 2016; Hosfield and Cole 2018; MacDonald 2018; Moncel *et al.* 2018), and all are considered, from a seasonal perspective, in the chapters that follow. The distribution patterns of Europe's Lower Palaeolithic sites and assemblages are therefore reviewed in greater depth below.

The Early Pleistocene

Southern Europe provides the vast majority of the evidence for an Early Pleistocene presence in Europe, from the Sima del Elefante, Gran Dolina and the Orce sites, to Monte Poggiolo and Pirro Nord (*e.g.* Falguères *et al.* 1999; Peretto 2006; Carbonell *et al.* 2008; Oms *et al.* 2011; Arzarello *et al.* 2015). The technological signal is consistent, with core and flake assemblages at all of these sites, but palaeoenvironmental records suggest that the earliest southern European occupations were not comfortable, or at least were only comfortable for a fraction of the time. Climate estimates from hominin and non-hominin sites suggest that the Early Pleistocene occupations were associated with, and constrained by, mild conditions (*i.e.* warm and wet; Agustí *et al.* 2009), while the high levels of carnivore competition implied by both site-specific data (*e.g.* Espigares *et al.* 2013) and modelling approaches (*e.g.* Rodríguez-Gómez *et al.* 2016) suggest significant dietary, and safety, challenges to Early Pleistocene *Homo*. These look especially significant given that, while evidence for dietary strategies at these very early sites is extremely limited, it is certainly not possible to confidently discuss hunting prior to *c.*

1 mya. The modest scale of the archaeological record in this earliest period underpins Dennell's (2003) suggestions that the 'long chronology' sites from the Mediterranean zone during the Early Pleistocene may reflect temporary dispersals rather than a sustained occupation of Europe (see also MacDonald *et al.* 2012).

However, there is also evidence of gradual change in the distribution, and perhaps scale, of Early Pleistocene occupations in Europe, particularly in the Galerian (*c.* 1.2–0.8 mya) at sites such as Gran Dolina (TD-3/4, and, especially, TD-6), Sima del Elefante, Vallparadís, Barranc de la Boella and Monte Poggiolo (Mosquera *et al.* 2013). A factor behind these changes may be found in broader patterns of mammal paleobiogeography (Rodríguez *et al.* 2013). Comparisons of fauna to the south and north of *c.* 45°N suggests that differences in the species' pools, and ecological structures, were at their peak in the *c.* Late Villafranchian (*c.* 1.6–1.2 mya) and much stronger than those of the present day. These differences then declined in the Galerian period however, with an increasing number of species present on both sides of *c.* 44–45°N (Fig. 2.6). This chimes broadly with the early northward expansions of *Homo* (*e.g.* at sites such as Happisburgh III,[21] Pont de Lavaud and Lunery-Rosières). It might also suggest a set of Early Pleistocene *Homo* adaptations that were relatively limited in scope and could not, at least prior to *c.* 1.2 mya, be extended beyond those Mediterranean tree savannah habitats that may have facilitated their initial range expansion into southern Europe. This appears to change after *c.* 1.2 mya, perhaps reflecting a combination of changing hominin behaviours and increasing ecological similarity between southern and northern Europe (although interestingly the results of Rodríguez's study contrast with the habitat reconstructions of Kahlke *et al.* 2011; see also Table 2.4). Changes in hominin taxa may also be a factor, with the first confirmed fossils of *H. antecessor* dating to the Galerian, although the current limitations of Europe's Early Pleistocene hominin fossil record do not permit this issue to be resolved.

The Early–early Middle Pleistocene transition

The question of whether or not European occupation was continuous in the south towards the end of the Early Pleistocene and in the early Middle Pleistocene also remains uncertain (MacDonald *et al.* 2012). Mosquera *et al.* (2013) have suggested a possible break in occupation in the early Middle Pleistocene between *c.* 800 and 600 kya, based both on the specifics of the Atapuerca sequence and the relative paucity of assemblages across Europe during this period (but *cf.* the evidence from Vallparadís and La Noira; Martínez *et al.* 2013b; Moncel *et al.* 2016). If correct such an abandonment might reflect specific ecological characteristics: Rodríguez-Gómez *et al.* (2014) have argued that the early Middle Pleistocene, based on the TD-8 portion of the Gran Dolina sequence at Atapuerca, was a period of higher intra-guild carnivore competition for resources and therefore more difficult for *Homo* given the seasonal pressures on food supplies. While this is possible, the study was based on a single site, and the absence of *Homo* in the TD-8 deposit may reflect very practical, local, concerns such as the dark, small entrance to the cave at this time and the extensive presence of hyena. A more recent, Europe-wide,

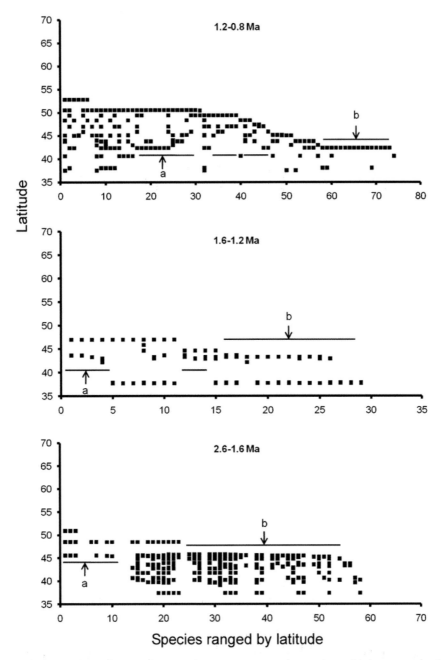

Figure 2.6: Latitudinal distribution of mammal species in Europe during the Early Pleistocene (Rodríguez et al. 2013, fig. 3). The lines mark the southern limit of those species restricted to the north (a) and the maximum northern range of those species restricted to the south (b). Note the increase in the number of species able to live on either side of the 45°N parallel during the 1.2–0.8 mya time interval (top panel).

study has challenged these results and suggested that, relative to the Early Pleistocene, competition was lower throughout the Middle Pleistocene (Rodriguez-Gomez *et al.* 2017).

There is clear evidence for a significant shift in both technology and hominins either side of the proposed 'gap' (Mosquera *et al.* 2013, table 3), and Mosquera *et al.* proposed a two-stage model for the post-800 kya period: a small-scale, ultimately failing, hominin dispersal, associated with the earliest Acheulean, between 800–500 kya; followed by a major *H. heidelbergensis* dispersal at around 500 kya. Yet the change in occupation intensity, or at least the duration of the gap, was perhaps over-stated. The period between *c.* 700–600 kya includes Notarchirico and Isernia la Pineta, Italy, La Noira and other sites in the Cher river valley, France, and Pakefield, England, while the period from *c.* 600–500 kya includes the earliest levels at Arago (P & Q) and rich artefact assemblages associated with the Bytham River in eastern England (Despriée *et al.* 2010; 2011; Parfitt *et al.* 2005; Pereira *et al.* 2015; Peretto *et al.* 2015; Falguères *et al.* 2015; Moncel *et al.* 2016; Davis *et al.* 2017). While these are not rich records in comparison with post-500 kya Europe, neither do they appear to be substantially more modest than the pre-*c.* 800 kya record. The two-stage model of Mosquera *et al.* (2013) therefore seems problematic. Instead I think it is more likely that the *H. heidelbergensis* dispersal occurred from *c.* 700 kya onwards, reflected in sites such as La Noira, with the 'lag-time' prior to its widespread distribution after 500–400 kya reflecting a period of wider acclimatisation to Europe (an acclimatisation which *H. antecessor* appears never to have fully managed: Messager *et al.* 2011; Agustí *et al.* 2015). Overall the sites and assemblages in western Europe from this initial post-1 mya period, while small, are perhaps suggestive of sustained occupations, albeit by changing hominin species and with a fluctuating geographical distribution. During cold stages the maintenance of this sustained presence would have been dependent on southern sites such as Notarchirico, which would have been the likeliest locales in which occupation could have been sustained while the north was periodically 'abandoned'.

The Middle Pleistocene

Whether the European early Middle Pleistocene, in particular, was or was not a period of possible 'abandonment' is of particular interest because the European record rapidly looks rather different soon after, with the first appearances of *H. heidelbergensis sensu lato* fossils at *c.* 500–600 kya (Table 2.8) and the first widespread appearance of Acheulean technologies in western Europe during the same period. As Mosquera *et al.* (2013) have argued, an occupation break strongly supports the notion of new hominins and new technologies entering Europe, most probably from the Near East, although their appearance would not necessarily require the prior disappearance of *H. antecessor* and is not followed by a simple or complete replacement of core and flake technologies (Villa 2001). This change in the occupation record may reflect the significant climatic, landscape and ecosystem changes associated with the Early–Middle Pleistocene Transition (*e.g.* Muttoni *et al.* 2018), combined with the adaptations, behaviours and environmental tolerances of *H. heidelbergensis* (MacDonald *et al.* 2012). An occupation break and new dispersal into Europe is also supported by the apparent paucity of sites tracking the emergence of

Acheulean technologies from core and flake traditions *within* Europe (Mosquera *et al.* 2016), with only very modest or controversial examples of Acheulean large cutting tools (LCTs) currently known from the Early Pleistocene (Jiménez-Arenas *et al.* 2011; Vallverdú *et al.* 2014; Mosquera *et al.* 2016). It is notable that this lack of continuity remains true whether the Acheulean is defined and detected on the basis of the presence/absence of LCTs or on the basis of a wider variety of technological traits (*e.g.* well-structured knapping, standardized cores, and diverse retouched flakes; Mosquera *et al.* 2016).

Overall the European Middle Pleistocene after *c.* 600 kya is characterised by a much more substantial hominin presence, albeit one still marked by cyclical northern extirpations, as has long been recognised (*e.g.* Roebroeks and van Kolfschoten 1994; Dennell and Roebroeks 1996; Hosfield and Cole 2018). What changes in Europe at this time? This has previously been explored from both palaeoclimatic (*e.g.* Candy *et al.* 2015; Candy and Alonso-Garcia 2018) and behavioural perspectives (*e.g.* Ashton 2015; Moncel *et al.* 2015). In behavioural terms emphasis has been placed on the timings of the first appearances of fire use evidence (Roebroeks and Villa 2011) and clear-cut hunting (Villa and Lenoir 2009), and changes in planning ability and landscape use (Moncel *et al.* 2015).

However, the sustained relatively 'mild' conditions in northwest Europe across MIS 15–13 (*c.* 620–480 kya; Candy and Alonso-Garcia 2018) may also have been significant. MIS 14 is the warmest of any cold stage of the last 1 myr, with the harshest conditions only lasting 9 kyr. While it is unclear whether conditions were generally more interglacial or interstadial in character, the latter would still favour a hominin dispersal, as many of the British sites from this period (*e.g.* Boxgrove, Happisburgh I, High Lodge) are associated with late warm stage or interstadial conditions (Table 2.3). A further key factor in the expanded hominin distributions may have been the north–south climate gradients of MIS 13. Candy *et al.* (2015) suggested a relatively modest difference of *c.* 4°C between Britain and Spain (based on sea surface temperatures from North Atlantic records), *c.* 2–3°C less than in later interglacials. These particular conditions may have enabled a detectable hominin range expansion,[22] in both spatial distribution and numbers, at this time.

The exact nature of these range expansions is inevitably a matter of some conjecture. Whether there was or was not overall population continuity in Europe from the Early to Middle Pleistocene (and beyond; Mosquera *et al.* 2013; Ollé *et al.* 2016), it seems logical that 'pre-dispersing' populations were primarily adapted to the particular conditions of the Mediterranean core areas, and perhaps also western Asia.[23] It thus seems likely that European range expansions from this southern 'core' would most easily have occurred through non-directional, territorial drift mapped onto familiar habitats, in response, at least in part, to local 'ecological events' such as shifting resource availability (*e.g.* Roebroeks 2006) or increased predatory threats. The extent and distribution of those 'familiar' habitats would fluctuate over time (*e.g.* reflecting warm stage vegetation successions), and range expansion into the north would probably have occurred incrementally, rather than through any large-scale dispersals into the unknown. In this model, the ability of the hominins to continue to meet those familiar challenges would seem to be the logical constraint on their range expansion. Comparably, range contraction might have involved an accumulation of relocations 'back' into core zones, in response to gradual climatic deterioration. The complication of the latter model however is that other hominin bands or communities may well have already been present in those core zones: hence the preference of Dennell *et al.* (2011) for extirpation in light of climatic deterioration.

But does this model of incremental range expansion fit with the evidence for a substantial expansion in the scale of the Lower Palaeolithic record in northern Europe, and elsewhere, after *c.* 600 kya? Population pressure in the southern core areas does not appear to be a factor (Mosquera *et al.* 2013). The 'mild' conditions of MIS 15–13 and the suggested low south–north climate gradient of MIS 13 (Candy *et al.* 2015; Candy and Alonso-Garcia 2018) may instead have been a critical 'trigger' factor, facilitating initial range expansion, subsequently reinforced by progressive adaptations to new conditions. This may well have operated in tandem with the various suite of behaviours associated with the early Acheulean phase (Moncel *et al.* 2015).

It is notable however that the post-600 kya Lower Palaeolithic occupation of the north was still broadly dominated by warm stage-only occupations. In northern France, for example, the primary context sites are mainly associated with optimal interglacial (*e.g.* La Celle), final warm stage (*e.g.* Soucy) or early warm stage conditions (*e.g.* Cagny-la-Garenne: Antoine *et al.* 2010; Limondin-Lozouet *et al.* 2015). Indeed, the apparent expansion in the scale of hominin activity in Europe, and especially northern Europe, immediately after MIS 12 may in part reflect the favourable conditions of MIS 11c (*c.* 425–398 kya, although its duration may be shorter in central and eastern Europe; Candy *et al.* 2014; Kousis *et al.* 2018). These conditions are especially well illustrated at La Celle in the Seine valley, where the optimal interglacial habitats were characterised by high temperatures, rainfall and Mediterranean plants (*e.g.* figs, box and hackberry) and mammals such as hippo and macaque (Antoine *et al.* 2010; Dabkowski *et al.* 2012). While derived handaxes are associated with cold climate deposits, these associations are uncertain and the most likely and parsimonious interpretation is that the north was effectively 'abandoned' (probably principally through extirpation, with perhaps some range contraction) during each glacial stage,[24] although occupations may have occurred during cooler transitional periods, as suggested at the Rue du Ménage site in the Somme valley, possibly also at the Carriere Carpentier (Antoine *et al.* 2015), and at various central European sites (Szymanek and Julien 2018). This suggests that any significant behavioural changes in the early Middle Pleistocene, such as fire control or hunting proficiency, only extended hominin tolerances so far.

The challenges of Europe during the Middle Pleistocene, exacerbated by the impacts of abrupt climate changes within both interglacials and warm stages (*e.g.* the climatic and habitat impacts of the OHO and the YHO during MIS 11; Candy *et al.* 2014), are therefore likely to have resulted in local group 'abandonments', and thus repeated expansions and contractions in hominin distributions and regional population sizes. Such 'abandonments' most likely occurred in the form of local extirpations in northern areas: Dennell *et al.* (2011) convincingly argued that the southern refugia were 'lifeboats' (in which local groups survived through glacial and warm stage periods) rather than 'arks' (to which northern groups retreated), and a similar argument has been made for the later Middle Palaeolithic (Hublin and Roebroeks 2009). Thus, regular northern extirpations, driven by stage and sub-stage climatic variations, were an integral component of European life in the Middle Pleistocene. The frequency of such extirpations might be further exacerbated by small group sizes (*e.g.* the Neanderthal-focused genetic studies of Lalueza-Fox *et al.* 2011; Prüfer *et al.* 2017): under such demographic conditions local communities would be particularly vulnerable to the impacts of individual deaths. This discontinuous nature of northern occupation has long been recognised: Stringer (2006) estimated that Britain was abandoned for roughly 80% of the period *c.* 500–12 kya. After allowing for absence throughout the duration of the cold stage intervals during this period (MIS 12, 10, 8, 6 & 4–2 span *c.* 273 kyr or 55%; stage boundaries as per Lisiecki and Raymo 2005), and acknowledging the impacts of a fluctuating North Sea/English Channel palaeogeography, this estimate also implies significant periods of absence within the warm stages. Highly dynamic and fragmented populations, both within as well as between warm stages, was thus a defining characteristic of the later Lower Palaeolithic and would result in differing levels of connectivity and isolation over time. This is in-keeping with the increasing recognition that Palaeolithic demography was characterised by regional and local variations, rather than a global pattern of slow growth and small populations (French 2016). Estimating population sizes is extremely difficult (see also Chap. 5: Box O) but it is likely that there were never more than a few thousand hominins in Europe at any one time.

Population dynamics may be reflected in the material culture signatures of Lower Palaeolithic Europe, for example the blend of handaxe and non-handaxe industries in Europe, especially *c.* 700–500 kya (Moncel 2010), or the repeated shifts between core and flake and Acheulean assemblages in the Italian peninsula around 600–700 kya (Peretto 2006), although other prosaic factors (*e.g.* raw materials, site function) may also have played roles. Population dynamics and fluctuations, in both space and time, may also explain the notable hominin variability during the later Middle Pleistocene (*e.g.* as evident in the fossils from Swanscombe, Arago, Ceprano, Mauer, and the Sima de los Huesos; Table 2.8), and the ongoing debates about the definition and membership of *H. heidelbergensis* in Europe (*e.g.* Rightmire 1998; Stringer 2012). While that documented hominin variability might be a consequence of multiple dispersals into Europe, it may also be a product of post-dispersal local divergences due to isolated populations, combined with repeated extirpations, and periodic re-blending in southern refugia. Dennell *et al.* (2011) in particular have stressed the cyclical inhospitality of the north, arguing that the history of Lower Palaeolithic Europe is one of repeated extirpations, followed by renewed occupation of the higher latitudes when conditions improved, stemming from southern refugia in Iberia, the Italian peninsula and the Balkans (and perhaps also areas in southern France and Germany). A repeated re-mixing of the biological and behavioural 'stock' in different regions of Europe should therefore be expected, as stochastic processes and 'historical' events led to groups going locally extinct in different places and at different times, different rates of genetic drift and change in isolated

northern populations and southern refugia (Skinner *et al.* 2016), and the specific character of range expansion (*i.e.* 'recolonisation') varying with each cyclical improvement
in conditions. There is increasing evidence for this in the hominin record. At the end
of the Lower Palaeolithic the mosaic appearance of Neanderthal traits has become ever
more apparent through the Sima de los Huesos fossils, whose morphological attributes
(Table 2.9) and genetic characteristics (Meyer *et al.* 2016) suggest that they are either
early Neanderthals or their close ancestors, at *c.* 430 kya. These Neanderthal affinities
are in notable contrast to broadly contemporary fossils (*e.g.* Arago and Ceprano), which
are characterised by different traits (*e.g.* with regards to dental morphology in the case
of Mauer and Arago; Martinón-Torres *et al.* 2012; Stringer 2012).

It is likely that such complexities may also have occurred during the Early Pleistocene
– but the limited fossil and archaeological record currently makes it impossible to detect
the impacts on hominin and material culture variability at that time. Speculatively, the
current restriction of *H. antecessor* to Atapuerca might be suggestive of similar local
variability in the early portion of the European Lower Palaeolithic.

Periods of abandonment?

To what extent may all of Europe have sometimes been empty of hominins during the
Middle Pleistocene? Dennell *et al.* (2011) suggested that all of Europe may have been
occasionally abandoned, with western Asia (the central area of dispersals of Eurasia
or CADE; Dennell *et al.* 2010) acting as a periodic source area. This is certainly possible,
given the severity of selected cold stages (*e.g.* MIS 12), and may in part explain the
morphological variability of Middle Pleistocene hominins. However, the potential for a
stable and persistent southern European population throughout the Lower Palaeolithic
(or at least from the late Early Pleistocene onwards) is perhaps supported by the
region's environmental characteristics. The glacials were certainly cool and dry, for
example at Notarchirico where the mammals and vegetation indicate a cold and open
climate in MIS 16 (Pereira *et al.* 2015), typical of the steppe-like conditions associated
with Italian Middle Pleistocene glacial stages (Combourieu-Nebout *et al.* 2015), and
hominins would have needed to adapt to more open conditions. However Moncel *et
al.* (2018) argued that the Mediterranean should be considered as a 'warm spot' even
during the glacials, a view supported by the evidence for cold stage occupations at
Notarchirico (Pereira *et al.* 2015) and Guado San Nicola (Orain *et al.* 2013), the diverse
range of habitats associated with Italian Middle Pleistocene sites (Orain *et al.* 2013), the
relatively modest temperature differences between 'glacial' and 'interglacial' stages at
Atapuerca (Blain *et al.* 2009; but *cf.* the MIS 12/11 pollen contrasts reported by Sánchez
Goñi *et al.* 2016), and the potential role of the Balkans as a tree refugia in MIS 12 (Kousis
et al. 2018). It is also notable that regional records, such as the Mediterranean planktonic
curve from Ocean Drilling Program (ODP) Site 975 (Lourens 2004), suggest that even
MIS 12 and MIS 16 may have had reduced impacts in the Mediterranean (although
the post-MIS 12 glacials were more comparable to the global marine and ice records).
Moreover, this was a region which contained glacial refugia with permanent moisture
(enabling the long-term survival of a Mediterranean vegetation which is constrained

by aridity not temperature), habitats to which hominins were presumably drawn at those times. In combination this evidence from the late Early Pleistocene onwards suggests that if they *did* occur, such Europe-wide 'abandonments' were the exception rather than the rule (as also argued by Dennell *et al.* 2011).

Geographical patterns

The other notable European Lower Palaeolithic pattern concerns the relatively limited evidence to the east of the Rhine, particularly prior to *c.* 500 kya. While the pattern is almost certainly partly taphonomic (Iovita *et al.* 2012; Romanowska 2012; Szymanek and Julien 2018), subtle climatic contrasts between the continental interior and the Atlantic West may also be a significant factor, with evidence for slightly colder, sub-freezing, winters and warmer summers in present-day north-central Europe (see also Chap. 1). There may well have been small but significant differences between the Pleistocene conditions and habitats of the interior and those inferred by Kahlke *et al.* for north-western Europe between 0.9–0.4 mya, during which period much of the latter region's Lower Palaeolithic archaeology accumulated:

> temperate climate, high precipitation and low seasonality, typical of oceanic mid-latitude Europe, supporting a diverse ecosystem dominated by forest but with productive open areas as well. (Kahlke *et al.* 2011, 1383)

However, when central European occupations occur, the sites show a preference for the early and late portions of warm stages (not the thermal maximums), associated with open-forest or forest-steppe environments. Moreover, there is occasional evidence for occupations in cool/cold and open conditions in these regions, such as at Korolevo VI and Kärlich H (Szymanek and Julien 2018, table 3), highlighting the potential range of these hominins' ecological tolerances and strategies.

Outstanding questions ... and a seasonal approach

The European Lower Palaeolithic record is therefore characterised by marked changes over time in the distribution, scale and permanence of hominin occupations. These changes are well documented and have been considered elsewhere with reference to climate changes, technological transformations, foraging strategies, and new hominin species. Yet they have rarely been considered from a seasonal perspective. This is a strange omission when investigating temperate Europe, with its annual fluctuations in temperature, precipitation, vegetation cover, day lengths and food supplies. To fully understand those large-scale changes I therefore think it is necessary to explore the short-term, seasonal challenges to survival, seek out the evidence for how the earliest Europeans coped with them, and consider whether long-term changes in seasonal strategies can be detected.

In investigating Early and Middle Pleistocene seasonality and its impacts on hominin behaviour in the European Lower Palaeolithic, the following chapters partly draw on data and examples from modern ecosystems, extant herbivores and carnivores, recent and living hunter-gatherers, and other periods of the Palaeolithic and discussions are necessarily speculative on occasions. The caveats associated with such an approach are fully acknowledged and should be kept in mind. The book's structure also introduces the danger of over-stressing possible differences between the seasons and/or artificially partitioning hominin activities into separate times of the year. That is not my intention, as it is very likely that the earliest Europeans were highly flexible in their foraging and other behaviours. At the same time, certain activities *were* more likely to have occurred at different times of year, whether because of the relative availability of, and/or need for, specific resources, the timing of events such as childbirth, or the daylight demands of specific tasks (see Table 1.1). Moreover, the scope for mixing tasks (*e.g.* collecting raw material while also foraging for food) would to some extent be limited by each individuals' carrying capacity (in weight and/or volume).

The chapters that follow therefore build a model of hominin activity through the seasons and across the year, with behaviours and strategies based on the available archaeological and palaeoenvironmental evidence of the climates, landscapes, flora, fauna and technologies of the Lower Palaeolithic, drawing on both primary context sites and disturbed assemblages. In particular the following chapters explore the themes of coping with winter cold (Chap. 3), the timings of hominin conception and birth (Chaps 4 & 5), the opportunities for learning in the long days of summer (Chap. 5), and the fluc-tuating availability of plant and animal foods (Chaps 3–6), in particular the potential to target herd aggregations in the autumn and possibly store food resources (Chap. 6). The final chapter re-evaluates the patterns in the long-term record, such as the punctuated long chronology, in light of the seasonal behaviours and strategies proposed below.

Notes

1. Thousand years ago.
2. Early and Middle Pleistocene interglacials and glacials are respectively odd- (*e.g.* MIS 13) and even-numbered (*e.g.* MIS 12) in the marine isotope stage (MIS) system (see Box A for further details).
3. Mesic habitats are characterised by a moderate or well-balanced supply of moisture.
4. This later temperate phase has been tentatively assigned to the 11a interstadial, although this correlation has been criticised as overly simplistic in light of the evidence for increasingly complex patterns of sub-stage variability (Candy *et al.* 2014).
5. Referred to as the Non-Arboreal Pollen or NAP phase in Britain (Candy *et al.* 2014).
6. Mesoclimatic refers to conditions at the scale of 10s and 100s of metres.
7. Defined by van der Made (2011) as the area to the west of the eastern borders of Germany, Austria and Italy.
8. Artiodactyla are even-toed ungulates such as deer.
9. However, late Middle and Late Pleistocene remains of *H. latidens* have recently been recovered at Schöningen and from the southern North Sea (Serangeli *et al.* 2015b).
10. Increased logging in the three Polish districts of the Białowieża forest, approved by the Polish government in 2016, was ruled against by the European Court of Justice in April 2018 (https://www.greenpeace.org/international/press-release/15961/european-court-of-justice-logging-in-bialowieza-forest-was-illegal/).
11. The taxonomy of the Sima de los Huesos fossils has been much debated (see the discussions in Stringer 2012 and Arsuaga *et al.* 2014 for example). Here I adopt the view that the material can

be seen as belonging to *H. heidelbergensis* as broadly defined (*i.e. sensu lato*) 'to include fossils with a generally more primitive morphology than the late Middle Pleistocene and Late Pleistocene Neandertals', even though the Sima fossils do exhibit some highly derived Neanderthal traits, especially in the dentition (Martinón-Torres *et al.* 2012; Arsuaga *et al.* 2014, 1362).

12. Although the emphasis on hunted foods varied markedly between different contributions in the Man the Hunter volume (Lee and DeVore 1968): *e.g.* compare Lee ('vegetable foods comprise from 60–80 per cent of the total [!Kung] diet by weight'; 1968, 33), Woodburn ('the Hadza rely mainly on wild vegetable matter for their foods'; 1968, 51), and Laughlin ('hunting is the master behaviour pattern of the human species'; 1968, 304).

13. Although access to the energy from carbohydrate-rich USOs has been argued to be limited in the absence of controlled fire use (Butterworth *et al.* 2016).

14. However wild plant use in modern Europe is widespread, with over 100 million EU citizens consuming wild foods (Schulp *et al.* 2014).

15. Key variables for estimating calorific demands include (Froehle and Churchill 2009):
    BMR: basal metabolic rate – the energy used by the body for maintenance and growth in the absence of activity or digestion.
    DEE: daily energy expenditure.
    MAT: mean annual temperature.
    PAL: physical activity levels – a coefficient for expressing DEE as a multiple of BMR.

16. For comparison, the UK's National Health Service currently recommends *c.* 2000 and 2500 kcal/day for women and men respectively, although these values vary according to factors such as age, metabolism and levels of physical activity (https://www.nhs.uk/common-health-questions/food-and-diet/what-should-my-daily-intake-of-calories-be/).

17. Theory of Mind is widely recognised to be an essential component for the understanding of verbal or visual language. While the level of intentionality achieved by chimpanzees has been much debated, 'mind-reading' (*i.e.* ToM) of other individuals' perceptual states (if not beliefs) has been suggested to be evident (Andrews 2017), indicating that ToM can be present without complex verbal language.

18. The hyoid bone is situated at the root of the tongue in the front of the neck and between the lower jaw and the largest cartilage of the larynx, or voice box. Its primary function is as an anchoring structure for the tongue.

19. A specimen from Castel di Guido (Italy; *c.* 300 kya) was originally interpreted as a hyoid bone and as evidence against pre-Middle Palaeolithic/MSA spoken language (Capasso *et al.* 2008). However, the identification has subsequently been revised, with the specimen now described as the posterior arch of the atlas (Capasso *et al.* 2016).

20. LCTs are defined here as incorporating bifacial handaxes, cleavers (bifacial and unifacial) and backed knives, typically between 10–30 cm in maximum dimension (after Kleindienst 1962).

21. Although the Happisburgh III age has been questioned (Westaway 2011), the presence of multiple Early Pleistocene sites to the north of the Alps from *c.* 1mya onwards suggests that, irrespective of the ages of individual sites, there is clear evidence for a small northwards expansion during the Galerian.

22. Such range expansions or dispersals would be very different in character to annual migrations (see Chap. 6).

23. The exact nature of the adaptations would vary depending of the dispersal routes that were 'followed' into Europe. For example, populations that dispersed into southern Iberia from north Africa would contrast with groups that spread through the Balkans/Danube or over the Caucasus and across the northern fringes of the Black Sea (*e.g.* MacDonald *et al.* 2012, fig. 4).

24. The behavioural implications of cool/cold-climate occupations highlights the significance of the various climate-driven models of river terrace formation (*e.g.* Bridgland 2000; Bridgland *et al.* 2006; Antoine *et al.* 2015).

Chapter 3

A winter wonderland?

It's grim in Europe: a winter challenge

The challenges presented by interglacial European winter conditions to Lower Palaeolithic hominins, inferred from site-specific evidence and modern parallels, were considerable, and will no doubt be familiar to many readers (see also Gamble 1987; Roebroeks 2006). The year's lowest temperatures (Table 3.1) would be further exacerbated by day/night contrasts and wind-chill. Mobility and the visibility of specific resources would sometimes have been hindered by snow-cover, although general visibility would be enhanced by the reduced vegetation. Days were shorter and the distribution, availability and quality of plant and animal foods would also have been reduced in contrast to the year's other seasons. The long dark hours of winter nights may also have been periods of heightened anxiety, fuelled by sounds (*e.g.* animal cries and other aural indicators of their presence), scents, and sights (*e.g.* animals' eyes glinting in the moonlight). Such challenges may also have become increasingly marked in the Middle Pleistocene, as seasonality increased (see also Chap. 2). In short, winter survival for Europe's earliest humans would frequently have been a major challenge, and probably the greatest threat in a Lower Palaeolithic year. An exploration of how hominins did survive, at least sometimes, therefore seems a logical place to begin a journey through the seasons.

In the broadest sense such challenging winters can be managed by one of two strategies: extensive annual mobility (*i.e.* migrations) or local residency (*i.e.* 'toughing it out').[1] Given the likely timings of any migratory movements (spring and autumn), the former will also be discussed in Chapters 4 and 6. Both strategies would have presented a number of difficulties, with several shared challenges. It is therefore appropriate to introduce them here in the context of 'toughing it out' and consider how they may have varied in different regions of Europe and at different times of the Lower Palaeolithic. Four European winter survival requirements are highlighted: firstly, securing and/or producing sufficient resources, whether shelter, clothing or fire, to ameliorate the effects of low temperatures; secondly, securing sufficient food resources to meet nutritional needs (see also Chap. 2); thirdly, managing group structure and composition; and finally, organising local mobility in a winter landscape. The winter problem, and how it was solved, is therefore central not only to the annual life cycle, but also to some of the big current questions in Lower Palaeolithic studies: how long-lasting, or successful, were continuous occupation phases in European, especially northern European, landscapes? What was the overall timing, duration and character of hominin range expansions into Europe, initially in the south (*e.g.* Agustí *et al.* 2009; Leroy *et al.* 2011; Arzarello *et al.* 2015; Muttoni *et al.* 2018), with subsequent expansions into the north (*e.g.* Roebroeks 2001; 2006; Dennell *et al.* 2011; Moncel *et al.* 2018; see also Chap. 2)? In recognition of the well-established chronological contrasts between the earliest occupation of southern and northern Europe (*e.g.*; Roebroeks and van Kolfschoten 1994; Dennell and Roebroeks 1996) this chapter therefore begins by assessing the similarities and differences between warm stage winter conditions in different regions and at different periods of the Early and Middle Pleistocene. Of particular interest are the potential contrasts in conditions either side of *c.* 44–45°N in western Europe, since both the archaeological and faunal records appear to vary to the north and south of this 'line' (see also Kahlke *et al.* 2011; Rodríguez *et al.* 2013; Ashton 2015).

Cold, dark and short days ... everywhere?

The general climatic description offered at the start of this chapter may be an overly north-centric view of European winters. Herpetofauna from Gran Dolina TD-6.2 in northern Spain suggest slightly milder winter temperatures (T_{min} [January]: 4.3±1.7°C), although the winter/summer contrasts are still marked (*c.* 18°C; Blain *et al.* 2013). Assessing regional contrasts in the European Lower Palaeolithic is difficult however, due to both the long-term climatic changes from the Early to the Middle Pleistocene (see also Chap. 2, Table 2.3 & Fig. 2.3) and the progressive climatic changes over the course of any individual warm stage. For example, an MIS 11 vegetation succession in Poland sees a change in winter conditions from -5°C at the beginning of the stage, to -1–0°C in the climatic optimum, returning to -4--5°C at the end of the warm stage (Szymanek 2017). At the other end of Europe, the Sima del Elefante sequence records a shift in winter temperatures from approximately 4°C (TE9c–TE13) to 2.5°C (TE14; accompanied by increased winter precipitation), which may reflect the climatic deterioration at *c.* 1.1–1.0 mya (Blain *et al.* 2010).

Nonetheless, comparison of site-specific winter temperature estimates from across western Europe (Tables 3.1 & 3.2) suggest that the major contrast is between the very earliest sites (*e.g.* Barranco León D and Fuente Nueva-3) and those of the late Early (*e.g.* Gran Dolina TD-6.2) and Middle Pleistocene (Blain *et al.* 2016). The former are characterised by very mild European winters (*i.e.* relatively warm and wet), and warm summers (Chap. 5), in marked contrast to the later sites (see also Sánchez Goñi *et al.* 2016). A comparison of winter temperature estimates for selected British, German and Spanish sites from the Early and Middle Pleistocene (Fig. 3.1) reveals an overlap between –1 and +6°C only (and no overlap at all when only Early Pleistocene sites are considered). This chronological difference may therefore be a key factor in understanding one of the long-standing chronological and geographical patterns in the presence and absence of hominins across Lower Palaeolithic Europe: Leroy *et al.* (2011, 1461) have suggested that *H. antecessor* (and possibly also *H. erectus*), present around the Mediterranean rim well before 1 million years ago, were only able to cope with a narrow winter temperature range of 0–+6°C. In short, a northern European winter in the Early Pleistocene, such as that at Happisburgh III, may have presented challenges which were frequently, although apparently not always, beyond the capabilities of the earliest European hominins.

For much of the Middle Pleistocene, winter temperatures contrast noticeably between the south and the north of western Europe, with values in the latter region fluctuating around freezing, while southern sites' winter conditions are typically a

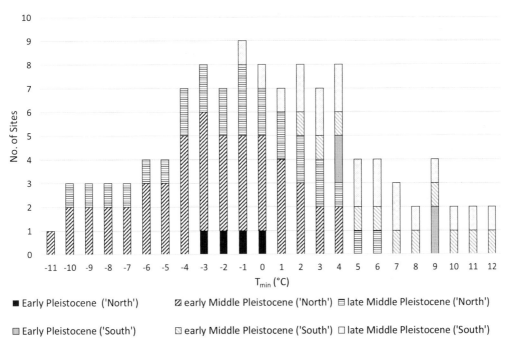

Figure 3.1: Comparison of winter temperature ranges for selected 'southern' (Spanish) and 'northern' (British and German) sites from the European Early Pleistocene, early Middle Pleistocene and late Middle Pleistocene. Number of sites in each temperature category calculated according to the temperature ranges for each site (e.g. eight sites have a T$_{min}$ range which includes –3°C). Spanish sites: Aridos 1; Barranca León D; Cúllar Baza 1; Fuente Nueva-3; Gran Dolina TD-6.2; Gran Dolina TD-10.1; Sima del Elefante; British & German sites: Bilzingsleben II; Boxgrove; Brooksby; Happisburgh I, Happisburgh III, High Lodge, Hoxne, Pakefield, Schöningen 13 II-4 (Sources and specific site levels as per Table 2.3).

few degrees above zero (although there is considerable variability in both samples; Fig. 3.1 & Table 3.1). At Ceprano and Vallo di Diano, for example, the occurrence of *Hedera* is noteworthy as Common European Ivy, *H. helix*, is sensitive to winter temperatures and water supply. Its significant presence in the pollen spectrum at those sites, reaching 10%, suggests mild winter conditions, with temperatures above −1.5°C in the coldest months (Margari *et al.* 2018). The northern data by contrast imply the possibility, and at times probability, of sub-freezing winter temperatures on those sites, although the presence of frost-sensitive insects on selected sites suggests that those sub-freezing temperatures were relatively short in duration (MacDonald 2018). Nonetheless in almost all cases the northern estimates fall below their present-day equivalents (Table 2.3), a difference that is exacerbated by the tendency of Pleistocene winter temperature estimates to be warmer than the reality.[2]

Comparisons between winter conditions in northwestern and north-central Europe are also of particular interest, since the marked difference in the scale of the Lower Palaeolithic record to the east and west of the Rhine is a long-standing archaeological pattern (Haidle and Pawlik 2010; Romanowska 2012). Unfortunately it is difficult to discuss specific temperature estimates that are directly associated with hominin occupations (excluding Bilzingsleben II and Schöningen 13 II-4; Table 3.1), both due to the relative paucity of sites to the east of the Rhine and the nature of the available evidence. However, although not directly related to hominin occupations, a number

Table 3.1: Selected winter temperature estimates (T_{min}) for European Lower Palaeolithic sites

Site	T_{min} (°C)
Early Pleistocene	
Barranco León D	+9.0
Fuente Nueva-3	+9.2
Sima del Elefante (Level TE9c)	+4.1
Happisburgh III (Bed E)	−3–0
Gran Dolina (TD-6.2)	+4.3
early Middle Pleistocene	
Pakefield (Bed Cii–Ciii)	−6–+4
Cúllar Baza 1	+2.5–+12.5
Boxgrove (Unit 4c & Freshwater Silt Bed ≈ Units 4b & 4c)	−4–+4
Happisburgh I (Organic Mud)	−11– -3
High Lodge (Bed C1)	−4 – +1
Brooksby (Redland's Brooksby Channel)	−10 – +2
later Middle Pleistocene	
Hoxne (Stratum D; HoIIIa)	−10 – +6
Bilzingsleben II	−0.5 – +3
Aridos I	+2 – +12
Gran Dolina (TD-10 [sub-level T1])	−0.5 – +7.5
Schöningen 13 II-4	−4 – -1

See Table 2.3 for summer temperatures, temperature data sources, site ages and references.

Table 3.2: Polish winter temperatures across the MIS 11 succession (Szymanek 2017). See also Table 5.2

MIS 11 Sub-stage	Vegetation Zone	January temperatures (°C)
Pre-optimum	Betula–Pinus	-5
	Picea–Alnus	-5 – -3
	Taxus	-1
Climatic optimum	Carpinus–Abies	-1 – 0
Post-optimum	Picea–Pinus–Pterocarya &	-3 – -2
End of warm stage	Pinus–Juniperus	-5 – -4

Table 3.3: Winter temperature estimates for selected non-hominin sites from central Europe (data from Szymanek and Julien 2018). See also Table 5.3

Site	Age (MIS)	Intra-stage phase	Proxies	January temperatures (°C)
Ferdynandów,	15	Not specified	Pollen, plant macrofossils	+1.5 – +3
Zdany &	14	Not specified		-12
Łuków (Poland)	13	Not specified		-2
Bilshausen	13	Beginning of warm stage	Pollen	-7
(Germany)		Interglacial maximum		-3 – -2
		End of warm stage		-8
Dethlingen,	11	Beginning of warm stage	Pollen, diatoms	-14
Hetendorf &		Interglacial maximum		-2 – +2
Munster-Breloh		End of warm stage		-10
(Germany)				
Ossówka,	11	Early warm stage	Pollen, diatoms	-5
Woskrzenice,		Interglacial maximum		-1 – 0
Kaliłów & Wilczyn		Late warm stage		-3 – -2
(Poland)		End of warm stage		-5 – -4

of sites from central Europe do provide indications of Early and Middle Pleistocene warm stage conditions, based on key palaeoenvironmental proxies (Szymanek and Julien 2018; Table 3.3). Compared to the Middle Pleistocene hominin sites (Table 3.1) winter conditions are perhaps slightly harsher in central Europe at the end of warm stages (*e.g.* compare Bilshausen [51°37′N] with Boxgrove [50°51′N] and High Lodge [52°20′N]), although peak interglacial winter conditions appear more comparable (*e.g.* compare the Dethlingen/Ossówka *et al.* data with Hoxne [Stratum D]).

Moreover, general descriptions of the conditions at the key archaeological sites in central northern Europe are available (Szymanek and Julien 2018). It is evident that hominin occupations were associated with both temperate woodlands and more open, steppic conditions, and with late warm stage periods (*e.g.* Stránská skála I and Miesenheim I; Table 3.4), suggesting that tolerances of continental conditions may have been relatively wide, and not the cause of the apparent west–east cline in hominin presence across northern Europe during the later Lower Palaeolithic. Such a breadth

Table 3.4: *Palaeoenvironmental conditions at selected Lower Palaeolithic sites from northern central and eastern Europe during the Middle Pleistocene (after Szymanek and Julien 2018, table 3)*

Site	Age (MIS)	Conditions
Stránská skála I (Czech Republic)	17–16	Forested and open environments, meadows and marsh-lands near cave, final part of warm stage
Mauer (Germany)	15	Warm, humid environments, forested areas with patches of open habitats, warm summers and mild winters
Korolevo VI (Ukraine)	14	Cool climate, steppe and birch forest, small lake in vicinity of site
Miesenheim I (Germany)	13	Sparse pine-birch forests, final part of the warm stage
Vértesszölös (Hungary)	13	Temperate climate, mosaic of woodland, shrubby and grassy areas
Steinheim-an-der-Murr (Germany)	11	Warm, Mediterranean habitats, forests with open landscape
Rusko (Poland)	11	Mixed forest, high proportion of coniferous trees

Table 3.5: *Precipitation data for selected European Lower Palaeolithic sites*

	Precipitation (mm)				
Site	Winter (Dec–Feb)	Spring (Mar–May)	Summer (Jun–Aug)	Autumn (Sep–Nov)	D_{max}[1]
Barranco León D	287.0	165.0	47.0	204.0	107
Fuente Nueva-3	314.0	189.0	49.0	223.0	115
Sima del Elefante (TE9c)	141.0	–	144.0	–	–
Gran Dolina (TD-6.2)	200.1	240.1	176.3	232.5	45
Aridos I	274.7	166.1	48.2	198.9	–

Seasonal figures calculated from monthly data in Blain *et al.* (2010; 2013; 2014; 2016). [1]D_{max}: difference between the driest month(s) and the wettest month(s).

of environmental tolerances is also supported by the inferred winter conditions at north-western European sites such as High Lodge and Happisburgh I (Table 3.1).

Winter precipitation levels also varied across Europe, both inter- and intra-regionally (Table 3.5). At Gran Dolina (TD-6.2) winters were relatively dry compared to autumn and spring, although slightly wetter than present-day Burgos, as were all the seasons (Blain *et al.* 2013). Similar conditions also prevailed at the adjacent, but earlier, site of Sima del Elefante (Blain *et al.* 2010). A further factor at the higher altitudes of Atapuerca (*c.* 1000 m asl), and elsewhere on the Spanish Meseta, would be occasional snowfall. By contrast, the winters at the very early sites of Barranco León D and Fuente Neuva-3 (MIS 43–49), in the southeast of Spain, were the wettest periods of the year, contrasting markedly with very dry summers, while conditions as a whole were drier than those of Gran Dolina (Blain *et al.* 2016). Such conditions are also apparent at the later Middle Pleistocene site of Aridos, near Madrid (Blain *et al.* 2014), with high levels of rainfall occurring during winter in combination with temperate conditions (a coldest month mean temperature of 8.8°C). In the northwest, a seasonal precipitation regime of cool, wet winters and warm, dry summers has been suggested for Pakefield in the UK (MIS 19 or 17; Candy *et al.* 2006). Finally, in the continental interior at the later site of Schöningen (site 13 II-4; MIS 9) relatively low levels of annual precipitation were suggested, with estimates varying between 400–450 mm and a minimum of 470 mm (Urban and Bigga 2015). While overall trends are unclear these varying winter conditions are significant because of the potentially damaging impacts of both cold/cool *and* wet conditions, in terms of human body temperature maintenance and the possible need for, and sustainability of, supporting cultural insulation technologies such as fire and/or clothing. Summer droughts would also impact significantly upon hominins (Chap. 5).

Overall these data, both direct and indirect, suggest that the initial Early Pleistocene occupations of southern Europe were associated with relatively mild and wet winter conditions (see also Leroy *et al.* 2011). By contrast the late Early and Middle Pleistocene occupations were associated with cool to cold northern winters characterised by temperatures at or around freezing, with evidence for milder, but still cool, winters in the south. Precipitation, snow and rainfall, would have varied on both continental and regional scales, although modern data suggests greater winter rainfall in the Atlantic West, and increasing snowfall (in depth and duration) along a southwest–northeast transect and in high altitude zones (Barron *et al.* 2003, figs 5.8 & 5.9; Chap. 1).

Interglacial-only?

The discussion above focused solely on warm stage European winters. However, it is likely that the expanded Lower Palaeolithic occupations of the later Middle Pleistocene were permanent, although sustained during the glacials by populations restricted to relatively favourable local conditions around the Mediterranean rim (see also Chap. 2). The likely importance of such micro-climates is highlighted by the general character of southern European cold stages, *e.g.* MIS 12, after the Early–Middle Pleistocene Transition: extremely dry and cold winters, with windy conditions and a predominantly semi-desert-type vegetation (Sánchez Goñi *et al.* 2016). Specific examples of favourable conditions are scarce, but the Gran Dolina amphibian and squamate sequence for TD-10 (spanning *c.* 400–300 kya [MIS 11–9], with a weighted mean age of 372±33 kya [late MIS 11]; Falguères *et al.* 1999), records minor differences in the mean temperatures of the coldest month of *c.* 1–1.5°C between 'cold' and 'warm' periods, with similar trends in the earlier levels of TD-5 and TD-6 (Blain *et al.* 2009). It is thus possible that aspects of the winter challenges of northern warm stages might be broadly comparable to those of southern glacials, at least in the case of favourable locations such as Gran Dolina.

Surviving winter

European winters are of particular interest because available site-specific reconstructions challenge possible assumptions about the tolerances and preferences of the earliest Europeans. Recent archaeological discoveries, mostly notably the work by

AHOB (*Ancient Human Occupation of Britain* project) and Leiden University at the Norfolk sites of Happisburgh I & III and the Suffolk site of Hoxne (Ashton *et al.* 2008a; Parfitt *et al.* 2010; Ashton and Lewis 2012; Lewis *et al.* 2019), have associated the hominin occupation evidence at these UK sites with later and/or cooler interglacial phases – and therefore with more challenging winters. The Happisburgh III conditions, for example, have been compared with southern Scandinavian-type environments, and thus seem to extend the range of environmental tolerances attributable to Early and Middle Pleistocene hominins in mid-latitude regions of the northern hemisphere. Late, post-temperate occupations in MIS 13 have also been suggested for High Lodge and the main archaeological deposits at Boxgrove (Ashton *et al.* 1992; Roberts and Parfitt 1999; Candy *et al.* 2015), while several sites in northern France have been associated with climatic transitions, such as Saint-Pierre-lès-Elbeuf at the MIS 11/10 boundary (Antoine *et al.* 2010). There are also a small number of northern sites that are even suggestive of MIS 14 and 12 cold stage occupations: Rue du Manège (Amiens) in the former, and Boxgrove (Eartham Formation), Kärlich H and Cagny la Garenne in the latter (Roberts and Parfitt 1999; Antoine *et al.* 2010; 2015; Haidle and Pawlik 2010), although some of these assemblage sizes are small and thus the duration of any hominin presence may have been very short. It is also very likely that the MIS 12 assemblages are associated with the early part of the glacial or 'milder' interstadial intervals within that stage, as Candy *et al.* (2015) estimated winter temperatures of –36 to –10°C in the early MIS 12 Ostend Arctic Freshwater Bed. These inferred environmental conditions, especially those of the post-temperate warm stages, have unsurprisingly led to discussions about possible adaptive hominin strategies, both behavioural (*e.g.* fire and sheltering technologies, migration) and physiological (Ashton and Lewis 2012; Ashton 2015). It is also important to consider that there may well have been differences between hominin habitat *preferences*, and habitat *tolerances*, as discussed by Cohen *et al.* (2012) with reference to Pakefield and Happisburgh III.

But before diving headlong into the question of winter survival we need to pause for a moment. How do we know hominins were present through these mid-latitude European winters? The honest answer is that we don't, due to an absence of seasonal indicators. But what we do know is that they were present at some point in the year, in order to make, use and discard the lithic artefacts recovered from sites and geological deposits across Europe. It is therefore worth evaluating the challenges of the various different winter scenarios (*e.g.* Early Pleistocene and Middle Pleistocene, southern and northern Europe), and the feasibility of the various possible winter-survival solutions available to hominins. That is the focus of this chapter (and also Chap. 6). While direct evidence for hominin solutions to the 'winter problem', such as fire, clothing and shelter, are frequently tenuous, possible archaeological indicators of those solutions are highlighted below, and the implications of their presence or absence are explored. Alternative 'winter' solutions such as migration are considered in the context of spring and autumn (Chaps 4 & 6), when any long-distance residential moves are likely to have occurred.

Built for the cold?

Before considering cultural insulation however, it is logical to consider purely biological solutions to the problem of low winter temperatures. Such solutions have long
been considered as explanations for the distinctively stocky and robust body shape
of the Neanderthals, interpreted as a form of cold-adaptation, typically with reference to modern, high latitude populations such as the Inuit (Ruff 1994; Holliday 1997;
Trinkaus *et al.* 1998). Such arguments have been difficult to extend to European Lower
Palaeolithic hominins, due to the relative paucity of post-cranial fossils until recently,
although contrasting body shape issues have been explored in low latitude archaic
hominins such as the Nariokotome Boy (Ruff 1991; 1994; Holliday 2012). However, the
robust tibia find from Boxgrove in the early 1990s enabled Trinkaus *et al.* (1999) to
model the body proportions and robusticity of a hominin individual generally assigned
to *H. heidelbergensis* (Roberts *et al.* 1994; Stringer *et al.* 1998). Taken in association with
the site's cool temperate conditions (late MIS 13), Trinkaus *et al.* suggested that the
Boxgrove hominins would have been reliant upon significant biological rather than
cultural solutions to the problems of conserving body heat.

At around the same time the substantial post-cranial evidence from the Sima de
los Huesos greatly expanded the available sample for Lower Palaeolithic hominins,
although the position of the Sima population relative to Neanderthals means that the
body shape data is not necessarily applicable to *H. heidelbergensis sensu lato* populations.
While slightly taller than the Neanderthal average (see also Chap. 2) Arsuaga *et al.*
(2015) suggested that the Sima hominins, like Neanderthals, had a larger costal[3] skeleton relative to their stature, in comparison with modern humans. While the sample
is much smaller, Carretero *et al.* (1999) tentatively concluded that *H. antecessor* may
have had limb proportion values close to those associated with sub-Saharan African
populations, and therefore a less effectively cold-adapted body. This is notable in
light of the slightly milder Early Pleistocene winters suggested at Gran Dolina TD-6.2
(Table 3.1), although Gómez-Olivencia *et al.* (2010) concluded that *H. antecessor* may
have had a broader thorax than living humans, as is also the case with Neanderthals.
Finally, Buck *et al.* (2018) have recently demonstrated ecogeographic patterns in the
skeletons of Japanese macaques that broadly follow the common trends for mammals
and hominins, *i.e.* Allen's and Bergman's rules (Ruff 1993).[4] It is thus possible that
Plio-Pleistocene hominins possessed a conserved primate pattern of skeletal plasticity.
The size and/or shape of skeletons of specific Lower Palaeolithic *Homo* taxon may
therefore have varied across Europe, by latitude if not also longitude, although as
noted above the available fossil record makes this difficult to test at present.

However, these cold-adaptation arguments, especially with reference to Neanderthals,
have been critiqued in recent years (Aiello and Wheeler 2003; see also Galway-Witham
et al. 2019 for a recent summary). Aiello and Wheeler's minimum sustainable temperature (MST) prediction[5] of 11.6°C for *H. erectus* suggested that purely biological solutions
would be of limited effectiveness in a European setting – and they reached a similar
conclusion for the robust and arctic body-proportioned Neanderthals. The Neanderthal

body shape has recently also been considered as being shaped instead by locomotion and woodland-based encounter and ambush hunting (Stewart *et al.* 2019), and the woodland habitats of much of the Lower Palaeolithic's warm stages would certainly have been conducive to encounter-based hunting. Adopting the Aiello and Wheeler (2003) methodology for *H. heidelbergensis*, based on the available Sima de los Huesos stature and body mass data (Arsuaga *et al.* 2015), also suggests a limited degree of cold adaptation (Table 3.6; comparable calculations were not possible for *H. antecessor* due to the current lack of body weight estimates). A similar conclusion was reached by MacDonald (2018) for the Sima hominins. With reference to the very earliest Europeans, *e.g.* as represented by the dental evidence from Barranco León (Toro-Moyano *et al.* 2013), the Aiello and Wheeler (2003) *H. erectus* data, based on the East African Nariokotome Boy skeleton, is clearly not an ideal proxy. However, the minimum sustainable temperature predictions for *H. erectus*, *H. heidelbergensis*, *H. neanderthalensis* and *H. sapiens* span just 8.0–11.6°C, suggesting that the values of *H. antecessor* and any other early European *Homo* would most likely fall somewhere within that range (Aiello and Wheeler 2003, table 9.2; Table 3.6). The narrow range of the modelled values representing those four species also suggests that the likely degree of any ecogeographic intra-species skeletal variability in Lower Palaeolithic Europe would still place demands on behavioural and technological adaptations (Buck *et al.* 2018).

However, the figures presented above assume a physiology and basal metabolic rate (BMR) comparable to modern humans. They also exclude other forms of biological insulation, such as increased muscle mass or subcutaneous fat, or body hair. Table 3.6 therefore provides alternative predictions for *H. heidelbergensis* and *H. erectus* (and Late Pleistocene *H. sapiens* for comparison) based on an elevated BMR, increased muscle mass and body hair, following the data, formula and equations from Aiello and Wheeler's original study. Their justification for an elevated BMR for Neanderthals lay in the observation that arctic-adapted people have significantly higher BMRs than tropical populations, apparently relating to a high animal fat and protein diet and also to the effects of temperature and day length on thyroid function. However, those conditions (persistent low temperatures and animal product-dominated diets) are less applicable and/or demonstrable for Lower Palaeolithic hominins, and so the enhanced lower critical temperature (LCT)[6] and minimum sustainable temperature values associated with elevated BMR in Table 3.6 may not be applicable. Subcutaneous fat is not included, in light of Aiello and Wheeler's previous observation that, for an 80 kg Neanderthal, the 3.2 cm of fat required to provide 1 clo[7] of insulation would weigh upwards of 52 kg and therefore seems unfeasible for an active Pleistocene forager.

These modified predictions reduce the minimum sustainable temperature for *H. heidelbergensis* to 4.9°C (for elevated BMR), 3.2°C (for elevated BMR + increased muscle mass), and -20.0°C (for elevated BMR + 1 clo of insulation). It is clear that without at least additional biological insulation (in the form of 1 clo of body hair: borrowing Aiello and Wheeler's estimation of 3.9 cm of relatively sparse hair covering the entire body), there are small discrepancies between these minimum sustainable temperature

Table 3.6: Lower critical and minimum sustainable ambient temperatures for H. heidelbergensis, H. erectus *and* H. sapiens *(after Aiello and Wheeler 2003, tables 9.1–9.3)*

Species	H. heidelbergensis[1]		H. erectus[2]		H. sapiens[3]	
	Kleiber BMR[4]	Elevated BMR[5]	Kleiber BMR[4]	Elevated BMR[5]	Kleiber BMR[4]	Elevated BMR[5]
Body Mass (kg)	69.1	69.1	68	68	70	70
Stature (cm)	163.6	163.6	185	185	177	177
BMR	81.487	93.710	80.512	92.589	82.282	94.624
Body surface area[6]	1.749	1.749	1.900	1.900	1.862	1.862
Human Conductance A[7]	5.0	5.0	5.0	5.0	5.0	5.0
Total Conductance A[8]	8.747	8.747	9.498	9.498	9.312	9.312
Lower Critical Temperature A (°C)[9]	27.7	26.3	28.5	27.3	28.2	26.8
Minimum Sustainable Temperature A (°C)[10]	9.1	4.9	11.6	7.8	10.5	6.5
Human Conductance B[11]	4.750	4.750	4.750	4.750	4.750	4.750
Total Conductance B	8.310	8.310	9.023	9.023	8.846	8.846
Lower Critical Temperature B (°C)[9]	27.2	25.7	28.1	26.7	27.7	26.3
Minimum Sustainable Temperature B (°C)[10]	7.6	3.2	10.2	6.2	9.1	4.9
Human Conductance C[12]	2.817	2.817	2.817	2.817	2.817	2.817
Total Conductance C	4.928	4.928	5.351	5.351	5.246	5.246
Lower Critical Temperature C (°C)[9]	20.5	18.0	22.0	19.7	21.3	19.0
Minimum Sustainable Temperature C (°C)[10]	-12.6	-20.0	-8.1	-14.9	-10.1	-17.1

[1]*H. heidelbergensis* data from Sima de los Huesos (Arsuaga *et al.* 2015); [2]*H. erectus* data from KNM-WT 15000 (Ruff 1994); [3]*H. sapiens* data from Předmost 3 & 9, Skhul 4 and Grotte des Enfants 4 (Ruff 1994); [4]BMR = 3.4 × mass (kg)$^{0.75}$ (Kleiber 1961); [5]Elevated BMR = BMR raised by 15% (after Aiello and Wheeler 2003, 150); [6]Body surface area (m^2) = 0.00718 × mass (kg)$^{0.425}$ × stature (cm)$^{0.725}$; [7]Typical human conductance = 5 W.m^{-2}.°C^{-1}; [8]Total conductance = typical human conductance × surface area (m^2); [9]Lower critical temperature (°C; 'the lower limit of the thermoneutral zone within which a mammal can regulate its core temperature solely by controlling its thermal conductance ... as the ambient temperature falls below this level, homeostasis can only be maintained by progressively increasing internal heat production, and incurring the additional energetic costs associated with this increase in heat production'; Aiello and Wheeler 2003, 148) = 37°C – (BMR/Total conductance); [10]Minimum sustainable ambient temperature (°C; 'the minimum temperature at which an animal can maintain normal body temperature by raising its basal metabolic rate to its maximum sustainable level, in humans usually about three times normal BMR'; White 2006, 568) = 37°C – ((3 × BMR)/ Total conductance); [11]Typical human conductance reduced by 5% to account for hominin muscularity (after Aiello and Wheeler 2003, 150); [12]Typical human conductance reduced by *c.* 44% to account for 1 clo of insulation (after Aiello and Wheeler 2003, 150). *H. heidelbergensis* (Sima de los Huesos) predictions differ slightly to those calculated by MacDonald (2018, table 3), due to small differences in the body mass and stature estimates used

predictions (3.2°C) and the site-based winter temperature estimates in northern Europe. However, this MST prediction does fall within the winter temperature ranges for southern European Middle Pleistocene sites (Table 3.1). This highlights that varying biological and/or behavioural adaptations would have been required for survival in different parts of Europe.

Although it is not possible to model *H. antecessor* at present, *if* its tolerances fell somewhere between *H. heidelbergensis* and *H. erectus*, then its minimum sustainable temperature prediction (without additional biological or cultural insulation; 3.2–6.2°C) would fall only at the upper end of the winter temperature range of the coldest of the Early Pleistocene sites in Table 3.1 (*e.g.* TD-6.2 and Sima del Elefante). All of these possible discrepancies between MST and winter temperatures become even more marked when wind-chill factors (using the Stage 3 Project models as a comparative data set; Van Andel and Davies 2003) and/or the impacts of rain are considered (MacDonald 2018).

Finally, and most importantly, these physiological predictions do not imply ecological viability (Aiello and Wheeler 2003): in short, maintaining internal heat production at the levels required by elevated BMR necessitates sustained, high levels of energy intake. As Mark White wrote:

> they [Neanderthals] could not have been constantly 'on the go', feeding as they went to fuel their energetic needs, and the problems of keeping warm during 'downtime' continues to force the issue. (White 2006, 558)

A hairy hominin?

Thus, some form of additional insulation, whether biological or cultural, would be needed. Would dense body hair or fur be a likely feature of Lower Palaeolithic life in Europe, at least for *H. heidelbergensis* if not also for *H. antecessor*? The all-over body hair option has frequently been discussed in a European context, both in writing and graphically (*e.g.* Stringer and Gamble 1993, pls 3 & 8; Moser 1998; Aiello and Wheeler 2003; White 2006, 557), with comparisons being made to various species of 'woolly' Pleistocene fauna (*e.g.* woolly mammoth; woolly rhinoceros: Guthrie 1990). The benefits are obvious: fur can significantly lower LCTs, *e.g.* to –30°C for reindeer, although hominin fur would probably be less dense, in-keeping with primates generally. However, the feasibility of the 'hairy hominin' option as a strategy in the European Early and Middle Pleistocene needs to be evaluated against the evolutionary background. It is likely that hair *loss* had previously been selected for, in association with the adoption of habitual bipedalism, mobile activities such as foraging, and improved sweating abilities, on the hot and open environments of the African Pliocene savannah (Wheeler 1996; Ruxton and Wilkinson 2011; Dávid-Barrett and Dunbar 2016), although the implications of surviving cooler savannah nights have received less attention (do Amaral 1996; Ashton 2015, 149; MacDonald 2018). Genetic data similarly suggests that hominins have been hairless for at least 1.2 myr (Rogers *et al.* 2004). There are also potential costs associated with body hair, some indicated by the evidence from other mammals, in particular the energy required to grow a winter coat[8] and a reduction in the efficiency of heat loss through sweating, although those reductions could be minimised by growing distinctive winter and summer coats.

A potential body hair strategy must also be seen against the geographical and chronological contexts of the European Lower Palaeolithic. Early Pleistocene climatic and archaeological records suggest a highly episodic presence, with frequent local extinctions, of *H. antecessor* and *H. erectus* in the higher latitudes of both Europe and Eurasia (Dennell 2003; Dennell *et al.* 2011). In light of this, a 'hairy' reversion, and other purely biological solutions, seem an unlikely evolutionary investment for brief, Early Pleistocene, forays into Europe – this extends the argument previously made by Ashton and Lewis (2012, 59–60) with specific reference to northern Europe.

Yet the apparent challenges of Middle Pleistocene winter conditions in both the south and the north of Europe (Table 3.1), combined with the more expanded archaeological record after *c.* 600 kya, may support the case for a sustained reversion to all-over body hair by *H. heidelbergensis* from the early Middle Pleistocene onwards. This strategy would therefore reflect specific selection pressures in those cooler conditions, and MacDonald (2018) has concluded that the costs of body hair were probably feasible (*e.g.* due to the relatively minor thermoregulatory pressures to keep cool in European summers), and may have been combined with subcutaneous fat, as a means of surviving winter conditions. The co-occurrence of macaques (*M. sylvanus*) with hominins at a number of later European Lower Palaeolithic sites (*e.g.* Swanscombe and Hoxne) might also be an indicator in favour of a body hair strategy for the latter, given the physiology of the former (Ashton 2015).

A further complication concerns the still sporadic nature of the northern European occupation during the Middle Pleistocene: the site and palaeoenvironmental record strongly suggests the cyclical 'abandonment' of the north, probably through widespread extirpation, at least on glacial/interglacial timescales and probably also on shorter stadial/interstadial timescales. Such a pattern would certainly have occurred during the later Middle Pleistocene, and very probably during the early Middle Pleistocene as well given the suggestion of a marked glacial–interglacial amplitude in the north-west prior to the Mid-Brunhes Event (Candy and Alonso-Garcia 2018). Set against repeated range expansions and contractions, would body hair have been maintained, or re-developed during each re-expansion into the north? Given the relatively minor climatic differences between southern and northern sites in the Middle Pleistocene (Table 3.1) the former would seem more likely, and southern conditions, both winter and summer, do not seem incompatible with body hair. Nonetheless Wells (2012) has recently argued that developmental plasticity can contribute to phenotypic change across generations (*e.g.* shifts in primate body proportions in populations exposed to novel environments) – and this raises questions as to the scope for phenotypic change related to body hair in European populations at the timescale of a peak interglacial or warm sub-stage.

A closely related issue concerns the possibility that Lower Palaeolithic hominins may have had distinctive or enhanced physiological adaptations that improved their ability to survive low temperatures without any additional cultural insulation. These arguments have been made for Neanderthals, with specific reference to the potential benefits of brown adipose tissue (BAT), enhanced vasoconstriction and localised vaso-dilation (Steegmann Jr *et al.* 2002; White 2006; MacDonald 2018). The specific benefits of BAT for children can be seen in modern populations: adipose fat is retained through early post-natal life and if needed it can be modified into ketones, which are the key neonate brain fuel, through oxidation of fatty acids (Hublin *et al.* 2015). More generally, the possibility of pre-modern human genetic adaptations to specific environments has been indirectly demonstrated with reference to the role of the hypoxia pathway gene *EPAS1* in the adaptation of Tibetans to their high-altitude plateaux: critically, the gene's distinctive structure has been linked to introgression from Denisovans or Denisovan-related individuals into *H. sapiens* (Huerta-Sánchez *et al.* 2014). However, as White (2006) has argued, biological buffering can be energetically costly (Steegmann Jr *et al.* 2002), and potentially maladaptive in the event of inadequate food supply. Moreover, the evidence for cold acclimatisation amongst modern populations as a whole is variable (Golant *et al.* 2008).

With specific reference to Lower Palaeolithic hominins, MacDonald (2018) reviewed the potential contributions of muscle insulation and enhanced BMR (both included in the calculations in Table 3.6), but also potential internal heat production by brown adipose tissues (BAT) and non-shivering thermogenesis (NST; via acclimatisation). MacDonald concluded that while adults might have survived at some northern sites, assuming temperatures at the top of the reconstructed ranges, they would have had to invest heavily in foraging and heat production and would have been vulnerable to wet weather and wind chill. At other, cooler, sites, they would have been at risk of hypothermia, especially at night. Finally, the vulnerability of infants would have been greater than that of adults as a consequence of their small body sizes: while studies have suggested that infants can greatly increase their heat production when subject to cold stress, this impacts on growth, so cannot have been a habitual response. Unsurprisingly, MacDonald (2018) concluded that any such physiological contributions would likely have been combined with other solutions, such as behavioural and cultural strategies.

One last option worth briefly considering is hibernation, for which there are a number of animal parallels. However, MacDonald (2018) concluded that hibernation was unlikely, due to the risks of starvation, vulnerability to predation and the need for a significant pre-winter weight gain (but see also the discussion of possible autumn 'stockpiling' in Chap. 6).

In summary, site-based temperature estimates, palaeoclimatic modelling, and predictions of *H. heidelbergensis*' hominin physiology together make a strong case that the winter temperatures recorded at known northern European occupation sites and, during the worst winters, at southern sites, were often, if not always, below the basic tolerances of a hairless, and non-culturally insulated, Middle Pleistocene hominin. Moreover, even when MST values exceeded (*i.e.* fell below) site temperatures, *e.g.* in southern Europe during the Early Pleistocene, the hominins would still have needed to secure significant food supplies to maintain internal heat production. All-over body hair, combined with sub-cutaneous fat and/or other physiological adaptations, is certainly one potential means of increasing climatic tolerances (MacDonald 2018), and was probably feasible for hominins across Europe given the relative similarities and partial overlaps of northern and southern conditions (Tables 2.3 & 3.1). Yet if a body hair reversion did occur in Europe, it is possible that it only appeared in the later Lower Palaeolithic, both in response to long-term climatic cooling trends and in association with the more sustained occupation evidence. But if not body hair, how else might Lower Palaeolithic hominins have survived winter climates?

Put the heating on ...

Fire, at least while humans work and/or rest in one place, is one of the three cultural methods of ameliorating harsh winter climates, the others being clothing and shelter. The benefits of fire would also have extended far beyond simply warmth in a Lower Palaeolithic context, spanning cooking, possible aids to tool-making, defence against predators, aiding and changing social interactions and extending the hominin day (Fluck 2007; Wrangham 2009; Barham 2013; Dunbar and Gowlett 2014; Barkai *et al.* 2017). The provision of artificial lighting would have enabled hearth-side tasks such as tool-making and, by allowing individuals to see each other's facial expressions and body language, the maintenance of social bonds (Wrangham 2009; Dunbar and Gowlett 2014). A relatively early origin for the controlled use of fire has also been proposed on the basis of our unique, universal 8-hour sleep pattern, which is distinctive from other diurnal primates in being decoupled from annual cycles of natural daylight and darkness (see also Table 1.1), and presumably evolved through firelight (*e.g.* Gowlett 2010; Twomey 2013). Yet the evidence for anthropogenic Lower Palaeolithic fires, as opposed to natural wildfires, has always been very limited, both within and outside of Europe (*e.g.* Roebroeks and Villa 2011; MacDonald 2018). This has long been an area of speculation and controversy for Palaeolithic archaeologists, with much debate over the claims of specific sites (*e.g.* James 1989; Rolland 2004).

Arguments for fire – the direct evidence

Recent years have seen a small expansion in both convincing, and more controversial, claims for hominin fires from the later stages of the European Lower Palaeolithic: Beeches Pit and Foxhall Road in the UK (White and Plunkett 2004; Preece *et al.* 2006), Menez-Dregan 1 and Terra Amata in France (De Lumley 1969; 2006; Mercier *et al.* 2004; Roebroeks and Villa 2011; Monnier *et al.* 2016), Schöningen and Bilzingsleben in Germany (Mania and Mania 2005; Thieme 2005), and Vértesszőlős in Hungary (Kretzoi and Vertes 1965; Roebroeks and Villa 2011; for other examples see also Rolland 2004, table 3). The controversies involve issues of site chronology (Menez-Dregan 1: Mercier *et al.* 2004), taphonomy (Bilzingsleben: Gamble 1999; Müller and Pasda 2011), and sedimentological traces (Schöningen: Stahlschmidt *et al.* 2015b), while variations in fieldwork methodologies can be a further complicating factor (*e.g.* Marquer *et al.*

Box G: Fire at the Schöningen lakeside?

Alongside its spectacular spear and butchered horse evidence, the German site of Schöningen was initially argued to have also made a significant addition to knowledge of Lower Palaeolithic fire. However, the four claimed hearths (*e.g.* Thieme 2005) have been strongly criticised recently (Stahlschmidt *et al.* 2015b). This later analysis has demonstrated that the localised reddening of the site's calcareous marl sediments (the 'hearths') in fact results from localised iron oxidation, rather than from the effects of heating. This has been further supported by a range of other studies: only very few dispersed and fragmented pieces of charred plant tissue have been identified, and those that have are herbaceous and likely to be from natural peat fires; there is no evidence for the alteration by heat of bone, mollusc and ostracod fragments; micromorphological studies show no difference in the structure and components of the samples taken within and outside the 'hearths'; there are no thermoluminescence differences within and outside the 'hearths'; and reflectance values on sediment and wood samples suggest humification not carbonisation (Stahlschmidt *et al.* 2015b).

Is this surprising? It depends on the nature of the kill-butchery site. If it was controlled by hominins, *i.e.* if other carnivores and scavengers could be repelled, enabling several hours or days of sustained butchery and consumption by an entire hominin group, then fire might well be expected here – and indeed might be central both to carnivore and insect repelling and perhaps also to cooking. Such a level of 'control' of the locality is suggested by the overall abundance of butchery evidence, the range of activities (*e.g.* skinning, defleshing and marrow extraction) and the limited evidence for carnivore gnawing (Voormolen 2008; van Kolfschoten *et al.* 2015a; see also Chap. 6). If on the other hand the kill-butchery site was a less secure place, one from which hominins sought to depart as quickly as possible, perhaps weighed down with meat and other animal products, then hearths should probably not be expected.

Stahlschmidt *et al.* (2015b) also criticised the Beeches Pit fire evidence (Gowlett 2006; Preece *et al.* 2006), on the grounds that the reported temperatures do not exclude natural fires,[1] and that the hearths were also identified on the basis of oxidised sediments. Renewed analysis of the Beeches Pit evidence is currently ongoing, with results pending. However, a rejection of the Beeches Pit claims would be more surprising than the changing views at Schöningen. This is because the Beeches Pit site setting in Beds 4 and 5, a closed deciduous woodland with tufa-forming springs (Preece *et al.* 2006), would seem a more likely venue for a maintained fire and associated sleeping site, if such behaviours occurred at all (see also Chap. 4: Box L). Moreover, the recurrence of localised burning traces over three of the site's beds (3b, 5 and 6) appears a strong argument against natural fires (Preece *et al.* 2006).

---

[1] Natural and anthropogenic fire temperatures can also be highly variable.

2012). The one chronological exception to the above sites is the Early Pleistocene Spanish site of Cueva Negra, where hearth-less fire tending, if not control, has been suggested on the basis of burning temperatures, inferred from modified stone, bone and sediment materials through microscopic examination of visual modifications and scanning electron microscopy and energy dispersive spectroscopy (SEM-EDS) (Rhodes *et al.* 2016; Walker *et al.* 2016).[9]

Irrespective of these individual cases two patterns can nonetheless be highlighted. First, the broader lack of generally accepted examples remains surprising, particularly in light of the climatic challenges of winter. This pattern may however be a by-product of the distinctive character of the Lower Palaeolithic archaeological record. Both the Beeches Pit and Bilzingsleben examples are from non-fluvial contexts which may have favoured the preservation of fire traces. The paucity of other examples of controlled fire use may well be a taphonomic consequence of the bias towards high energy river floodplain sediment traps in the European Lower Palaeolithic record (*e.g.* Roebroeks and van Kolfschoten 1995; see also Chap. 2: Box F). Such environments would be unfavourable to both the preservation of fire traces and perhaps also to hominins' inclination to make fires in such locations, where they would be both exposed and vulnerable to other predators (see also Box G). The size of fires may also be a key factor in the issue of archaeological visibility. Gowlett (2006, fig 3) suggested that the hearths at Beeches Pit were ovoid in shape and *c.* 1 m across, although the suggestion of a knapper sitting at least *c.* 1 m away, on the basis of the spatial distribution of lithic artefacts, is potentially indicative of the degree of heat, and light, generated. Similar dimensions were suggested for the charred bone areas at Vértesszőlős (Kretzoi and Vertes 1965). Nonetheless, as Pettitt and White have argued:

> Most unstructured hearths would probably have consisted of little more than a few twigs and branches around which people huddle very closely. It is extremely unlikely that such features will survive intact in anything other than the most exceptional preservational environments. (Pettitt and White 2012, 196)

However, there is also a wider issue. The Lower Palaeolithic fire debate has typically been focused around the quality of the evidence (*e.g.* Preece *et al.* 2006; Stahlschmidt *et al.* 2015b) and less around the need for, and costs of, fire, whether in winter or at other times of the year. Taking the latter approach the absence of evidence for fire in Lower Palaeolithic cave sites, such as Gran Dolina (occupations are associated with levels TD-6.2 and TD-10) and the early levels at Arago, is noteworthy, as it suggests that unfavourable site settings and preservation conditions may not be the only reason for a limited fire record (Roebroeks and Villa 2011). It might even suggest that sheltering in caves removed a need for fire. Secondly, it is notable that the majority of those Lower Palaeolithic European fire sites which are known are located to the north of the Pyrenees and date to MIS 13 or later, while the earliest combustion structures in southern Europe are dated to MIS 7c (*c.* 230 kya) at Bolomor Cave, in Valencia, Spain (Fernández Peris *et al.* 2012). While the current distribution is certainly a partial one, and is almost certainly spatially and temporally biased, these existing geographical and chronological patterns suggest that managed fire may have been a relatively late innovation in Europe, and in the Lower Palaeolithic was primarily utilised as a source of heat, rather than for other purposes such as cooking.

Arguments for habitual fire – a fuel-rich environment

Pollen records from interglacial and warm stage sites certainly highlight a wide range of potential wood fuels for starting and maintaining fire: hornbeam, alder, juniper and pine are all good fuels for example, the latter two because of their resin content (Bigga *et al.* 2015). Indeed Théry-Parisot and Meignen (2000) have suggested that there is sufficient, locally available, dry standing wood and deadwood in dense forest habitats to support sustained fire burning. Specifically, they proposed that an area of no more than 1 km radius around a hypothetical site would support four fires, burnt 24 hours a day, for 6 months (Henry 2017). Animal dung and/or bone would also be alternative options, although ongoing experiments (Alex Pryor, pers. comm.) suggest that heat transfer from flaming to unlit portions of a bone may be relatively inefficient. In terms of kindling, birch bark is widely recognised as an effective material by bushcraft professionals (*e.g.* http://paulkirtley.co.uk/) and has been frequently documented ethnographically (*e.g.* Holloway and Alexander 1990), although a variety of other barks (*e.g.* lime) and materials can also be used. Seed fluff or fleece (*e.g.* the seed heads of marsh thistle and cattail) sparks very effectively, and are available in autumn and winter, while lichens such as Old Man's Beard (*U. barbata*) and the resin-rich bark of pine and spruce can also help with fire starting (Bigga *et al.* 2015). Many species of polypore fungi are also effective tinder materials (Peintner *et al.* 1998), including King Alfred's cake (*D. concentrica*) and the tinder fungus (*F. fomentarius*), the latter of which was identified at the Middle Palaeolithic site of Salzgitter-Lebenstedt (Tyldesley and Bahn 1983).

Since fuel can be bulky to carry, it seems likely that fuel collecting was at least sometimes a dedicated task, although it could also be acquired by unsuccessful hunters. Henry *et al.* (2018) noted that these tasks are often undertaken by women and children in recent hunter-gatherer communities. Collecting such static resources, with no risk of fight or flight, might well favour the demographic extremities in a hominin group (*i.e.* the young and the old), with elder individuals potentially teaching, or children independently learning, how to identify 'good' and 'bad' wood. An obvious example of the 'bad', namely saturated wood, would be especially common from late autumn to early spring, although this issue would not have been exclusively limited to those times of year. Fresh wood and rotten wood would also likely be avoided where possible (Henry 2017).

Arguments for habitual fire – a social-encephalisation perspective

The issue of wet or damp fuel and tinder highlights that often the starting of a fire is a greater problem than the maintaining of one, and thus 'fire minding' may have been a key social role in the winter months, fuels stocks permitting. This has implications for day-to-day tasks and the composition of local hominin foraging groups, as group-wide benefits would arise from individuals being responsible for remaining at, and maintaining, domestic fires, either through refuelling or allowing it to smoulder during the day. Those individuals might be elders, young infants, or individuals of limited mobility, for whom engagement in food or other resource-getting tasks was difficult, but whose contribution to group survival may still have been critical[10] (see also Pettitt 2000). Given the amounts of timber required for fire maintenance, and the demands of fuel-getting in the winter months, combining fuel-getting with food-getting might not have been feasible in terms of how much individuals could carry and the relatively short daylight hours available for foraging. Fuel-getting might therefore also have been a task for those 'left' around the fire, although not necessarily women and children (*cf.* Henry 2017).

Indirectly, the need for these social 'fire roles' can be framed as possible support for habitual fire use, through reference to the evidence for marked hominin encephalisation in the Middle Pleistocene. If European winter climates, in the form of low temperatures and high precipitation, required artificial heating and drying during the day as well as the night, then significant cognitive demands are implied.[11] Those cognitive costs would result from regular, perhaps daily, group divisions, with different individuals or groups foraging and maintaining the fires respectively, as suggested above. The consequences of such regular group sub-divisions and re-formations, also known as fission and fusion (see Chap. 4: Box K), include the need for reliable food-sharing and robust social bonds. In short, if members of a group are staying in or close to camp to tend a fire at a time of limited resource availability then they need to be able to rely on the foragers returning, bringing food with them, and sharing it. The marked encephalisation associated with *H. heidelbergensis* from around 600,000 years ago is noteworthy in this context, especially when connected to the chronology and geography of the fire sites (cool and wet winter conditions are likely to have been especially prevalent in the Atlantic West, where the majority of Lower Palaeolithic sites tend to cluster). By contrast, the mildest conditions of the southern Early Pleistocene sites (Table 3.1) may not have required habitual winter fires, or the associated cognitive costs.

Arguments against habitual fire – what about the costs?

Given the abundance of fuel within the various warm stage forests and woodlands of the Pleistocene, why might fire not be a habitual winter technology, and therefore more commonplace in the archaeological record (taphonomic bias notwithstanding)? One reason may stem from our tendency when discussing fire to accentuate the positives: heat, light, cooking and protection (Henry 2017; Henry *et al.* 2018). Yet there are costs to fire technology too: the challenges of locating dry fuel, especially tinder, in the wetter months, acquiring, retaining and passing on the knowledge and skills required to reliably make and maintain a fire, potentially becoming increasingly visible and vulnerable to other groups, and the social requirements of keeping a fire alight (Barham 2013, 170).

A further risk, rather than benefit, of fire has been emphasised by Hardy *et al.* (2016) with reference to enclosed environments and the potential for fire smoke to act both as an irritant (*e.g.* to the eyes and as a cause of coughing) and as a source of serious health problems, *e.g.* to the lungs. The presence and size of the micro-charcoal fragments in the dental calculus samples from the Qesem Cave in Israel suggested their accidental inhalation and a sooty atmosphere in the cave, as does the detection of soot in cave speleothems (Vandevelde *et al.* 2018). While enclosed cave and rockshelter habitats appear to have been less prevalent in the European Lower Palaeolithic than in later periods (and indeed smoke hazards might explain the absence of fire evidence at sites such as Arago), the principles of smoke management would still apply to some extent within temporary shelters exploiting vegetation, animal hides or tree-throw pits.

These and other demands and costs have recently been considered with regards to the potentially sporadic evidence for Neanderthal fire use in the Middle Palaeolithic (Sandgathe *et al.* 2011; Dibble *et al.* 2017; Henry 2017; but see also Sorensen 2017). Focusing more widely on early hominins, Henry *et al.* (2018) emphasised the process of fuel gathering and, while using a limited cost/benefit model (costs = fuel gathering; benefits = enhanced energy from cooked foods), made a strong case against obligate fire use. Thus, while open-air site taphonomy and preservation factors are certainly a major complicating factor, a seasonally-structured cost and benefits-type approach can help to clarify patterns in the Lower Palaeolithic record.

Arguments against habitual winter fire – the challenges of finding fuel

A particular winter cost for Lower Palaeolithic hominins would have concerned the difficulty of gathering the necessary raw materials for fire at those times of the year when, from a purely insulation perspective, it was most typically needed. Fuel-getting at any time is a significant day-to-day task. The recent experiments of Henry *et al.* (2018) burned 19 kg of fuel on a continuously flaming fire, *c.* 30 × 50 cm in size, for approximately 3.5 hours, while a 2018 experiment at Barnham by Davis, Hogue and Scott (Rebecca Scott, pers. comm.) burned 35 kg of seasoned wood on a *c.* 60 cm diameter fire over 2 hours (temperatures returned to pre-burn levels after *c.* 12 hours). Both experiments highlight the investments in time and effort that would be required to maintain fires over several hours or days, while Palaeolithic fire performances would be further complicated by factors such as variable fuel condition. The greater prevalence of wet fuel in the winter months would increase carried weights and/or, if dry wood was preferentially sought, extend search times (Henry *et al* 2018), which combined with short winter day lengths, and the implications of the latter for the length, scale and focus of foraging activities, might be a significant challenge to the maintenance of winter campfires over several days. The experiments by Henry *et al.* (2018) unsurprisingly indicated greater search efforts in open habitats (in terms of distances covered and fuel weight gathered, over a set time), but also a tendency for wetter, lower quality wood from forests. This highlights the importance of mosaic landscapes and ecotones, often indicated by the available archaeological habitat data

(Chap. 2), in facilitating access to both open and closed habitats, for sighting game and gathering fuel respectively. Overall, these experiments highlighted that winter fire maintenance might be difficult in those wooded environments that appear to dominate the Lower Palaeolithic record, depending on moisture variations in the available dead-wood.

A possible mechanism for reducing the costs of fuel gathering might be in the stockpiling of dead wood during other times of the year. However, this implies the existence of sustained local territories and/or residential spaces (see also Box I), and evidence of planning ahead across weeks or months. The available lithic transfer data is suggestive of relatively small-scale mobility (Wilson 1988; Féblot-Augustins 1999; see also Chap. 5), although the duration of a group's stay at a particular location, or even within a local area, is unknown. While residential sites have been identified (Chap. 4: Box L), the relative paucity of cave and/or rockshelter occupations, or artificial shelters, further complicates any fuel stockpiling – although wood could perhaps have been gradually dried and/or kept dry in the crooks of trees. Finally, fuel depletion may well have been a significant problem in more open habitats, such as early/late warm stage environments. This has recently been argued by Pryor *et al.* (2016) with reference to Gravettian groups in the harsh environments of the mid-Upper Palaeolithic, where deliberate tree killing in advance of need has been proposed as a fuel management strategy.

An alternative is that fires were made where and when natural dead wood supplies permitted fire-making at relatively low 'economic' cost and/or when the benefits of fires were most strongly needed. This is perhaps more in-keeping with the highly sporadic evidence for fires in the Lower Palaeolithic archaeological record. Such benefits need not be limited to cooking and/or food preservation: fire as a source of warming and drying may have been equally critical during cold, wet winters. Whether fires were episodic or habitual, the choice of winter landscapes may well have been guided by accessible fuel stocks, whether natural or pre-prepared by hominins. Such choices may have been guided and aided by a symbiotic, albeit perhaps rather parasitic, relationship with beavers. Thompson *et al.* (2016) have noted that beavers create significant quantities of dead-wood in contemporary boreal forests, as damning-created inundation results in the widespread die-off of trees in the flood zone. Such a relationship might also have extended to the exploitation of beavers for their fur and/or fatty tails.

Arguments against habitual fire – the challenges of fire-starting?

A further question concerns how fires were started, particularly in light of the suggestion by Dibble *et al.* (2017) that Neanderthals were unable to generate fire at will during the Middle Palaeolithic. While there are a wide range of ethnographic methods (*e.g.* bow-drills, spindle and fire board, flint and tinder; Barham 2013), archaeological evidence for fire-lighting in the Palaeolithic, as in later prehistory, is sporadic. Sorensen *et al.* (2014; 2018) have, however, identified possible strike-a-lights from the Middle and Upper Palaeolithic records. Polish and striations on late Middle Palaeolithic bifaces, combined with modern experiments, suggest the percussion of pyrite fragments against the face of the artefact to produce sparks (Sorensen *et al.* 2018). Comparable examples from the Lower Palaeolithic may yet be found – the incipient percussion cones on handaxes from South Woodford are intriguing (Pettitt and White 2012, 166–167 & fig 4.11), while Wymer (1968, fig. 32) noted localised battering on several handaxes from Swanscombe. However, there are no unambiguous examples yet known from the Lower Palaeolithic record, raising the question as to whether the fires at sites such as Beeches Pit, Menez-Dregan and, on the margins of Europe, Qesem Cave (Karkanas *et al.* 2007; Shahack-Gross *et al.* 2014), were generated on-demand or exploited natural wildfire events? This issue has been explored in Gowlett's (2016) proposal that hominin use of fire can be seen to have evolved through various different stages: from opportunistic (*i.e.* an attraction towards natural fire in order to benefit from it and/or other, associated, resources) to limited conservation (*i.e.* maintaining the duration of naturally acquired fire beyond the burn time of the original source) to kindled (*i.e.* modern fire use). Would a limited conservation-type strategy be feasible in Lower Palaeolithic Europe? With reference to natural fire sources Sorensen (2017) stressed that lightning flashes, ground strikes and wild fire rates are by no means comparable: while the Dordogne receives *c.* 3 lightning flashes/km^2/year today, about 75% of those flashes are restricted to the clouds, and only 1–4% of the remaining cloud-to-ground strikes ignite a fire (based on sub-boreal forest environments; even lower rates are suggested for temperate deciduous forests). Sorensen concluded that between 0.0075–0.03 fires/km^2/yr could be expected in the Dordogne. Might such low rates explain the relative paucity of Lower Palaeolithic fire evidence? However, Sorensen also highlighted that fire can be detected over significant distances:

> Depending on weather conditions and terrain, smoke plumes are potentially visible up to 50 km away...Using the average single day foraging distance for modern hunter-gatherers of around 15 km (round-trip), one could reasonably expect to encounter between 1.3 and 5.3 natural fires yr^{-1} (of any size) in the Dordogne today within this 176.7 km^2 daily foraging area surrounding a site...Logs left smouldering can potentially burn for days or weeks...meaning if the desire were great enough, any fire within visible range could be reached within a few days and may still be exploitable long after the flaming fire front has been extinguished. (Sorensen 2017, 119)

In this scenario fire could be acquired from the margins of wildfires, presumably in their late stages, or from smouldering trees, and curated, as burning or glowing brands, until site fires could be lit. Peintner *et al.* (1998) suggested that dried and partly hollowed specimens of the polypore fungi species *F. fomentarius* could also have been used to transport fire from place to place, due to their ability to smoulder for several hours. Moreover, if hominins moved *to* natural fires (and assuming a smoke plume visibility over a 50 km range), then *c.* 59–236 fires/year could potentially be exploited (these estimates are slightly lower than those suggested by Sorensen [2017]), although the largest of such moves would be exceptional in the context of typical raw material transfer distances at this time (Féblot-Augustins 1999). Still-burning or smouldering tree trunks, in particular, might have been an attractive fire resource. Wildfire evidence is certainly known from Europe at this time. Lebreton *et al.* (2018) documented wildfire traces in the Boiano Basin, southern Italy, during the Middle Pleistocene, from micro-charcoal analysis (although analysis of the archaeological sites of Isernia la Pineta and Gaudo San Nicola in the same study was unable to demonstrate fire use by hominins).

However modern forest fires show clear seasonal patterns. Data from the European Forest Fire Information System (EFFIS; http://effis.jrc.ec.europa.eu/) highlights April–September as the key period for Germany, and May/June–September for Spain, Italy and France, although occasional winter fires also occur (UCJRC 2011; data: 2006–2010). While modern habitats, and causes of forest fires, are clearly much more complex than their Pleistocene equivalents and include some non-comparable factors (*e.g.* agricultural burning), the data strongly suggest that a winter reliance on natural fire would have been a high-risk strategy for Lower Palaeolithic hominins. Thus, a failure to be able to produce fire on demand may have been one of the key threats to winter survival, especially in the cooler, highly seasonal winters of the Middle Pleistocene when the energetic costs and risks of whole group moves to the nearest wildfire source would be greater. The exploitation of natural wildfire as an explanation is also brought into question by those sites where fire use appears more sustained and is perhaps therefore a relied-upon strategy rather than the product of fortuitous opportunities (*e.g.* the multiple-level hearths at Menez-Dregan, France and, especially, Qesem Cave, Israel). Since those sites are located at specific points in the landscape the catchment of available wildfire sources (*i.e.* accessible through daily round trips) is reduced, suggesting an ability to produce fire on demand in those cases.

In light of these potential sources Twomey (2013) has proposed a model of fire access rather than production during the initial stages of controlled use of fire. Access is suggested to have occurred either through stealing, by stealth or confrontation, or through negotiation with other groups. However, I find this unlikely given the low population densities suggested by both archaeological and carnivore models (Grove 2009; Churchill *et al.* 2016; Chap. 5: Box O). Moreover, I am not convinced that such approaches would be the least cognitively demanding, as Twomey argued, in comparison to simple stone-on-stone sparking. Nonetheless, Twomey (2013) highlighted other important issues, including the potential time budget conflicts between fire behaviours (*e.g.* fuel gathering) and food foraging, and the potential importance of sharing and communicating fuel knowledge (*e.g.* locations). Twomey also emphasised the vulnerability of small fires to heavy/sustained rainfall, although the impact of this would be reduced under forest canopies. Twomey's suggested cognitive requirements, in particular anticipatory planning and future-directed group-level cooperation, remain applicable to fire lighting as well as fire access models, particularly with reference to fuel gathering in anticipation of ongoing needs (although it is less clear whether this would operate at the scale of hours, days or even weeks). This would probably be especially relevant in a seasonal environment where the condition of fuel varies through the year, and where the problem of damp winter fuel and tinder would be a persistent issue.

Arguments against habitual fire – fire for cooking?

The Qesem fires are noteworthy for their apparent association with meat roasting and cooking (Barkai *et al.* 2017). Yet from a European winter perspective, this relationship may not be so straightforward. Henry *et al.* (2018) highlighted the relatively modest differences in energy between a sample of selected raw and cooked foods and suggested that the costs of fuel gathering for a short-lived fire were difficult to offset through the energy gains of cooked food, particular plant foods. Moreover, energy gains can be made without cooking, by reducing masticatory effort through the pounding of foods such as underground storage organs (Zink and Lieberman 2016), using the percussion tools known from sites such as Barranco León D and Fuente Nueva-3 (Barsky *et al.* 2015; Titton *et al.* 2018). Pounding can also remove toxins from plant foods, although in low latitude environments this has often been in combination with other methods such as grating, leaching in water and cooking (Rowland 2002; Harris 2006). Rotted meats similarly reduce the necessary masticatory efforts and might also reduce the need for storage by smoking or drying (Speth 2017). Henry *et al.* (2018) suggested that the most likely foods associated with a cooking strategy would be larger game, as their size would enable them to be cooked on a single fire but then consumed over several days. The nature and availability of winter game, *e.g.* their reduced body fat and generally poorer condition, would presumably reduce the degree of such benefits (as would a predominantly scavenging rather than hunting-based strategy for accessing animal foods; see also Chap. 4), further complicating the cost/benefit balance during the coldest months of the year.

While the overall sample size of Lower Palaeolithic fire makes any chronological trends highly tentative, it is notable that the absence of earlier cooking at Atapuerca during the Early Pleistocene is also suggested by buccal dental microwear evidence from Gran Dolina TD-6 (*H. antecessor*) and Sima del Elefante (*Homo* sp.). The density of scratches is greater than in later hominin samples, and has been interpreted as indicative of harder, tougher foods that were not thermally processed (Pérez-Pérez *et al.* 2017). In the context of possible later changes in diet and/or food processing methods, it is perhaps noteworthy that the *H. heidelbergensis* samples from this study were more comparable to Neanderthals than to *H. antecessor*, although this is clearly not unequivocal evidence for widespread fire use and cooking in the later Middle Pleistocene.

Lower Palaeolithic fire – an episodic occurrence?

Overall the evidence, both direct and indirect, for habitual fire is ambiguous. In light of an early hominin presence in Europe, especially northern Europe, that was episodic and fragmented (*e.g.* Dennell *et al.* 2011), fire would therefore seem to have been an important, but probably not habitual, part of the hominin arsenal of survival strategies. Nonetheless the winter temperatures in both the south and the north certainly suggest the seasonal importance of fire in both regions, particularly during the Middle Pleistocene (Table 3.1), and the drying and warming benefits of fire (but not necessarily cooking) may have been especially important in those areas with cold, wet winters, such as the Atlantic west. It is therefore noteworthy that some of the most convincing and sustained fire evidence currently stems from Menez-Dregan in Brittany (Monnier *et al.* 2016).

Clothing, or I can't feel my fingers …

A key limitation of fire, and shelters, concerns their stationary character. Given the need for hunter-gatherers to keep warm while foraging for food and other resources, some form of mobile insulation would be needed. Numerous reconstructions of Lower Palaeolithic hominins, and their Neanderthal descendants, have portrayed those insulation 'strategies' as either all-over body hair, very localised skin/fur coverings or simply going naked, the latter frequently combined with specific postures to respect modern sensibilities, with seemingly little thought given to the meeting of genuine survival needs (Figs 3.2–3.5). The all-over body hair option was reviewed above and considered to be a possibility during the more sustained European occupations of the Middle Pleistocene. Without it, the need for clothing becomes paramount. As John Wymer wrote:

> There seems no reason to think that a species which evolved in the tropical or Mediterranean climates of Africa would have been any more equipped to cope with the British climate during interglacials than ourselves. Even if they were hardier and hairier it would seem unthinkable that they would not have sought or made means of protecting themselves from the cold and wet of summers, let alone winters, or during periglacial conditions. (Wymer 199, 36)

The specific winter risks, as anyone who has worked outside without gloves in winter or just stood on a cold football terrace for 90 minutes will testify, are frostbite ('injury to body tissues caused by exposure to extreme cold, typically affecting the nose, fingers, or toes and often resulting in gangrene'; Oxford Dictionaries 2018) and hypothermia ('having an abnormally [typically dangerously] low body temperature'; Oxford Dictionaries 2018). Modern medical advice states that mild hypothermia spans core body temperatures of 32–35°C,[12] with moderate to severe hypothermia occurring below body temperatures of 32°C,[13] and leading rapidly to death (Parsons 1993; Golant *et al.* 2008). Modern populations as a whole typically rely on food, shelter and/ or clothing to prevent both hypothermia and frostbite (Golant *et al.* 2008; Pocock *et al.* 2018). The specific environmental conditions associated with hypothermia vary depending on the age and/or mobility of the individual, but external temperatures above 16–20°C, and appropriate clothing, are recommended to avoid its occurrence

Figure 3.2: Reconstruction of unclothed (and body hair-less) Lower Palaeolithic hominins at Boxgrove (Roberts and Parfitt 1999, rear cover; © Peter Dunn, English Heritage Graphics Team).

Figure 3.3: Lower Palaeolithic hominin reconstruction, with torso-covering clothing and limited body hair (© José Antonio Peñas).

in babies and immobile adults (NHS 2017: https://www.nhs.uk/conditions/hypothermia/).

The available site-based temperature reconstructions suggest that exposure to hypothermia could potentially occur throughout the continent. Such risks would apply both during daily, local foraging (assuming momentarily that 'domestic' spaces included the warming benefits of fires) and during any longer journeys, such as residential moves or seasonal migrations. They might also apply while hominins were at rest, if fire use was sporadic, especially in areas where diurnal temperature variations were more marked. Thermal stress is also a particular issue for infants and young children, due to their small body sizes, high surface area/weight ratios,

Figure 3.4: 'Hairy' H. erectus *(although relatively short hair lengths are suggested) (by Zdenek Burian © 1972).*

and higher thermoneutral zones[14] (Mateos *et al.* 2014; MacDonald 2018). Its avoidance in the Lower Palaeolithic would therefore place significant demands on, at the least, the abilities of Early and Middle Pleistocene hominins to acquire adequate resources (food, and possibly, fuel) and to balance the needs for, against the risks of, high levels of daily movement. It might also impact on the timings of any seasonal migrations (see also Chaps 3 & 5).

In light of the earlier discussions of biological insulation (Table 3.6), and especially if all-over body hair was absent, clothing may have been *the* critical requirement in the Lower Palaeolithic (Gilligan 2017). The complex issue of clothing in the earlier Palaeolithic can be broken down into a number of separate questions. What do we mean by 'clothing'? Did it exist at all? If it did, what was it like? Which raw materials, tools and techniques were used to produce it? Did 'clothing' vary across the year, or in different locations, and if so how? When did it first appear, and was it consistently used after this point? The working definition of clothing as used here is as follows:

Figure 3.5: Neanderthal 'loin-cloth'-type clothing (by Zdenek Burian © 1950).

any and all additional materials, covering all or part of the body, for the purposes of protecting exposed skin/hair from immediate weather conditions (*i.e.* air/surface temperature, precipitation, sunshine, wind). Those materials can be animal or vegetable, may be tailored or untailored, and can encompass worn clothes (covering any part of the body), coverings used when sitting or sleeping, and carriers (*e.g.* slings) for infants (Taylor 1997).

So was such clothing a component of Lower Palaeolithic lives? The available archaeological evidence is scarce. Unsurprisingly there is no direct evidence for clothing, in the form of preserved hides or furs. Moreover, there is none of the indirect evidence that has formed the basis of clothing reconstructions for the European Upper Palaeolithic, such as bead distributions on buried bodies (*e.g.* Sunghir: Trinkaus and Buzhilova 2012), bone needles (Gilligan 2010), textile imprints and other traces (*e.g.* Adovasio *et al.* 1996; Soffer 2004), and occasional representations of clothing on figurines (*e.g.* Soffer *et al.* 2000). Yet indirectly, evidence in favour of some sort of clothing comes in the form of palaeoenvironmental reconstructions (Table 3.1) and hominin models (Table 3.6), which suggest a possible need for cultural insulation. Site-based palaeoenvironmental data suggests that while Early Pleistocene winter conditions in the occupied south were milder, Middle Pleistocene interglacial winter temperatures were typically at or just below freezing in the north (and just above freezing in the south), although these obviously varied across the duration of a warm stage. Since the occupation of Europe was a punctuated process, limited to the Mediterranean until *c.* 1 mya, and only expanding substantially into northern Europe after *c.* 700–600 kya, it is therefore possible that the adoption of clothing *sensu lato* was a gradual, multi-staged process, along the lines of Gowlett's (2016) conceptual approach to fire.

As summarised in Table 3.6, minimum sustainable temperature predictions for *H. heidelbergensis* (and probably *H. antecessor*) do suggest limitations for non-hairy biological solutions (*e.g.* body shape, muscle mass or elevated BMR) to the challenges of winter temperatures, particularly in northern Europe and particularly during the Middle Pleistocene. This becomes especially true if elevated BMR is not permitted. If one clo or more of cultural insulation is permitted, then higher summer temperatures are broadly comparable with the modelled lower critical temperatures (18.0 or 20.5°C for *H. heidelbergensis* [Kleiber and elevated BMR respectively], while minimal sustainable ambient temperature estimates (-12.6 or -20.0°C) exceed typical winter conditions. Even if additional physiological advantages were also present, such as elevated brown adipose tissue levels (brown fats generate body heat and help to maintain core body temperature), and enhanced vasoconstriction and localized vasodilation (Steegmann Jr *et al.* 2002), extra cultural insulation would still markedly reduce the energy intake demands of such physiological adaptations and/or of maintaining internal heat production at three or more times BMR (MacDonald 2018).

The need for additional insulation is supported by Wales' (2012) modelling of likely clothing coverage for Neanderthals, based on an ethnographic survey of 245 modern and historical hunter-gatherer groups. Even set against the favourable Pleistocene extremes of MIS-5e (a time of hippopotamus and lions beneath present-day Trafalgar Square in London; Franks 1960), and using southern Britain as a 'point'-specific example, Wales (2012, figs 3, 5, 8 & 9) suggests minimum–maximum ranges of 20/30–40/60% (of the body covered by clothing), 25–75% (the probability that hands were covered by clothing) and 25–50% (the probability that feet were covered by clothing). Across Europe, those predictions vary along a broadly northeast–southwest transect, with reduced clothing needs in the Atlantic west and increased needs for hands and feet to be covered in high relief areas such as the Alps and the Carpathians. Moreover, for the warm phases of MIS-4 and 3, the southern Britain figures change, respectively, to 30/40–70/80%, 50–90%, and 75–90%. Although the majority of Lower Palaeolithic occupations are likely to have been in conditions closer to those of MIS 5e (*e.g.* Table 2.3), the latter set of figures is particularly interesting with regards to the southern Scandinavian-type conditions of Happisburgh III (Parfitt *et al.* 2010) and the late MIS 11 occupation of Hoxne (Ashton *et al.* 2008a).

So what form might such clothing have taken? The absence of formal bone needles in the pre-Upper Palaeolithic archaeological record has often led to assumptions of very simple clothing (if at all), such as loincloths, wraps or capes – although most commonly illustrated on the body of Neanderthals (Fig. 3.5). The limitations of loincloths, namely the leaving of a large percentage of the body exposed, are obvious. Capes could provide the space for infants to shelter beneath them, especially during periods of travelling, and potentially benefit from skin-to-skin contact (MacDonald 2018; see also Fig. 5.2). However, untailored wraps or capes covering the majority of the body present their own problems, principally the pinning of the arms within capes and the absence of effective air-capture systems that reduce the replacement of warm air by cold air and permit the venting of excess heat and moisture (White 2006). There is though some disagreement as to the need for tailored clothing: MacDonald's (2018) review of ethnographic examples suggests that draped, singe-layer clothes would offer 1–2 clo of insulation and would be adequate for winter survival, if combined with additional energy inputs (*i.e.* food), although clothing performance would vary depending on local conditions (*e.g.* windchill and rainfall).[15]

Producing tailored clothing would require cutting tools (*e.g.* flakes, handaxes), piercing tools (bone splinters?; Gilligan 2010), scraping tools, and bindings (vegetation or thin strips of animal hide?). In terms of technology and access to materials these are all familiar elements of the Lower Palaeolithic world, and some of the flake tools from Schöningen and, slightly further afield, Rahuma (Israel) are intriguing candidates as possible piercing tools (Thieme 2005, fig. 8.4; Ronen 2006; Serangeli and Conard 2015), although the hide-working use-wear from Schöningen appears to be linked to butchering activities (Rots *et al.* 2015). Selected artefacts such as bifacial handaxes and the Schöningen spears also demonstrate a sophisticated awareness of 3-dimensional forms, which might be applicable to the production of effective clothing. An undergraduate dissertation at Reading University (Piet 2018) recently demonstrated that a closed 'dress' garment can be produced from deer hide, a simple backed knife, and binding (Fig. 3.6). One challenge is that, unlike for the Middle Palaeolithic, there is no convincing independent evidence for composite technologies in the archaeology of the earliest Europeans (the possible hafting evidence from Schöningen is intriguing but unconfirmed; Rots *et al.* 2015). This does not necessarily rule out an early form of tailored clothing (by definition a composite technology), but it is a notable disconnect. An alternate option might be the wearing of layers of untailored hides as a means of enhancing air-capture and reducing warm air replacement, although the problems of arm-pinning and poor venting would likely be exacerbated.

Whether tailored or not, the cleaning and processing (even if not necessarily piercing) of animal hides for clothing production also makes particular demands on the available stone tools. Robust, but relatively blunt, tool edges would be most appropriate: in short, scrapers. The visibility of such artefacts in the Lower Palaeolithic record is mixed, but it is likely that this reflects archaeologists' collecting habits and other biases in the record, alongside the technological practices and preferences of past hominins. In particular, the collection of a large proportion of the artefact record from active gravel pits has tended to favour the recovery of the larger and more visible handaxes (for which antiquarians and archaeologists would often pay more to the gravel workers: *e.g.* Hosfield 2009; Roe 1981; see also Chap. 2: Box F). The relative dominance of river banks, floodplains and other waterside settings in the archaeological record may also be a factor favouring handaxes over flake tools. 'Urgent' butchery tasks with handaxes rather than the lengthier (and arduous) processing of already

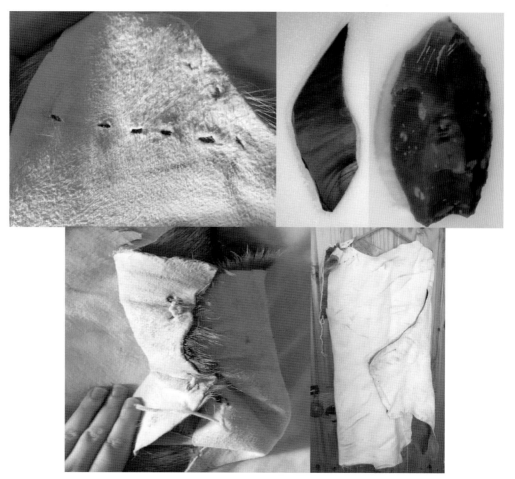

Figure 3.6: Experimental reconstruction of early Palaeolithic 'dress'-type clothing (bottom right). Incisions (top left) made with naturally-backed knife (top centre) and backed-knife (top right), binding with synthetic material (bottom left; Piet 2018, figs 6–9 & 11).

skinned animal hides would seem more likely in such exposed habitats. Yet despite these potential biases there are still examples of scrapers and other flake tools, at sites including High Lodge (Ashton *et al.* 1992), Hoxne (Singer *et al.* 1993) and La Noira (B. Hardy *et al.* 2018). The uses of flake tools for hide-scraping, and other activities, were suggested at Hoxne on the basis of use-wear traces (Keeley 1980; 1993), although some of the original identifications were revised in the later publication. Nonetheless numerous experimental activities have confirmed the suitability of scrapers for such tasks. However, these tasks are time-intensive and physically demanding (Shaw *et al.* 2012), as are the hide tanning processes which would have been necessary for clothing items to have any degree of longevity. Thus, while clothing may have been a requirement for winter survival, its production may well have occurred at other times of year (Chap. 6).

Shifting away from the core of the body, what of the extremities? Frostbite can occur through exposure to temperatures below -0.55°C and can be further exacerbated by wind chill and by high levels of heat loss (NHS 2017: https://www.nhs.uk/conditions/frostbite/). The latter can occur as a consequence of various factors potentially relevant to Palaeolithic hominins, including wet clothing, over-exercise, and unreplaced calories, while the impacts of exposure and wind chill are an argument in favour of sleeping- or camp-sites within relatively closed woodland. Given the importance of manual dexterity, and the potential for frostbite in the likely winter conditions (Table 3.1, with the variable addition of snow cover and wind chill), hand and foot coverings would seem likely strategies for the avoidance of frostbite to fingers and toes, especially in the north and east, as argued by Wales (2012). While significant question marks have been raised regarding the potential for complex, tailored clothing, hand and foot 'wraps' would not seem to be beyond the capability, and certainly not beyond the winter needs, of Lower Palaeolithic hominins (Fig. 3.7), although they would likely reduce the dexterity and mobility of their wearers, and hand wraps especially might need to be removed for specific tasks. Scraped pieces of hide could also be tied to each foot, with hide strips, rather than wrapped around them. Hairy, 'hobbit'-like feet, in a Tolkien-sense rather than a Flores one, might also be an option.

All too often discussions and visual reconstructions of Palaeolithic clothing, at least for pre-modern humans, have focused on the torso and the waist – although that may reflect residual Victorian sensibilities. But what of the head? Vreeman and Carroll (2008) have argued that the head does not lose disproportionately high levels of body heat but, rather, that any uncovered part of the body loses heat, reducing core body temperature in proportion to the body part's surface area. However, Pawłowski (2005) has stressed the particular risk to hominin infants of heat loss

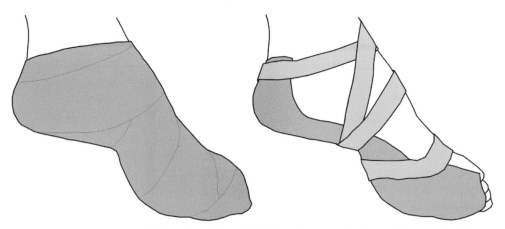

Figure 3.7: Foot 'wrapping' (left) and 'tied hide' (right) techniques which could be used with strips of animal hide/fur.

through the head, exacerbated by the larger, encephalised heads of later *Homo*. This occurs when human newborns spend time at temperatures below a 31–33°C optimum, and is especially significant for humans, as our head: body surface area ratio is relatively large, in comparison to chimpanzees for example. It is true that as the ambient temperature falls then the metabolic rate increases. But this can only work for so long. Pawłowski (2005) suggests that a neonate can maintain thermal balance in extreme conditions for no longer than 60 minutes, and that it only works down to an ambient temperature of 22–24°C. Below 22°C the energy for tissue growth will be spent on thermoregulation and replacing lost heat instead. Moreover, while cradling and skin-to-skin contact reduces body heat loss, it is much less effective when it comes to the loss of heat through the head (MacDonald 2018). Avoiding such heat loss would be a serious challenge in the climates of mid-latitude Europe, due to both seasonal variations and diurnal contrasts. Sheltering under cape-type coverings would provide a degree of insulation (as documented amongst the Yahgan at the southern tip of South America, for example; see MacDonald 2018 for other examples), as could untailored head coverings consisting of animal furs or even vegetation. A further possibility is that hominin infant survival selected for a covering of hair, presumably reasonably thick, on the head. Finally, there may have been other cultural insulation behaviours: MacDonald (2018) highlights the potential benefits of animal fat smeared onto the face and/or body, and the possibility of insulated carriers for infants (*e.g.* hide sacks lined with furs and feathers), although the latter has implications for tailoring.

Cold conditions would have been particularly problematic for all when hominins were relatively, or completely (*e.g.* at night) immobile. During the daytime a degree of body warmth may be internally generated through exercise (*e.g.* foraging activities), although this has knock-on implications for reliable energy intake. In terms of night-time warmth (or maintaining body warmth for temporarily immobile individuals, such as the very young or the injured), Pleistocene 'duvets' in the form of treated, but un-tailored, animal hides from large ungulates (horse or bison?) may have been especially important.

The practicalities of clothing production, and evidence for skinned fauna, are discussed further in Chapter 6, since it is likely that the requirements of hide condition and availability, and the time required for production, would favour a pre-winter 'scheduling' of this activity. In the meantime, the combination of palaeoenvironmental reconstructions, hominin physiological tolerances and available technologies suggest that clothing was a potential component of Lower Palaeolithic lives. But if so, how common was it? Palaeotemperature estimates from Early and Middle Pleistocene sites across Europe highlight broad trends in both regional and seasonal variations (Tables 2.3 & 3.1) that are comparable to the present (Chap. 1): *i.e.* increasingly warm summers and milder winters from north to south, and increasingly warm summers and colder winters from west to east. However, these trends are very generalised, and local habitats and micro-habitats would also have significant impacts. Nonetheless, none of these winters, whether in Boxgrove or Burgos (Atapuerca; Table 3.1), would seem to permit a clothing-free existence. Diurnal variations would be a further challenge throughout much of the year and across the continent, with more marked daily contrasts in the south and during the summers, but lower overall temperatures in the north and during winters (Chap. 1). Given the limited evidence for cave occupations in the European Lower Palaeolithic, keeping warm at night would have been a notable challenge, with infants particularly vulnerable to the lower, night-time, temperatures. In this context untailored, although presumably tanned or greased (see also Chap. 6), hides would have been an important source of night-time warmth. There would also be a further added benefit, since while clothing production generates significant costs (*e.g.* hide acquisition, processing), these are 'one-off' costs, in a way that a reliance on fire for warmth (*e.g.* fuel getting, monitoring) is not.[16]

As to when clothing may have appeared, the available evidence is admittedly scant. On the basis of the earliest skinning traces (Gran Dolina TD-6.2 and, albeit in very small quantities, Gran Dolina TDW4 and Sima del Elefante TE9–TE14; Huguet *et al.* 2013), the earliest clothing might even date back as far as the late Early Pleistocene, although the purposes of the skinning are uncertain. More large-scale skinning evidence is known from the late Lower Palaeolithic (*e.g.* Boxgrove, Arago, Schöningen and Gran Dolina TD-10.2: Parfitt and Roberts 1999; van Kolfschoten *et al.* 2015a; Lebreton *et al.* 2017; Rodriguez-Hidalgo *et al.* 2017). This would broadly tally with the apparent transformation in the scale of the European archaeological record during and after MIS 13–11, particularly to the north of the Alps (Hosfield and Cole 2018). However, Early Pleistocene and early Middle Pleistocene winter conditions to the south of the Alps (Table 3.1) were still challenging for hominins, initially *H. antecessor* and latterly *H. heidelbergensis*, with temperatures at or around their un-insulated minimum sustainable temperatures (MSTs). This favours the presence of some form of clothing during those Early and early Middle Pleistocene occupations as well, at least during the cooler climatic intervals and/or seasons (while genetic studies of clothing lice suggest an appearance between *c.* 83–170 kya, this only relates to the modern human population expansion from Africa, and is uninformative about archaic clothing use; Toups *et al.* 2010). A further, speculative question is whether enhanced hide processing in the form of tanning, and therefore more durable clothing and other types of hide insulation (*e.g.* shelters and bedding), were critical to the more sustained northern expansions that appear to be associated with the latter Middle Pleistocene? Without it hominins might be more reliant on favourable climates (and reliable, high-quality food supplies) – and hominin sites such as Hoxne (Ashton *et al.* 2008a), Kärlich H (Haidle and Pawlik 2010) and Korolevo VI (Szymanek and Julien 2018) do not appear to meet those climatic requirements.

It therefore seems likely that fur and fat, or clothing, or both (Fig. 3.8), supported by selected physiological advantages and additional energy intake, was key to the avoidance of hypothermia in European Lower Palaeolithic winters (MacDonald 2018). I favour the use of clothing, although it is likely that the amounts, and perhaps 'styles',

Figure 3.8: 'Survival gear', H. heidelbergensis-style (left; © Mark Gridley), contrasted with the unclothed 'Boxgrove man' reconstruction (right; © Peter Dunn, English Heritage Graphics Team). The left-hand reconstruction includes a 'cape'-type garment, attached at the neck with an organic binding or tie (e.g. nettle cordage or sinew): it could be wrapped around the body when required, but also 'pushed' back behind the shoulders when greater freedom of movement of the arms and/ or legs was needed.

of clothing varied markedly across the year, and possibly also between regions and between the Early and Middle Pleistocene. Hide acquisition, and clothing production may have been a significant autumnal activity (Chap. 6), with hominins drawn towards key resources by indicators such as red deer stag bellowing during the rut.

Shelters?

Clothing and fire, in their various guises, offer sources of mobile and static winter insulation. But what of shelters, whether artificial or natural? There is limited evidence for cave or rockshelter occupation during the European Lower Palaeolithic. Many of the artefact assemblages recovered from within caves have been transported, with or without sediments, from their original points of discard at or beyond the mouth of the cave by a variety of natural processes. Examples include Kent's Cavern, Brixham Cave (both UK), and Sima del Elefante (Cook and Jacobi 1998; O'Connor 2000; Proctor *et al.* 2005; Huguet *et al.* 2017). Arago Cave, Gran Dolina (TD-6.2 and TD-10) and Menez-Dregan, Combe Grenal, Pech de l'Azé II and La Micoque, all in France and Spain, are amongst the few examples of genuine cave and rockshelter occupation in the European Lower Palaeolithic, with the majority dating to the later Middle Pleistocene (Gamble 1986; de Lumley *et al.* 2004; Saladié *et al.* 2014; Rodríguez-Hidalgo *et al.* 2015; Monnier *et al.* 2016).[17] This pattern of rare cave occupation inevitably raises questions: is the pattern genuine? If so why? What alternatives did Lower Palaeolithic hominins utilise for shelter?

The pattern of cave occupation is dictated partly by geology: where are habitable caves available? While karstic rocks are widely distributed in Europe, they are by no means continuous (Chen *et al.* 2017; Fig. 3.9). Yet areas with significant Middle and/or Upper Palaeolithic cave occupation records, such as the French Périgord, show little evidence of Lower Palaeolithic activity within caves. One reason for this may be the taphonomic complexity of caves, such as the repeated cycles of sedimentation, frost shattering and re-mobilization events documented at Kent's Cavern (Lundberg

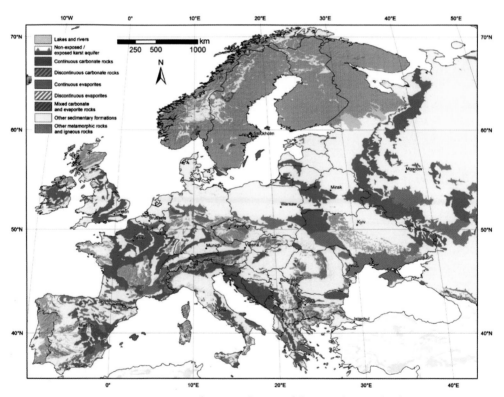

Figure 3.9: Karst aquifer map of Europe (Chen et al. 2017, fig 9).

and McFarlane 2007) or the cyclical flushing out of Middle Pleistocene sediments demonstrated in Cantabrian caves (Butzer 1981; 2008), which may remove Lower Palaeolithic-age occupation evidence (or at least disturb or destroy the specific contexts). Stating that caves were not, or only rarely, utilised by hominins in the Middle (or Early) Pleistocene is thus a dangerous assumption (Butzer 1981).

However, this apparent settlement pattern may also reflect a limited pyrotechnology amongst hominins, since artificial light would be a key requirement for cave occupation, particularly during the shorter daylight hours of late autumn–early spring. As Roebroeks and Villa (2011) have observed, the hominin occupations at six caves, including Arago and Gran Dolina, appear to occur without fire. If this pattern is genuine, it also challenges the suggestion that cave use *began* to prevent hominin fires from being extinguished by heavy rainfall and other weather conditions (Twomey 2013). Cave occupation, at least in the near-mouth zones of caves, might therefore be more feasible in the summer months, due to the longer daylight hours and generally stronger light levels. However, cave occupation might also be less important at such times of the year, given the generally improved weather conditions, although the degree of protection offered by caves would still be significant. A further factor behind the limited evidence for cave occupation might be the competition from other carnivores and potential threats, such as Deninger's bear (*U. deningeri*), which used Kent's Cavern as a hibernaculum and breeding site (Proctor *et al.* 2005), while the Gran Dolina bear (*U. dolinensis*) and spotted hyena (*C. crocuta*) are recorded at Gran Dolina TD-6.2 (Rodríguez *et al.* 2011).

Moving to the outdoors there is very little evidence for artificial structural elements, such as branches or tree trunks being deliberately selected and positioned. While this is predominantly a consequence of preservation (since the majority of archaeology from this period is derived and in secondary context; Chap. 2: Box F), the more occasional primary context sites have nonetheless failed to provide examples of settings for branches or alternative materials such as large mammal bones. This is in contrast to occasional Neanderthal examples, such as the Combe Grenal post-hole from south-western France (Mellars 1996) or the suggested windbreaks at Molodova, Ukraine (Chu 2009; Pettitt 1997). The only possible Lower Palaeolithic examples in Europe are the suggested shelters at Bilzingsleben (Mania and Mania 2003; 2005). These were inferred on the basis of semi-circular concentrations of large stone and bone elements, *c.* 3–4 m across, although their exact architecture is unclear and varies significantly between different visual representations (Fig. 3.10). They have also been much debated, due to the complex taphonomy of the site, in particular the evidence for lake flooding which has modified the spatial distributions of *c.* 80% of the site's material (Mania and Mania 2005) and inevitably raised doubts over the suggested interpretation of living structures (Gamble 1999, fig. 4.29). Either way, such evidence is exceptionally rare for the period, and the use of natural, rather than artificial, shelters seems the most likely hominin behaviour. There is ample evidence in the present for other animals' use of natural open-air shelters, such as red deer use of

Figure 3.10: The Bilzingsleben shelters. Above: Past and present views of huts in organic materials, with the exit controlled (above left: plan of Bilzingsleben; above right: wigwam from Wisconsin: Otte 2012, fig. 10); Below: The campsite at Bilzingsleben (© Landesamt für Denkmalpflege und Archäologie Sachsen-Anhalt, Karol Schauer).

macro and micro-topography and vegetation to reduce heat loss through exposure (*e.g.* Grace and Easterbee 1979), and the surface denning behaviour of wolves (*e.g.* under roots of fallen trees) and badgers (*e.g.* dense vegetation and hollow trees; Theuerkauf *et al.* 2003; Kowalczyk *et al.* 2004). There seems no reason to assume that their Pleistocene equivalents would not have behaved in a similar manner, and it seems equally unlikely that hominins able to observe animal behaviours and utilise that knowledge for the purposes of hunting (see also Chap. 6) could not extend this into other spheres.

But what exactly do I mean by natural shelters, from a hominin perspective? Trees, both standing and fallen, could be used as structural elements within organic shelters, supporting additional material such as covering hides, pieces of dead wood (Mears 2009), and/or leafy vegetation if necessary. The tree would most likely be fundamentally unmodified; however, I suggest that the overall structure becomes artificial if other material is added (or perhaps an augmented natural shelter). While amounts and types of winter snow are uncertain, they would be unlikely to have been

Figure 3.11: Potential natural shelters (clockwise from top left: rock outcrop; undercut river bank; pine (P. sylvestris) with low-hanging branches; tree root base and shallow tree throw pit; hollow tree (Sources & copyrights: details in Fig. acknowledgements).

sufficient for use as a structural element in the building of snow houses during warm climate stages (*e.g.* Kershaw *et al.* 1996), at least as a regular and widespread strategy. However, layers of snowfall accumulating on vegetation or hide coverings would add a valuable insulating element, further enhancing the benefits of the shelter (25 cm of fresh snow provides the insulatory equivalent of 15 cm of fibre-glass; Yankielun 2007). The problem is that such shelters would have little or no archaeological footprint.

Further possible categories of natural shelters include tree throw pits, hollow trees, the spaces beneath trees with low-hanging branches (*e.g.* pine and other conifers), small rock outcrops and/or rockshelters, and even undercut banks (Fig. 3.11). Some of these natural and modified shelters may also have been key for maintaining small fires (in some cases there is also the potential danger for accidentally setting the shelter alight), although the fire-extinguishing impacts of rainfall and high winds would be reduced generally in dense woodlands. In the case of tree throws the space behind the root base would provide shelter from the prevailing wind, while the pit itself would also provide shelter if sufficiently deep (see also Otte 2012). Such pits would be too small for the sizes of local bands estimated by the social brain hypothesis (Chap. 2), although they might have been adequate for smaller foraging groups or intimate networks (Gamble *et al.* 2014). They could also be wet, depending on the local soil and weather conditions. Rockshelters and undercut banks would potentially accommodate more individuals, and offer significant shelter in at least one orientation, although the degree of protection would vary depending on wind directions.

While nomadic or semi-nomadic lifestyles would favour portable, flexible dwelling options (Otte 2012), mobility might be reduced during winter (assuming that winter territories were mapped onto key prey species' winter feeding grounds and/or the availability of plant foods), and dwellings might show greater evidence of a structural investment and a longer presence. Beyond Bilzingsleben (as interpreted by Mania and Mania 2005) we have little evidence of this however. Overall there is limited evidence for significant environmental modifications for the purpose of shelter – the evidence, or more accurately the lack of it, favours the use of natural, outdoor shelter such as tree pits, fallen trees, tree 'under-spaces', rock outcrops/shelters and undercut banks. Since many other animals also make use of bedding material (*e.g.* badgers), it is likely that shelters were adorned where necessary and possible with leafy vegetation (*e.g. Carex* [sedges] or rushes; Bigga *et al.* 2015), feathers, and/or animal hides,[18] and perhaps also with fire. Bedding and/or clothing, probably combined with huddling, may have been especially critical as a means of reducing heat loss while sleeping in the winter months (MacDonald 2018).

Winter is coming, the deer aren't getting fat ...

Keeping warm and dry were by no means the only overwintering problems in Pleistocene Europe. The challenges of finding winter food were essentially threefold: 1. finding sufficient quantities of foods; 2. finding sufficient quantities of the right foods (*i.e.* those which met, sufficiently if not optimally, nutritional requirements); and 3. minimising foraging times and distances where possible, to reduce the exposure to low temperatures and wind-chill that would be especially problematic if hominins were reliant on only minimally tailored clothing to keep warm. Dietary requirements were discussed in Chapter 2, and a mixed omnivorous diet will be assumed for the following discussions. It is also argued here, on the basis of site-specific evidence (see Chap. 6 for details), that European Lower Palaeolithic hominins were effective and skillful hunters and butchers since at least *c.* 1 mya (*e.g.* Parfitt and Roberts 1999; Voormolen 2008; Saladié *et al.* 2011; Huguet *et al.* 2013; Rodríguez-Hidalgo *et al.* 2015; van Kolfschoten *et al.* 2015a), although it is notable that some of the strongest evidence dates from the later Middle Pleistocene.

Despite this, meeting the first of the challenges outlined above is notably prob-lematic in the mid-latitudes. The sizes of mammal territories, especially those of carnivores, increase at higher latitudes, due to resources becoming increasingly aggregated in space and constrained in seasonal availability from the equator to the Arctic (Kelly 1995; Roebroeks 2001; 2006). These trends would also apply to homin-ins. The relative aggregation or disaggregation of resources can be assessed from a human perspective by measuring the average distances covered by hunter-gatherer groups as they move their camps between resources (*i.e.* residential mobility). These distances, and resource aggregation, increase with decreasing effective temperature (Kelly 1995, fig. 4.7 & Fig. 3.12). These patterns in resource distribution, territory size and mobility would be especially challenging to hominins during the short, cold days of winter, not least because of the risks of winter mobility. In that regard it is interesting that there is no evidence for late winter mortalities at one of the few sites with seasonality indicators (the Gran Dolina TD-10.2 'bison bone bed'). This is in marked contrast to the evidence for concentrated deaths in spring/early summer and autumn at the site (Rodríguez-Hidalgo *et al.* 2016; Chaps 4 & 6). This may reflect the local disappearance of the bison during winter as a result of migration, or perhaps limitations in hominin hunting ability and/or success at this time of year, but either way it highlights the challenges of winter.

However, animal territories are also likely to have varied in accordance with food supply. In the case of the European wolf for example, its modern territory varies between 100 and 1000 km² (Macdonald and Barrett 1993, 92). Moreover, the overall

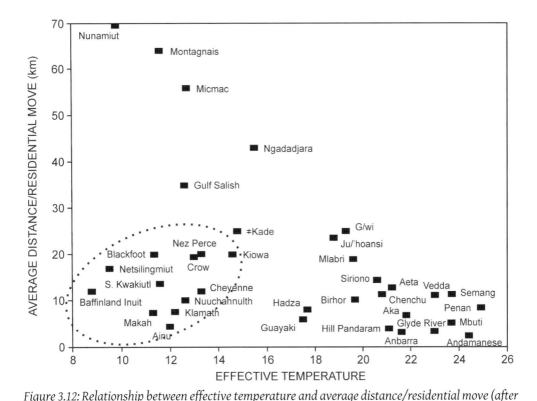

Figure 3.12: Relationship between effective temperature and average distance/residential move (after Kelly 1995, fig 4-7). Note the examples (circled) for groups making relatively short mean residential moves in low effective temperature environments (see Kelly 1995, 128–130 for details). Effective temperature (ET) is derived from the mean temperatures (°C) of the warmest and coldest months $\left(W \text{ and } C; \text{ where } ET = \frac{18W - 10C}{(W - C) + 8} \right)$, *and its value varies from 26 (equator) to 8 (poles). High ET values are associated with tropical, non-seasonal environments (in terms of temperature, not precipitation) with long growing seasons. Low ET values are associated with cold, seasonal environments with short growing seasons (Kelly 1995, 66–69).*

trend for increasing resource segregation at lower temperatures ignores the evidence for groups hunting local, dispersed resources in low temperature environments. For example, the Netsilingmiut and the Baffinland Inuit are able to make relatively short average residential moves (Fig. 3.12) by relying on solitary animals, freshwater fish and seals, the latter hunted through their breathing holes in the winter ice (Kelly 1995, 129).

While I am certainly not proposing either fishing or seal hunting in the Lower Palaeolithic (although there would clearly be potential for the exploitation of beached seals and other marine mammals), the potential for utilising local resources in colder, seasonal environments becomes relevant when considering the apparent evidence for distinctive, mixed landscapes in Pleistocene Europe. This is based on the repeated associations of woodland and grassland fauna, evident for example at Swanscombe, Hoxne and Boxgrove (see also Chap. 2). Combinations of fauna from river floodplains (*e.g. C. fiber*), open grasslands (*e.g. E. ferus, P. antiquus*) and temperate deciduous wood-land (*e.g. D. dama, S. scrofa*) have been observed at all three sites. Stewart *et al.* (2003) have previously questioned whether faunal communities from MIS 3 are genuine associations or palimpsest effects,[19] an issue that is further complicated by potential ecological differences between Pleistocene species and their present-day equivalents (Box H). However, if these Lower Palaeolithic examples are genuine then such locally diverse habitats and resources could result in lessened herd mobility and/or migration distances. This is because where food clumping is reduced and the even distribution of food is increased, then range sizes decrease (Clutton-Brock *et al.* 1982, 245). Such circumstances would favour foraging over a smaller area and, therefore, partly reduce the difficulties of residential winter survival strategies.

Moreover, there are herd animals on Pleistocene sites that may have been char-acterised by generally lower levels of mobility (*i.e.* essentially residential species), the exploitation of which would thus also reduce hominin mobility and foraging times during the winter. While Gamble (1987, 87) has previously highlighted musk ox and possibly rhinoceros, a number of smaller ungulate species recorded on Lower Palaeolithic sites are today characterised by relatively minor summer/winter migrations and/or by reduced winter territoriality (Table 3.7 & Box H). In a study of non-migratory red deer in the BPF, adult female winter ranges, although at their largest relative to other seasons, averaged 7.1 km^2 during the winter months (Kamler *et al.* 2008). Average adult male home ranges, although slightly larger than those of the females (as they were throughout the year in the BPF), were also at their smallest (11.6 km^2) during the winter. Horses' specific water requirements may have made them a favourable prey species in European winters, as the ranges of semi-wild modern representatives tend to reduce in size at this time of year (thus reducing hominin search costs), in part because water is more easily available (Corbet and Harris 1991). Although based on a Late Pleistocene sample, Julien *et al.* (2012) have also argued for non-migratory behaviour in steppe bison (*B. priscus*) on the basis of tooth enamel isotopes, in contrast to previous assumptions. However, the distributions and behav-iours of even non-migratory species or populations present their own challenges to

Table 3.7: Modern home range, density and mobility data for key deer species, documented on selected Early and Middle Pleistocene sites

Species (extinct Pleistocene taxa underlined)	Home range[5]	Density[5]	Mobility[5]	Site examples
C. capreolus (Roe deer) C. priscus C. suessenbornensis	0.05–1 km²	15–25/km²; Solitary/small groups in closed woodland[9]	Reduced territoriality in winter & congregation (herds up to 30)	Boxgrove[13] Bilzinsleben[6] Gran Dolina TD-10.1[10] Hoxne[15] Soucy[4] Swanscombe (LL)[12]
D. dama (Fallow deer)[8] D. dama clactoniana D. vallonetensis	0.5–2.5 km²	12(?)/km²; Small groups (<7/8) in closed/open woodland[9], larger temporary aggregations in open ground with good feeding	Habitat use shifts seasonally (e.g. summer: open habitats; autumn → spring: woodlands[9])	Arago (levels F–G & J)[2] Barnham[7] Bilzingsleben[6] Gran Dolina TD-6.2[11] & TD-10.1[10] High Lodge?[14] Hoxne[15] Sima del Elefante (TE9c)[3] Swanscombe (LG)[12]
C. elaphus (Red deer)	0.5–8 km²; Smaller upper limits also suggested[1]	5–45/km²; Small groups (1–3) in closed woodland[9]	Summer → winter range migrations up to 6 km (e.g. lowland woodlands → open uplands [UK])	Arago (levels F–G, J & L)[2] Barnham[7] Bilzingsleben[6] Boxgrove[8] Gran Dolina TD-6.2[11] & TD-10.1[10] High Lodge?[14] Hoxne[15] Schöningen 13-I & 13 II-4[16–18] Soucy[4] Swanscombe (LL)[12]

[1]Clutton-Brock *et al.* (1982); [2]de Lumley *et al.* (2004); [3]Huguet *et al.* (2017); [4]Lhomme (2007); [5]Macdonald and Barrett (1993; modern European data – it is fully acknowledged that Early and Middle Pleistocene species' ecology would not have been identical to their modern equivalents: see Box H); [6]Mania and Mania (2005); [7]Parfitt (1998); [8]Parfitt (1999a; the fallow deer's late rut results in males' poor condition during winter; there are also significant size differences between modern fallow deer and *D. dama clactoniana*); [9]Putman (1988); [10]Rodríguez-Hidalgo *et al.* (2015); [11]Saladié *et al.* (2011); [12]Schreve (1996); [13]Smith (2013); [14]Stuart (1992); [15]Stuart *et al.* (1993); [16]Thieme (2005); [17]Voormolen (2008); [18]van Kolfschoten *et al.* (2015a). Site units: Swanscombe (LL): Lower Loam; Swanscombe (LG): Lower Gravels

the task of finding animal foods in winter. The relationships between animal range size and habitat 'quality' (*e.g.* for deer; Putman 1988) would be a further complicating factor for example, as wider ranging prey would place greater demands on hominin foraging mobility and energy budgets during the short, cold days of winter, unless food supplies were buffered by significant autumn 'stockpiling' (Chap. 6).

Box H: Can we understand Pleistocene animals from their modern counterparts?

Modern animal behavioural data are rightly considered to be potentially problematic when applied to Pleistocene ecosystems. It is true that many modern European mammals evolved during the Quaternary, and their adaptations and distributions were therefore shaped by those environments (Lister 2004, 237–238). However modern species and populations have also undergone various modifications. In the specific case of red deer for example, Geist (1998, 185) highlights that populations have been bred and mixed, and that present-day populations have been subject to considerable taxonomic noise due to hybridization and bottlenecking.

Nonetheless there is potential to infer aspects of species' behaviours from the conditions at Lower Palaeolithic sites. The site-specific occurrences of different ungulate species highlight their Pleistocene habitat preferences and, frequently, their adaptable nature. Lister (1984) has demonstrated that red deer were associated with both optimal interglacial woodlands, such as the mixed oak forests at Clacton, but also with more open conditions, such as the late Anglian levels at Hoxne, characterised by scrub/herb communities and only scattered clusters of birch trees, and the damp open grasslands of the Upper Middle Gravels at Swanscombe. Red deer would thus be likely to be available to hominins during cooler, as well as optimal, phases of the interglacials. While present-day distributions are at least in part a product of human activity, key limiting factors appear to be snowfall (40–50 cm annually) and exposure. The red deer heat loss resulting from exposure to cold winter winds can be reduced through favourable topography and woodland shelter – thus increasing the possibility for human/red deer co-presence in winter woods. Modern faecal pellet data have also suggested that red deer have a strong preference for open thicket habitats within forests and avoid closed, dense canopy areas (Stevens *et al.* 2006), a behaviour which might have impacted on specific hunting strategies.

Similarly, while the natural ecology of fallow deer is more difficult to deduce,[1] its presence in both the Lower and Middle Gravels at Swanscombe, and in the Hoxne Upper Sequence, suggests a tolerance both for deciduous forest and for more scattered, and largely coniferous, tree cover (Lister 1984). A diet of grasses, tree browse, herbaceous plants and fruits also supports a distribution in woodlands combined with access to open ground. A broadly similar set of tolerances have been documented for roe deer, although coniferous forest is used only if a leafy understory is present.

Snow depth (<50 cm annually) is argued to be a more important constraint on distribution than vegetation type but animal availability and abundance at different times of the year would likely also be linked to specific plant communities: for example, the association of horse with fresh spring grass or reindeer with winter lichens (Sturdy and Webley 1988). Finally, two other factors are of potential significance to hominins. First, roe deer need regular fresh water – a potential means of predicting their location. Secondly, they may spend nearly the full 24 hours feeding in the winter, when food is scarce, to meet their high energy requirements – might this make them vulnerable to predators?

However, there is also clear evidence both for the non-analogue nature of Quaternary environments, and for both specialist and flexible adaptations amongst different mammal species. Non-analogue environments have been most notably emphasised by Stewart (2005) for MIS 3 at the end of the Middle Palaeolithic period and the start of the Upper Palaeolithic, with reference to the combinations of mammals living together (which are not found in the present) and the extinct faunal elements. Lister (2004, 224) has highlighted the combination, within red deer, of fixed but broad-use adaptations, behavioural flexibility, and ecophenotypic[2] plasticity, resulting in a species which is not only broadly distributed across varied habitats in the present, but also persisted through varied Quaternary habitats. By contrast, roe deer cannot consume larger quantities of grass and have both restricted present-day ranges and were limited to wooded phases of the Quaternary. Lister (2004, 225–226) also suggests that fine-scale spatial and temporal variability, via migratory movements and seasonality, can also lead to the evolution of flexible adaptations. Finally, recent studies of dental wear and isotopes (Pushkina *et al.* 2014; Rivals *et al.* 2008; 2015; Rivals and Ziegler 2018) have highlighted that many of our long-standing assumptions about Pleistocene animals' diets, mobility and habitat preferences may be, at least partially, incorrect (see also Chap. 2).

In light of this, a cautionary usage of modern behavioural data is clearly appropriate. However the data are used here (*e.g.* Tables 3.7 & 3.8) solely to demonstrate both the documented presence of specific animals at Lower Palaeolithic sites (although those species were not necessarily exploited by hominins), and *relative* contrasts between the different species' likely distributions and potential availability for exploitation.

---

[1] The marked size differences between living *D. dama* and the Middle Pleistocene sub-species *D. dama clactoniana*, the latter being *c.* 20% larger (Kurten 2017), are a further complicating factor, although Schreve (1996) argued that it is not unreasonable to assume a temperate woodland preference.

[2] Variation in an animal or plant's observable characteristics (its phenotype), as a result of its specific environment: hence ecophenotypic.

Modern animal data (Table 3.7) therefore suggests that at least some ungulate species might have been locally available to hominins during the winter months, with opportunities to hunt and/or scavenge influenced by the specifics of animal behaviours, short day lengths and weather conditions, and the settlement/mobility strategy adopted by the hominins (see also Box I). The 8-month weaning period of red deer for example (Stevens *et al.* 2006), after a birthing season in May–June, suggests that mother–calf dyads would be available through the early winter months as a potential food source. This, combined with the larger winter ranges of adult males and the impacts on search times, may have made dyads a preferable prey, although males are larger in size. However a related red deer behaviour with potential further impact on any hunting strategies concerns the tendency of the calves to 'hide' (a consequent behaviour is their ability to recognise their mothers' calls: Vaňková *et al.* 1997) – thus any hunting of calves, throughout the latter half of the year, would presumably have required relatively high search times, although these might be reduced in more sparsely vegetated winter landscapes, and perhaps also specific searching/stalking strategies. Animal cues would have been critical and both wide-ranging and seasonally variable. Modern British red deer, for example, consume bark from tree trunks and exposed roots (in winter and spring) and from fallen branches in spring (Corbet and Harris 1991). This behaviour produces wounds on trunks (typically *c.* 80 cm from the ground) and, while this specific behaviour may not be directly comparable to Pleistocene populations it highlights the potential for visual indicators.

Some of these ungulates' relatively small body sizes (*e.g.* 16–35 kg and 32–80 kg for modern fallow and roe deer; Macdonald and Barrett 1993) may have presented further challenges in terms of the frequency of hunting required and the hominin group sizes that could be sustained, although Pleistocene populations do vary (*e.g.* the Boxgrove roe deer sample is noticeably larger than those from other British sites, including Thatcham in the early Holocene: Parfitt 1999a, fig. 169).

However, finding prey is only part of the European winter challenge. Modern studies repeatedly document the poor condition of many potential winter prey species, with animal fat sources (*e.g.* bone marrow) subject to depletion, particularly in late winter and spring (*e.g.* Jochim 1981, fig. 3.1 for roe deer; Spiess 1979, fig. 2.2 for caribou). In the BPF for example, red deer calves are characterised by an extremely low autumn–winter fat content (Okarma *et al.* 1997), with mean marrow fat content in the femur measuring 66% in October–January, compared to 27% in February–March (measured between 1996–1999: Jędrzejewski *et al.* 2002). This is a problem because the issues around lean animal-dominated winter diets have been well-established ethnographically (*e.g.* Speth and Spielmann 1983). In short (see also Chap. 2), excessive lean animal meat carries the risk of over-dosing on protein (occurring at levels above *c.* 50%; the so-called 'rabbit fever'): this is a particular problem for pregnant females (Speth 1990; 1991b; Hockett 2012). Its intake therefore needs to be balanced against other macro-nutrients, and thus successful access to fats, marrows, carbohydrates and other vitamins and minerals is thus a critical requirement for the winter resident.[20]

Speth and Spielmann (1983) suggest three broad strategies for meeting these nutri-tional winter challenges: first, targeting animals with higher winter and spring fat con-tent (*e.g.* beaver, bear and waterfowl, the latter of which have up to 70% fat content by edible body weight: Jochim 1981); secondly, internal and/or external fat storage, either through body fat reserves or through the rendering of bone grease; and/or thirdly, targeting alternative sources of carbohydrates and other minerals and vitamins. The latter could include ungulate stomachs (Buck and Stringer 2014a; but *cf.* Fediuk *et al.* 2002 with regards to caribou stomachs and sources of vitamin C) and/or winter plants (*e.g.* seaweed/kelp, which can provide significant levels of vitamin C; Fediuk *et al.* 2002).

There would therefore have been various animals of potential 'winter value', which could have provided sources of fat, as well as, in many cases, being likely winter resi-dents (Table 3.8). Modern beaver tail, for example, increases its fat content from *c.* 7% (late spring) to *c.* 60% in autumn and early winter (Coles 2006, 55). Guthrie (1990, 247) has highlighted that more sedentary species of modern northern ungulates tend to put on larger quantities of winter fat than the long-distance migrants. Larger species (*e.g.* rhinoceros) would also offer a particularly valuable additional winter resource in the form of bone as a potential fuel source for fires (and the burning of which would result in fragmentation and their relative 'invisibility' in the archaeological record). A variety of modern-day migratory water birds also have a documented presence in northern Europe (Peterson *et al.* 1993), including Whooper swan (*c.* 8–11 kg), Greylag goose (*c.* 2–4 kg) and the now-extinct Great Auk (*c.* 5 kg). All three species have been recorded at Boxgrove (units 4, 4d and 4c respectively; Harrison and Stewart 1999). The particular value of such birds as food sources has been demonstrated for Canada goose, which provides enhanced levels of protein, zinc and iron, with the organs of particular value to the young and the pregnant (Belinsky and Kuhnlein 2000). Further examples of the range of specific nutritional benefits which can be gained from discrete animal body parts have been documented more widely (albeit with a specific, non-European, geographical focus) in a series of Arctic hunter-gatherer-focused ethnographic studies (*e.g.* Fediuk *et al.* 2002; Kuhnlein *et al.* 2006; Hidiroglou *et al.* 2008; Fig. 3.13). It would seem reasonable that at least some of these sources (or their equivalents) were known to Lower Palaeolithic hominins, through behavioural selection via trial and error. However it is also accepted here that some of the other food types in Figure 3.13 were presumably not available (at least not as a regular or semi-regular, as opposed to occa-sionally scavenged, food sources) due to the technological demands associated with their acquisition (*e.g.* marine mammals and fish or freshwater fish).

The faunal lists from a variety of Early and Middle Pleistocene sites therefore highlight a range of species that could potentially be exploited to assist hominins in meeting the specific survival demands of a Pleistocene winter (Tables 3.7 & 3.8). This is true both in regards to their lean season fat reserves (and other nutritional benefits) and their coats (see also Chap. 6), and in terms of their territorial sizes and ranging behaviours. However, a clear disconnect occurs when direct evidence for the exploitation of some of these animals is sought: Harrison and Stewart (1999, 193) explicitly note,

Table 3.8: Fat-bearing and/or residential winter animals, with modern distribution data for comparison, documented on selected Early and Middle Pleistocene sites

Species (extinct Pleistocene taxa underlined)	Home range[7]	Density[7]	Mobility[7]	Site examples
C. fiber (European beaver)	500 m–5.5 km (along river)	1.0–1.8/km²	Family movements within territory	Arago cave (levels D, F–J & Q)[5] Bilzingsleben[8] Boxgrove[10] Hoxne[15] Sima del Elefante (TE9c)[3] Soucy[6] Swanscombe (LL)[13]
S. scrofa (Wild boar)	2–20 km²	ND	Sedentary (if stable environment); ♀ Small herds; ♂ Solitary	Barnham[9] Bilzingsleben[8] Gran Dolina TD-6.2[12] Sima del Elefante (TE9c)[3] Soucy[6]
U. arctos (Brown bear) U. dolinensis U. deningeri U. deningeri-spelaeus	150–4000 km²	1–190/ 10,000 km²	Solitary; Travel 2–3.5 km/day; Hibernation (with accumulated fat)[4]	Arago (levels F–G, J & L)[1] Barnham[9] Bilzingsleben[8] Boxgrove[10] Gran Dolina TD-6.2[12] Hoxne[15] Sima del Elefante (TE9c)[3] Soucy[6] Swanscombe (LL/LG)[13]
D. bicornis (Black rhinoceros)[2] S. etruscus S. cf. hemitoechus S. hundsheimensis S. kirchbergensis	Few ha–75 sq. km	ND	♀ + young; ♂ Solitary; Resident & local (if resources sufficient)	Arago (levels F–G, J & L)[1] Barnham[9] Bilzingsleben[8] Boxgrove[10] Gran Dolina TD-6.2[12] & TD-10.1[11] Hoxne[15] High Lodge[14] Sima del Elefante (TE9c)[3] Soucy[6] Swanscombe (LG)[13]

[1]de Lumley *et al.* (2004); [2]Haltenorth and Diller (1980; modern African data – it is fully acknowledged that Early and Middle Pleistocene species' ecology would not have been identical to their modern equivalents: see Box H); [3]Huguet *et al.* (2017); [4]Jochim (1981); [5]Lebreton *et al.* (2017); [6]Lhomme (2007); [7]Macdonald and Barrett (1993; modern European data — it is fully acknowledged that Early and Middle Pleistocene species' ecology would not have been identical to their modern equivalents: see Box H); [8]Mania and Mania (2005); [9]Parfitt (1998); [10]Parfitt (1999a); [11]Rodríguez-Hidalgo *et al.* (2015); [12]Saladié *et al.* (2011); [13]Schreve (1996); [14]Stuart (1992); [15]Stuart *et al.* (1993). Site units: Swanscombe (LL): Lower Loam; Swanscombe (LG): Lower Gravels.

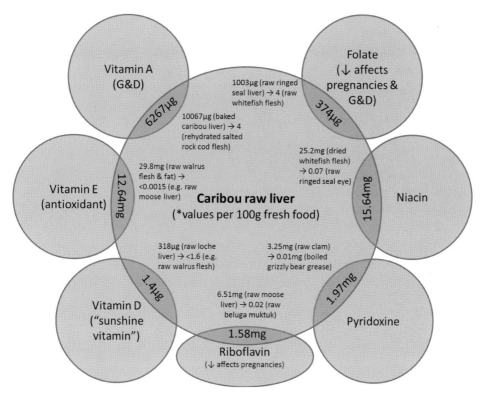

Figure 3.13: Selected sources of vitamins in Arctic hunter-gatherer diets (Hosfield 2016, fig. 9; data from Kuhnlein et al. 2006; Hidiroglou et al. 2008). Values per 100 g of fresh raw caribou liver (e.g. 1.58 mg for Riboflavin) compared against alternative food sources (e.g. raw moose liver [6.51 mg] and raw beluga muktuk [0.02 mg] for Riboflavin). G&D: growth and development.

for example, the lack of cut-marks associated with bird bones at Boxgrove, although they acknowledge that cut-marked bird bones are generally rare in prehistory. Another claim for bird exploitation in the European Lower Palaeolithic is similarly speculative: Thieme (2005) has drawn a tentative connection between the 'throwing stick' and goose bones at Schöningen 13 II-4 and the hunting of birds in flight by Australian aborigines. Instead the first relatively widespread evidence for bird exploitation in the European Palaeolithic comes with the Neanderthals, for both decorative and dietary purposes (*e.g.* Finlayson *et al.* 2012; Blasco *et al.* 2014; Negro *et al.* 2016). As a general rule, these animals are present on Lower Palaeolithic hominin sites, but there are only very occasional indications of their exploitation (*e.g.* at the Sima del Elefante: Huguet *et al.* 2013).

There are nonetheless occasional glimpses of the exploitation of residential, fat and/or fur-bearing animals (Table 3.9). Butchered bear remains have been recorded at Boxgrove (with skinning inferred from cut marks on a skull of *U. deningeri*; Parfitt and Roberts 1999, 402–403) and Bilzingsleben (Mania and Mania 2005, 108), while there is also evidence for the exploitation of rhinoceros and, at Bilzingsleben and

Arago cave, beaver (Lebreton *et al.* 2017). With regards to a possible winter focus on residential or locally mobile species, it is notable that wild pig, felids and roe deer were all exploited occasionally at Bilzingsleben. Interestingly, Mania and Mania (2005, 108) explain this low-level usage as a consequence of inefficiency (roe deer) or danger (wild pig and felids). However an alternative interpretation is that these species were primarily only exploited during the winter, due to their relatively residential behaviour and/or valuable resources (Tables 3.7, 3.8 & 6.6), at a time of the year when those attributes outweighed issues of danger and inefficiency. With the exception of Hoxne and Gran Dolina TD-10.2 (red deer and bison remains respectively; Table 3.9), all of these archaeological occurrences lack seasonality data. However, the presence of bear (and perhaps also the roe deer, felids and wild pig) at Bilzingsleben is particularly intriguing, given the admittedly controversial claims for fires, shelters and a sustained occupation at this site (Mania and Mania 2005; but *cf.* Gamble 1999, 153–172).

An alternative potential solution to periods of winter scarcity centres on internal storage and the accumulation of reserves of body fat immediately prior to winter. This has been linked with the buffalo gorging documented amongst Native American groups, although the duration of the benefits was not specified (Speth and Spielmann 1983, 19). In this regard the quantities of butchered animals at specific sites are intriguing. These quantities are further emphasised if we consider the potentially small sizes of day-to-day groups in the Lower Palaeolithic (see also Chap. 4: Box K). These small group sizes have been suggested as part of the social brain hypothesis (Chap. 2), on the basis of possible 'residential' sites such as Beeches Pit (Preece *et al.* 2006) and Bilzingsleben (Mania and Mania 2005), and have also recently been argued on genetic grounds for the Neanderthal descendants of *H. heidelbergensis* (Lalueza-Fox *et al.* 2011; Prüfer *et al.* 2017). The rich Schöningen horse remains (see Chap. 6 for details) are now argued to represent a series of hunting episodes (Voormolen 2008, 128). However just one stallion and 2–6 mares and their foals (*i.e.* a horse family group) would still yield approximately 400–1000 kg of meat. This can be compared with White's (2006, 563) estimate of one reindeer-sized animal (120 kg) every 3 days or one horse-sized animal (250 kg) every 6 days to sustain a party of ten adult Neanderthals. Nutritionally-beneficial seasonal gorging by a small group of hominins, as opposed to waste (at least for the hominins) or the cognitively-demanding collaborations of multiple groups, seems a parsimonious explanation of such zooarchaeological assemblages and one with benefits in terms of aiding winter survival.

External storage of animal fats, through the production of pemmican,[21] has also been widely documented in the case of recent hunter-gatherers (Speth and Spielmann 1983, 19). It is commonly produced by boiling bones to extract grease, and there is no evidence at all for cooking vessels in the Lower Palaeolithic (although water can be successfully boiled in filled perishable containers over open flames; Speth 2015). However alternative methods for its production also exist. One involves heating rocks in a fire, transferring them to a perishable container (*e.g.* an animal skin drawn together to form an unsewn bag) with broken-up bones, and then mixing the rendered fat with lean meat. Storage through dehydration (*i.e.* drying and/or smoking; Stopp 2002) or, perhaps more feasibly,

Table 3.9: Butchery by species and technique, from selected Lower Palaeolithic sites

Species	Butchery evidence	Sites
C. fiber (European beaver)	Skinning; Disarticulation; Defleshing	Arago cave[4]
Bos or *Bison* sp., incl. *B. schoetensacki* (Wild cattle or Bison)[a]	Marrow extraction & cut-marks (filleting?); Filleting; Cut-marks; bone breakage; Skinning, disarticulation/dismemberment, filleting, defleshing, evisceration & marrow bone breakage; Cut-marks, defleshing and marrow bone breakage; Dismembering, filleting, defleshing & marrow bone breakage	Barnham[5] Boxgrove[6] Gran Dolina TD-10.1[7] Gran Dolina TD-10.2[8] Happisburgh I[2] Schöningen 13 II-4[10,11]
C. capreolus & *C. priscus* (Roe deer)	Cut-marks; Defleshing; Cut-marks	Boxgrove[3] Happisburgh I[2] Gran Dolina TD-10.1[7]
C. elaphus (Red deer)[a,b,c]	Skinning, dismemberment, filleting & marrow bone breakage; Cut-marks; bone breakage; Marrow bone breakage & cut-marks (seasonality data: late Summer → Spring); Skinning, dismemberment & filleting; Cut-mark	Boxgrove[6] Gran Dolina TD-10.1[7] Hoxne[9] Schöningen 13 II-4[10,11] Westbury[1]
E. ferus & *Equus*. sp. indet. (Wild horse)[d]	Disarticulation, filleting & marrow bone breakage; Cut-marks; bone breakage Marrow bone breakage & cut-marks; Dismemberment, filleting, boning, defleshing & marrow bone breakage	Boxgrove[6] Gran Dolina TD-10.1[7] Hoxne[9] Schöningen 13 II-4[10,11]
S. hundsheimensis & *S.* cf. *hemitoechus* (Rhinoceros)	Disarticulation & filleting; Cut-marks; bone breakage; Disarticulation	Boxgrove[6] Gran Dolina TD-10.1[7] Happisburgh I[2]
U. deningeri (Deninger's bear)	Skinning	Boxgrove[6]

[1]Andrews and Ghaleb (1999); [2]Ashton *et al.* (2008b); [3]Bello *et al.* (2009); [4]Lebreton *et al.* (2017); [5]Parfitt (1998); [6]Parfitt and Roberts (1999); [7]Rodríguez-Hidalgo *et al.* (2015): the following processes were identified in the sample as a whole: skinning, defleshing, disarticulation, dismembering, evisceration and periosteum removal; [8]Rodriguez-Hidalgo *et al.* (2017); [9]Stopp (1993); [10]van Kolfschoten *et al.* (2015a); [11]Voormolen (2008). In other cases reported butchery processes were not distinguished between species: [a]*Bison* sp. and *C. elaphus* were butchered at Sima del Elefante, where processes included skinning, dismembering, defleshing and bone breakage (Huguet *et al.* 2013); *C. elaphus*, *S. etruscus*, *E. stenonis* & *E. giulii* were butchered at Gran Dolina TDW4, where processes included skinning, dismembering, defleshing, evisceration and bone breakage (Huguet *et al.* 2013); [b]The deer remains at Gran Dolina TD-10.1 were described as *Cervus/Dama* sp. indet. (Rodríguez-Hidalgo *et al.* 2015); [c]Cut-marks at Gran Dolina TD-6.2 were predominantly found on deer (species included red and fallow deer) and *Homo*, but other butchered species included: *B. voigtstedtensis*, *Equus* sp., *S. etruscus* & *U. dolinensis*; butchery processes included skinning, defleshing, disarticulation, dismembering, evisceration, periosteum removal and possible tendon removal (Saladié *et al.*, 2011); [d]*E. caballus mosbachensis* was butchered at Arago (level G), with evidence for cut-marks and bone breakage (Bellai 1995); a wide range of other species are present at Arago, including *B. priscus*, *C. elaphus*, *D. dama clactoniana*, *H. bonali*, *O. antiqua*, *P. priscus*, *R. tarandus* & *S. hemitoechus*, with frequencies varying by level and climate (de Lumley *et al.* 2004)

rotting (Speth 2017) are both worthy of consideration as means of managing hunting surpluses (after Bailey 1981). This could potentially ensure at least one reliable food supply through the winter months, although such a strategy has significant implications in terms of the need for limited residential mobility (Box I), possibly controlled fire-use, and/or the carrying capacity of individual hominins.

Winter plants?

Wild plant foods are an often ignored but, nonetheless, potentially key part of any solution to the nutritional challenges of winter survival. While plant availability is inevitably less during mid-latitude winters, Mabey (2012) and Mears and Hillman (2007) illustrate the potential of winter plants: roots and rhizomes (starch), mushrooms (vitamins B and, if exposed to sunlight, D) and various leafs and shoots (*e.g.* nettle) are all presently available in the UK during the winter months. The nutritional importance of plant foods in earlier Palaeolithic diets has received widespread recent discussion (see also Chap. 2), with particular reference to Neanderthals (Hardy 2010; 2018; Hardy and Moncel 2011; Hardy *et al.* 2011; 2012; Weyrich *et al.* 2017), while Karen Hardy (2015) has stressed the importance of carbohydrates in human evolution generally. Bruce Hardy (2010) has given particular emphasis to the underground storage organs (USOs) of species such as reed mace (cattails) and wild carrot, noting their widespread distribution (albeit across the Neanderthal range), the peaks in their energy storage in late autumn and winter, and the presence of visible, above ground winter vegetation for selected species. Occasional winter mushrooms and fungi may also have been a potential resource (see also Chap. 5 & Table 5.5).

The identification of plants to species-level can be problematic on Early and Middle Pleistocene sites, as it is in later periods too. Nonetheless, comparison of plant family data from Hoxne (Mullenders 1993) and from Godwin's (1975) history of British flora against modern species available to winter foragers (Mears and Hillman 2007; Mabey 2012) offers insight into the range of potential resources available (Tables 3.10 & 3.11). The nutritional value of seaweed/kelp (*e.g.* vitamin C; Fediuk *et al.* 2002, table 1 and fig. 2) and sea beet also emphasises the possible appeal of coastal sites and landscapes to early hominins, although in the case of the UK Mabey suggests that seaweeds are in their poorest condition during the winter. Selected discoveries at Pakefield (the oldest artefact was from the upper levels of estuarine silts: Parfitt *et al.* 2005), Happisburgh 3 (the river channel deposits were close to an estuary and salt marsh: Parfitt *et al.* 2010) and Boxgrove (Slindon Silt deposits: Roberts 1999a) are potentially intriguing in this respect (see also Ashton and Lewis 2012; Cohen *et al.* 2012; Ashton 2015), although the direct evidence from these sites suggests the exploitation of other, terrestrial, resources, *e.g.* the GTP 17 (Unit 4b) horse butchery episode at Boxgrove (Roberts 1999b).

The strongest direct evidence of plant food availability, if not unambiguous exploitation, has come from Schöningen. Bigga *et al.* (2015) have specifically highlighted the high density of submerged USOs of sedges and cattails along rivers and lakeshores: their rhizomes are starch-rich, while the roots of the Common reed

Table 3.10: Plant families identified at Hoxne, with comparison to modern plant species available to winter foragers

Family/Species identified at Hoxne[3]	Modern winter foraging species[1,2]		
	Species	Habitat	Key nutrients
Caryophyllaceae	Common chickweed (*Stellaria media*)	Woodland fringe	Vitamins A, D, B complex, C, and Rutin
	Common mouse-ear chickweed (*Cerastium holosteoides*)	Grassland	–
Brassicaceae (previously Cruciferae)	Garlic mustard or Jack-by-the-hedge (*Alliaria petiolata*)	Woodland fringe	Vitamins A, C & E
Ericaceae	Cowberry (*Vaccinium vitis-idaea*)	Pine forest	Vitamins A, B & C
Apiaceae (or Umbelliferae)	Wild parsnip (*Pastinaca sativa*)	Grassland	Potassium
T. latifolia	Reed mace/Bulrush (*Typha latifolia*)	Wetland	Protein & carbohydrate
Urticaceae	Stinging nettle (*Urtica dioica*)	Woodland & river valley	Protein and vitamin C

[1]Mabey (2012); [2]Mears and Hillman (2007); [3]Mullenders (1993, table 6.3 & figs 6.1–6.3). None of the above information should be used as a 'safety guide' with regards to the picking and consumption of wild plant foods. Any readers wishing to do so are strongly recommended to consult an appropriate, dedicated guidebook, such as Mabey (2012)

Table 3.11: Pleistocene records of potential winter plant foods (Mabey 2012 & Table 3.10), after Godwin (1975)

Species	Common name	Pleistocene?	Earliest record[1]
A. petiolata	Jack-by-the-Hedge	No	Roman
C. holosteoides	Common mouse-ear chickweed	Yes	Hoxnian
F. sylvatica	Beech	Yes	Cromerian
P. sativa	Wild parsnip	Yes	Cromerian
S. media	Common chickweed	Yes	Cromerian
T. latifolia	Reed mace	Yes	Cromerian
U. dioica	Stinging nettle	Yes	Cromerian
V. vitis-idaea	Cowberry	Yes	Hoxnian

[1]In light of Godwin's (1975, table 1) climate stage model (including the following sequence: Beestonian > Cromerian > Anglian > Hoxnian > Wolstonian > Ipswichian > Weichselian), 'Hoxnian' is cautiously interpreted as MIS 11 or MIS 9, and 'Cromerian' as spanning the early Middle Pleistocene. None of the above information should be used as a 'safety guide' with regards to the picking and consumption of wild plant foods. Any readers wishing to do so are strongly recommended to consult an appropriate, dedicated guidebook, such as Mabey (2012)

(*Phragmites australis*) contain starch and sugar. Consumption of starchy plant foods has been suggested at the Sima del Elefante (*c.* 1.2 mya) on the basis of dental calculus (Hardy *et al.* 2017), although that data lack a seasonal dimension. The berries of common bearberry (present at Schöningen) remain on the plant throughout the winter, even under snow, while the stems and leaves from common marestail are also available. At the very end of winter the catkins and young leaves of selected trees (willow, alder and birch) could also be eaten (Bigga *et al.* 2015).

Access to some of these potential winter foods might have clear technological requirements, not least because Shea (2015) has recently criticised the functionality of stone artefacts as digging tools. Moreover Bigga *et al.* (2015) have noted that while plant foods tend to come with low gathering risks, this can be offset by the high efforts required, *e.g.* in digging for USOs (although digging up the submerged USOs of sedges and cattails from muddy sediments is not difficult, and would be unlikely to require specific tools). There is little convincing evidence for Lower Palaeolithic digging sticks (although Middle Palaeolithic examples have been identified; Aranguren *et al.* 2018; Rios-Garaizar *et al.* 2018): Schoch *et al.* (2015) have emphasised that the double-pointed wooden artefact at Schöningen is shorter than Hadza digging stick examples (while the site's spears are longer, and lack appropriate use-wear traces), although the Hadza artefacts grew shorter with use. Ashton (2015) noted the associated processing demands of this resource (*i.e.* soaking or cooking to remove toxins), and the Palaeolithic role of USOs has been more widely critiqued by Ben-Dor *et al.* (2011). Nonetheless Speth and Spielmann (1983, 20) have previously emphasised the higher essential fatty acid contents of many plant foods, and their potential as a means of building up storable carbohydrate reserves during the autumn (see also Chap. 6). Critically, recent developments in dental calculus analysis have begun to suggest plant food consumptions on specific Lower Palaeolithic sites, although the accumulation of particles in calculus is taphonomically complex and can occur through a number of different pathways (K. Hardy *et al.* 2018). Finally, Buck and Stringer (2014a, 164–165) have emphasised herbivore stomachs as a further source of plant food intake, with reference to the widespread ethnographic evidence for the practice, on the grounds on taste and culture alongside nutrition. Among the Greenland Inuit for example, reindeer stomachs offer the best source of dietary carbohydrates, with the exception of more seasonally available berries.

Scavenging as a strategy: shovelling revisited ...

In the late 1980s Gamble (1987) suggested that frozen winter carcass might provide a unique foraging opportunity for a technologically-assisted (in the form of fire and wooden snow probes) and socially-organised hominin. While to some extent Gamble's paper was overtaken by the Lower Palaeolithic hunting evidence discovered from the 1990s onwards, particularly the faunal assemblages at Boxgrove, Schöningen and Gran Dolina, its emphasis on winter carcass resources remains highly pertinent. How common might such carcasses have been? Gamble suggested biomass densities of 3360 kg/km² for contemporary temperate grasslands (perhaps increasing to values as high as those from present-day East African savannah systems [23–31,000 kg/km²] for the Eurasian steppe-tundra of the Pleistocene), while Zimov *et al.* (2012) has similarly suggested that animal density on the mammoth steppe was comparable to that of the African savannah. However, these estimates are based on environments that were not typical of Lower Palaeolithic occupations. For the optimal interglacial forests of Europe, with which many of the northern sites are associated, the Białowieża Primeval Forest (BPF) offers some interesting data.

In the BPF today wild boar mortality from starvation and/or disease is strongly influenced by the degree of snow cover (as deep snow and/or frozen ground make it hard for them to root), and especially by the quality of the acorn crop in the previous year. Interestingly this is because high yields result in an increase in the number of boar, many of which then die in the following year, because a poor or nil acorn crop tends to follow a mast year (Okarma *et al.* 1995; in the BPF oak produces heavy seed crops at 6–9 year intervals). While such annual patterns cannot be applied directly to Pleistocene forests, this does highlight the potential variability of habitats, and carcass accumulations, over the short term. In modern British red deer populations there is similarly a significant winter variability in calf mortality, varying according to a range of factors including initial birth weight, population density, heavy rainfall during the preceding autumn, low late-winter temperatures and prolonged snowfall (Corbet and Harris 1991). Adult mortality is concentrated in late winter and is also influenced by low winter temperatures. In the BPF red deer mortality is increased by very deep snow, which causes their death from starvation and inanition (exhaustion caused by lack of nourishment). It is therefore unsurprising that in the harsh winter of 1969–70, when snow depths of 70 cm+ persisted for over 2 months, nearly 30% of the BPF ungulate populations were found dead – with the exception of bison (Okarma *et al.* 1995). Thus while scavenging can be an all-year round strategy, it is likely that it was most significant during the winter and early spring, when the condition and abundance of live prey was also at its poorest (Selva *et al.* 2003). This intra-annual pattern has been documented amongst the wolves of the Białowieża Primeval Forest, although their scavenging rates as a whole were low (Jędrzejewski *et al.* 2002).

It is therefore likely that Pleistocene winters, especially if there was deep and sustained snow cover, would generate a significant dead animal biomass within the interglacial forests. This highlights the complexity of labels such as a 'harsh' winter – while it would bring many challenges for hominins there were also opportunities, *if* carcasses could be reached before other scavengers did so, and then effectively exploited. Gamble (1987) argued that an ability to produce fire on demand would enable Middle Pleistocene hominins to defrost carcasses and, therefore, access a frozen resource that would be unavailable to other carnivores. It was also argued that the use of wooden technology, the so-called snow probes, would allow hominins to access hidden carcasses, *e.g.* those buried in deep snow drifts. However, Selva *et al.* (2003) observed that the rate of bison carcass depletion in the BPF was not significantly impacted by snow cover (or precipitation), and that the carcasses were mainly utilised by carnivores at temperatures between +15°C and –15°C: below the latter temperature they were simply too frozen to be accessed. Yet winter temperatures below –15°C would definitely have presented a suite of other problems for Early and Middle Pleistocene hominins, most probably before they reached any carcasses. It is therefore likely that any carcass access by hominins occurred in the context of competition, either real or potential, with other carnivores, and may not have always required the aid of fire technology (perhaps thankfully, given the mixed evidence for fire production).

The BPF data also highlights that the consumption behaviours of different predators will markedly adjust the scale and duration of any scavenging opportunities. In the BPF Jędrzejewski *et al.* (2000) noted that wolves consume their kills quickly and comprehensively: 57% were completely eaten on the first day after the kill, and the wolves utilised 91% of the edible parts of the prey, on average. By contrast lynx feed for several days (*c.* 8 hours–8 days: Okarma *et al.* 1997) and were often observed moving a carcass over short distances from kill sites to a safer location (*e.g.* within dense vegetation or under a log) to protect it from scavengers. They would also cover carcasses with materials (*e.g.* soil, leaves, moss, deer hair, snow) for the same purpose (Jędrzejewski *et al.* 1993). There are implications here for the length of any hominin carcass defence: the likely numbers in hominin foraging groups are clearly more comparable to wolf packs than the solitary lynx, suggesting that the defence of any single carcass by hominins would only need to occur over a relatively short period, as beyond this the carcass would have been butchered, and then consumed and/or transported elsewhere. The several hours suggested by Roberts (1999b) for the comprehensive horse butchery by multiple individuals at Boxgrove GTP 17 is in-keeping with those estimates and highlights the advantages of the processing speeds available to tool-assisted hominins.

The question of scavenging always raises the question of how quickly hominins were able to reach the carcass. While relatively little is known about the olfactory abilities of Middle Pleistocene hominins (genetic analysis suggests differences in the functional olfactory genes between modern humans, Neanderthals and Denisovans; Hughes *et al.* 2014), there are other indicators of the presence of a carcass in the landscape – for example the cries and calls of predator and prey, and the visual and/or auditory signals of scavenging birds such as vultures or ravens (Blumenschine and Cavallo 1992). What is clear from the BPF is the range of species that would likely be attracted to carcasses. Selva *et al.* (2003) recorded 13 species at bison carcasses in the BPF, the most frequent of which were raven, red fox, wolf, common buzzard, racoon dog and white-tailed eagle. While complicated by Pleistocene preservation issues, wolf and red fox occur persistently on various Early and Middle Pleistocene sites, which also include a wider range of larger mammal carnivores (Table 3.12). Selva *et al.* (2003) also noted that while the European temperate forests of the present-day northern hemisphere are characterised by a relatively unspecialised scavenger guild, numerous birds and mammals rely on carcasses in critical periods (*e.g.* pine marten, red fox, ravens and bears) – it seems unlikely that hominins would actively exclude themselves from this portion of the ecosystem.

What may have been critical is that thick, hard-skinned carcasses can be a challenge to many scavengers, especially when frozen. In the case of the BPF bison the other scavengers sometimes had to wait until wolves had opened up the carcass

Table 3.12: Comparison of scavengers in the Białowieża Primeval Forest (Selva et al. 2003) with selected Early and Middle Pleistocene sites

Białowieża Primeval Forest scavengers	Selected Early & Middle Pleistocene sites				
	Gran Dolina (TD-6.2)[1]	Arago (Level G)[2]	Boxgrove[3]	Gran Dolina (TD-10.2)[4]	Schöningen (13 II-4)[5]
Raven	✗	✗	✗	✗	✗
Red fox	✓ *(V. praeglacialis)*	✗	✗	✓	✓
Wolf	✓ *(C. mosbachensis)*	✓ *(Canids)*	✓	✓	✓
Common buzzard	✗	✗	✗	✗	✗
Raccoon dog	✗	✗	✗	✗	✗
White-tailed eagle	✗	✗	✗	✗	✗

Sources: Parfitt (1999b); de Lumley *et al.* (2004); Saladié *et al.* (2014); Starkovich and Conard (2015); Rodriguez-Hidalgo *et al.* (2017). [1]Fauna represented at TD-6.2 (other potential scavengers include *C. crocuta*, *U. dolinensis* & *Lynx* sp.); [2]Fauna represented in level G (other potential scavengers include *L. spelaea*; *P. Leo* & Ursids); [3]Fauna represented in the lagoonal Slindon Silts and the palaeosol and associated deposits (other potential scavengers include *C. crocuta*, *F. sylvestris*, *Meles* sp., *M. erminea*, *M. lutreola*, *M. nivalis*, *P. leo* & *U. deningeri*); [4]Fauna represented at TD-10.2 ('bison bone bed'; other potential scavengers include *C. alpinus europaeus*, *M. meles*, *M. putorius*, *P. leo spelaea* & *Lynx* sp. *pardinus* cf.); [5]Fauna represented at Schöningen 13 II-4 (other potential scavengers include *M. erminea* & *M. nivalis*)

(and then provided progressive access at later stages in the process) – wolves thus acted as a critical keystone species in the BPF, as hyaena do for vultures in Africa (Selva *et al.* 2003). Their howling could be a potential, albeit risky, guide to the locations of carcasses, as Harrington and Mech (1979) observed in the Superior National Forest (Minnesota) that throughout the year the howling reply rate was significantly higher among all packs and lone wolves attending prey kills (and the more food that was remaining at a kill, the higher the reply rate).[22] Stone-tool assisted hominins would surely be a fellow potential candidate, although not the only one, for opening up the carcasses of the Pleistocene. Faunal lists from key Lower Palaeolithic sites highlight the presence of a range of fellow potential 'carcass opening' scavengers: *e.g.* cave lion (*P. spelaea*), lion (*P. leo*) and spotted hyaena (*C. crocuta*) at sites such as Bilzingsleben, Barnham and Swanscombe (Schreve 1996; Parfitt 1998; Mania and Mania 2005). At Early Pleistocene sites such as Barranco León D and Fuente Neuva-3 in the Orce Basin the list of scavengers was even more exotic, including the bone-cracking hyaena *P. brevirostris* and the jackal-sized *C. mosbachensis* (Rodríguez-Gómez *et al.* 2016). The opening up of carcasses has been seen in the BPF to reduce competition and allow more species to feed together: observations at the bison carcasses in the BPF revealed that the avian scavengers would mainly eat the intestines, the stomach, the viscera and the muscles, medium-sized scavengers the viscera, muscles and smaller/softer bones such as the sternum and ribs, while wolves would eat it all, including the hard skin and the larger bones, with gnawing of the head of the leg bones for marrow a key indicator of their presence (Selva *et al.* 2003). It is an intriguing thought whether or not hominins might also have been welcome at such a table, or indeed whether this was a table they would have wished to join. While modelling work has placed hominins within the preferred prey size ranges of all the major Early and Middle Pleistocene carnivores (Rodriguez-Gomez *et al.* 2017), it is also possible that a small[23] group of tool-assisted adult hominins could have rapidly exploited targeted parts of the carcass and then left the scene, even if they were intimidated by other carnivores.

Nonetheless the potential presence and size of other secondary consumers highlights the risks of scavenging (Blumenschine 1991), and raises the question of who would have been involved? If searching for carcasses was an active strategy, perhaps in response to environmental cues, then the benefits of a larger searching group (more eyes, ears and noses and more ground covered) are easily apparent, and the latter point especially might be important during shorter days. However, would this outweigh the costs? Winter scavenging would expose individuals to cold conditions, potentially for several hours, without the benefits of shelter, fire and/or huddling together (although clothing could, and perhaps would, still be an option). Moreover, the carcass itself would likely require a division of tasks between carcass processing and monitoring fellow carnivores. Adding vulnerable individuals (*i.e.* infants and younger children) to the active scavenging group therefore seems unlikely, although older children could perhaps have participated (and could certainly have been active searchers; see also Chap. 5). Younger children could be carried by adult searchers, although this would add to the energy demands on the latter and reduce their walking speeds, particularly in the absence of slings or other carrying aids (Wall-Scheffler *et al.* 2007). Once the carcass was found they could be 'crèched' (perhaps by older adults, late pregnancy females or older siblings) away from the immediate vicinity of the carcass. The other alternative is that only some adults and older children were involved in scavenging, with food returned to and shared with the remainder of the group (*e.g.* carers, very young infants and perhaps also late pregnancy females). This raises issues as to how food was transported, with knock-on impacts for separate foraging groups, scales of mobility, food provision and/or sharing across the entire hominin group (see also Chap. 4), especially given the potential state of a scavenged carcass.

Finally, the bison carcasses of the BPF also highlight the potential duration of scavenging opportunities where large animals are concerned. The bison were utilised for 106±61 days, until over 80% consumed, with a maximum range of 22–239 days – although the mean daily consumption of the carcass was highest over the first 2 weeks, and utilisation times depended on the degree of carcass openness, the number of wolf visits, time of year (utilisation times were reduced with the advance of winter, as less live food became available) and habitat type (Selva *et al.* 2003). While these times may be exceptionally long due to the absence of brown bear from the BPF since the 19th century and are the product of a non-Pleistocene scavenger guild, they nevertheless highlight the potential role of carcasses in enhancing predator fitness, especially in winter.

'Every mile is two in winter …' (George Herbert, 1592–1633)

The above discussions of fire, clothing, shelter and food lead naturally onto the question of the organisation and degree of mobility of hominin groups during European winters (see also Box I). Grove (2009, fig. 3) indirectly highlights the potential conundrum facing European Lower Palaeolithic hominins during winter: foraging areas are likely to be seasonally larger (reflecting poor resource availability[24]), and would increase further, as local resources were progressively exhausted, if residential occupations were long in duration. Yet the latter might be appealing, if possible, to avoid the energetic costs and risks of residential movements in low temperature winter landscapes, which might be a particular threat to the young, old, sick and pregnant.

Low mobility might be especially likely if hominin conception followed the patterns suggested by Mussi (2007). Mussi proposed a concentration of conception in late summer, reflecting peak periods of body condition in the seasonal environments of Europe. In such a model, the second and third trimester would fall in winter/early spring, possibly resulting in differences in levels of mobility and types of activities between pregnant females and other adults and older children (see also Chap. 4). Reduced mobility for the group as a whole might also be favoured by the need for more frequent and close monitoring of infant health (in light of lower temperatures and/or reduced food supply), while the relatively poor condition of ungulates might necessitate more careful observations of wolves, hyenas and the many other Early and Middle Pleistocene predators (*i.e.* a 'period of vigilance'; after Bliege Bird *et al.* 2009). Thus, winter landscapes might encourage semi-nomadic rather than nomadic living (Gamble 1991, table 1), although the nature of those residential sites is currently uncertain (see also Chap. 4: Box L). However episodic residential movements might also be enforced by the accumulation of hunting/scavenging debris and the potential for such food waste to attract social carnivores (Grove 2009).

It also seems likely that larger, community-level aggregations (*c.* 120–130 individuals, as predicted by Dunbar's social brain hypothesis; Chap. 2) did not occur in winter. This is suggested for two principal reasons: first, the movements involved in forming social aggregations would incur additional energetic costs; and secondly, since foraging areas increase with group size, band-, or even foraging group-, level social units seem a more likely scenario. Large aggregations at other times of year, perhaps linked to hunting and/or mating opportunities in summer and autumn (Chaps 4 & 5), would be much more feasible, since conditions for movements between foraging areas would generally be favourable, and enhanced habitat quality (*i.e.* resources) would likely partially offset the impacts of larger groups on foraging area sizes.

The challenge of coping effectively with short, cold days and dispersed resources is brought into particular focus by the mobility behaviour of another European social predator. Okarma *et al.* (1998) observed a marked increase in the upper limits of wolf home range sizes in the Polish portion of the BPF, where wolves are protected, between spring–summer (May–September: 141–168 km²) and autumn–winter (October–April: 99–271 km²). Jędrzejewski *et al.* (2001) similarly observed that daily movement distances were longest in the autumn–winter. This resulted in rapid increases in the cumulative areas covered by wolves on consecutive days: in the latter study the wolves covered nearly 70% of their whole territory in 8 days, on average (Jędrzejewski *et al.* 2001). These data suggest that a wolf-type model of diet and mobility would make significant demands on the abilities of hominins to move effectively through winter landscapes. The alternative would be a reliance on stored foods (Chap. 6), a wider range of more locally available foods, and/or a more effective hunting performance.

However, wolves' use of their home ranges is rather more complex than is suggested by the figures above. While overall observations of wolves in the Polish BPF indicated a home range of 173–294 km² (based on 100% of recorded wolf locations), there was also a significantly smaller core area within this of just 11–23 km², or 5–13% of the total home range (based on 50% of recorded locations; Okarma *et al.* 1998). This core area was home to the majority of the wolves' diurnal resting sites, although the pack hunted in both the core and peripheral parts of their range (Okarma *et al.* 1998). While there are obviously caveats to wolf/hominin comparisons, these data suggest that the latter's core areas may have been significantly smaller than an overall territory. This might have been especially true in winter, potentially reducing although not removing the challenges of harsh conditions, although some longer-range mobility may still have been required to secure mobile prey food (as with the wolves). These core dwelling areas can perhaps be thought of as a local foraging radius (Grove 2009; see also Box I).

Given the particular challenges and costs of winter mobility and foraging (scarce resources, short days, low temperatures and possible snow cover: Kelly 1995, 124), the hominins' choice of a local or regional landscape in which to survive the winter months would be critical. This issue remains relevant whether the annual cycle included bi-annual migrations (Chap. 6) or all year-round occupation of a single region, although local micro-climate habitats such as sheltered valleys or dense forest patches would provide improved palaeoenvironmental conditions without the high 'costs' of long-distance migrations. Key criteria at a local or intra-regional scale would probably have included low altitude, sheltered settings (offering temperature, snowfall and exposure/wind chill benefits), availability of natural shelter opportunities, and nearby access to fresh, flowing water (reducing mobility costs), key prey species' winter feeding grounds, winter plant foods (possibly including coastal resources such as seaweed), likely scavenging opportunities, and (possibly) access to fuel supplies. Awareness of the territories and movements of other predators (*e.g.* wolves) might also be significant at this time of year, when competition for prey and/or the aggression of other predators might be at their peak. This is a complex suite of considerations and suggests that ideal (or at least good enough) winter refuges, once found, would not be swiftly abandoned, and thus a model of longer residences in specific places seems

Box I: Mobility in the Palaeolithic

One of the most interesting aspects of thinking about Lower Palaeolithic hominins, and indeed any human groups, concerns mobility. What do we mean by this? Human mobility spans a range of types, functions and scales. We can see this most obviously by reflecting on our own lives, from the commute to our places of work or study (daily), to the trip to the local shops (weekly?), to more occasional longer-distance trips (the annual holiday). More formally, as Gamble (1991, 5) highlighted: 'mobility is the means by which population is mapped onto the environment thereby averting shortage and conflict'. But what was mobility like for the earliest Europeans?

Binford (1980) usefully distinguishes between two types of mobility: residential (the movements of all members of a camp from one location to another) and logistic (movements of individuals or small groups from a residential location, for 1 or more days, typically to undertake specific tasks). These movements, and the concept of a camp or home base, are ubiquitous amongst extant and recent hunter-gatherers, and have underpinned numerous other studies (Kelly 1983). But are they applicable to the Lower Palaeolithic?

The notions of home bases, residential and logistical mobility, and related concepts such as central places and resource catchments, assume the concept of a 'site' – a place that an individual plans to return to. Is that likely or is our thinking over-influenced by our own preconceptions and the importance of 'home' in our own lives? Would groups have lived a continuously mobile lifestyle instead, spending each night in a new location? This seems unlikely for three inter-linked reasons: first, the nutritional requirements of large-bodied, large-brained and altricial hominins likely required a high quality and high quantity diet of both plant and animal foods in a mid-latitude environment (Bogin and Smith 1996; Cordain *et al.* 2000; Aiello and Key 2002). Secondly, the seasonal and mid-latitude character of Europe would result in time-delimited and spatially segregated resources, and prey animals with relatively large territories (Kelly 1995). Ecology suggests that 'sites' (*i.e.* dense patches of materials) should be expected in the high latitudes, due to these resource characteristics, although the resulting archaeological signature can be further blurred by individual resourcing trips (Gamble 1991). Thirdly, the fundamental differences between plant and animal foods (stationary and mobile), combined with points 1 and 2, meant that sufficient resources would be most efficiently and safely accessed (*i.e.* minimising risk of failure and reducing energetic costs and dangers to vulnerable individuals) by multiple, differently focused, foraging groups. Thus I propose that day-to-day living involved logistical mobility[1] from a home base or central place (to access both categories of food resources but with the majority of longer forays associated with mobile animal foods), overlain onto periodic residential mobility (as resources within the local foraging radius were exhausted). In the terminology of Lieberman and Shea (1994) this model therefore combines both circulating and radiating mobility (Fig. I.1).

It is also helpful to consider mobility and residency through the concept of specific hominin resource needs: water (for drinking, and perhaps, in the case of open wounds, cleansing), plant and animal foods, raw materials for tool-making (potentially including fuel for fire), natural shelters, security, and favourable micro (or macro)-climates. Supplies of some of these resources (water, food and potentially firewood) would need to be renewed on a daily basis (especially if, *e.g.,* food storage was not practised), and would gradually be locally exhausted over time, and there is an important division between static (*e.g.* shelter, water sources) and mobile (*e.g.* animal prey) resources. The concept of a temporary residence (whether termed a home base, sleeping site, camp site or other label) therefore

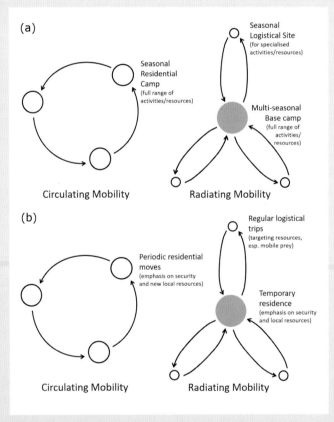

Figure I.1: *Circulating and radiating mobility models (a; Lieberman and Shea 1994, fig. 4), and modified for the Lower Palaeolithic (b).*

seems likely as it provides security and access to local resources through logistical mobility, combined with residential mobility as resources became exhausted.

Other factors which are likely to have influenced residential moving decisions
include the distance to the next food patch, the locations of suitable residential
locales, terrain difficulty, attraction of carnivores to food residues at the existing
residence, and short-term weather (Kelly 1995; Grove 2009). Residential moves
may also have been triggered by animal, plant and temporal cues (*e.g.* the end of
hibernations, the shedding or regrowth of an animals' winter coat, spring blossom, fruiting, the turning of the leaves, deer ruts, bird and mammal migrations,
longer and shorter days and changes in air temperature).

Grove (2009) proposed that habitat quality (using precipitation and effective
temperature indices) is the best predictor of foraging radii and therefore of
distances moved between camps. Unsurprisingly the foraging radius is small in
high quality habitats and increases as quality declines: as the radius increases so
do the average distances moved between camps, to avoid exploiting previously
depleted areas. This also has implications for the number of moves made over
the course of a year, since this has a negative relationship with average distances
moved (*i.e.* varying from few, long moves to many, short moves).

The nature and preservation of the archaeological record does not help, but
the suggested home bases at Gran Dolina (TD-6.2 and TD-10), Arago and, more
controversially, Bilzingsleben are supportive of residential mobility models, as are
other 'fixed location' sites on and beyond the fringes of Europe (*e.g.* Qesem cave).
The leading European candidates (*e.g.* Gran Dolina, Arago Cave, Beeches Pit and
Bilzingsleben) are all adjacent to fresh water sources, supporting the notion that
much of the day-to-day mobility associated with food/water resourcing was fundamentally local, concentrated in what Binford (1982) referred to as the foraging
radius.[2] Critically the available evidence, while limited, suggests temporary and
repeated use of such sites (*i.e.* residential mobility). At Gran Dolina TD-10.2 faunal
dental wear indicates short, seasonal occupations, while differences in the distribution of anthropogenic and carnivore marks on fauna (the former only occur on
bison, the latter on a wide range of taxa) suggest independence of hominins and
other carnivores in the formation of the animal bone assemblage, and therefore
that there were periods of hominin absence (Rodriguez-Hidalgo *et al.* 2017). While
the use of Lower Palaeolithic open-air residences is more difficult to prove due
to their very ephemeral 'footprints', the temporary use of dens and rendezvous
sites by other social predators (*e.g.* wolves) would seem to be strongly suggestive
of the existence of such sites, albeit probably characterised by relatively short-
lived occupations (Theuerkauf *et al.* 2003; Schmidt *et al.* 2008).

Evidence for the activity sites that fell within the local foraging radius is perhaps more abundant, since carcass-strewn kill-butchery sites such as Boxgrove and
Schöningen seem unlikely locations for multi-day residences (although carcass

butchery may have gone on for several hours). The evidence at certain sites for the introduction of artefacts from elsewhere, and specific tasks being undertaken (*e.g.* imported tools, and on-site roe deer and bison butchery, at Happisburgh 1; Lewis *et al.* 2019) is perhaps also suggestive of logistical trips, as are occasional raw material 'quarry' sites (Chap. 5). Repeat visits are evident at a number of these sites, presumably reflecting their resource potential, although the time-depth of the returns to specific locations is less certain (every few days? seasonal? annual?). Potential resource exhaustion and prey alertness would likely have been factors influencing the use and re-use of specific locations, and the frequency of residential moves to new core foraging areas. A sustained presence has been suggested for Soucy (Soucy 3: level P), on the basis of the ages of individual animals, which cover multiple species (Lhomme 2007). However, this seems surprising given the amount of accumulated fauna (over 20,532 remains were recovered from a 700 m² horizon [a gravel hillock slope on a river edge]) and the presence of other carnivores (wolf). If correct Lhomme's interpretation strongly suggests 'habitat control' by the hominins, although I suspect that a regular, but not continuous, presence is more likely.

The concept of a foraging radius also raises the question of what is 'local' or 'close' in a Lower Palaeolithic world. Classic studies of Palaeolithic site catchment analysis (*e.g.* Vita-Finzi *et al.* 1970) have employed measures such as 1 or 2 hours' travel, originally derived from ethnographic observations but subject to further modifications in light of local terrain conditions (Bailey and Davidson 1983). It is inevitably difficult to discuss specific examples, given the major transformations to many landscapes since the Lower Palaeolithic. However sinuous but relatively flat river terrace landforms might have facilitated movement, in contrast to more potentially constraining boundaries such as deep, wide rivers, snowlines, or steep scree slopes or cliffs (Sturdy and Webley 1988). Other contextual issues such as day length, weather conditions and the degree of threat (from carnivores and/or other hominins) would also impact upon perceptions of 'nearby' or 'far away'. Any trip that involved an 'overnight stay' might well be crossing a 'security rubicon', as it involved individuals' sleeping 'out', away from the remainder of the group (in the case of logistical groups), or exposing the more vulnerable members of the group to temporary sleeping site conditions (in the case of residential mobility). Even in the case of a residential group foraging close to 'home', remaining within-sight of other members of the group would be unlikely in the majority of peak interglacial landscapes, given the palynological evidence for significant tree cover, although remaining within ear-shot might not be. While hominin notions of risk are ultimately unknowable, and would have varied from one particular situation to the next, the available lithic transfer data (Table 5.9) suggests that the majority of individual trips may only have covered a few kilometres. Throughout this book I have chosen to define local mobility as any movement which can be completed within a single day.

A final factor concerns how mobility, at any scale but perhaps especially at local scales and in the context of day-to-day logistic mobility, may have varied between different group members. Lugli *et al.* (2017) have demonstrated local mobility in the case of one Lower Palaeolithic hominin individual from the Italian site of Isernia la Pineta, on the basis of strontium isotope data, calibrated against other faunal dental data (from rodents, bison and rhinoceros) and local modern plants. In this instance local mobility is most likely within 15 km of the Isernia site, and tallies with the definition of local offered above. But was this local living applicable to all at Isernia? It is noteworthy that the hominin data is from a deciduous incisor which, as a result of forming within the uterus, is informative about the mobility of a pregnant adult female. By contrast, the dental isotope data for rhinoceros and, to a lesser extent, bison, from Isernia suggest larger landscape movements, perhaps seasonal migrations, of up to 50 km. From a food-getting perspective this raises the interesting possibility of temporary or daily group fission/fusion (hereafter dF-F; see also Chap. 4: Box K), in which certain individuals, including pregnant females, were more locally focused than other members of the group. This shows interesting parallels with two very different species: chimpanzees in Gombe National Park, amongst whom pregnant females spend less time travelling than other females (Murray *et al.* 2009); and wolves in the Białowieża Primeval Forest, where pregnant females reduced their mobility by half in the ten days prior to parturition (Schmidt *et al.* 2008).

---

[1] Although the general absence of task-specific Lower Palaeolithic sites highlights that this is almost certainly not a logistical approach to landscape exploitation of the type described by Binford (1980).

[2] Binford (1982, 7) defined this as an area that could be searched and exploited by work parties in a single day, ending in a return to the residential camp.

likely over those months.[25] If artificial or naturally-modified shelters were used, then a residential strategy would also avoid the costs of transporting structural elements such as hides (MacDonald 2018). Testing this model against the available evidence is challenging due to the limited availability of seasonality data but selected primary context sites (Table 3.13) suggest that these criteria may have been successfully met in a diverse range of settings and habitats.

Present-day micro-climate contrasts in the UK include upland/lowland (*e.g.* temperature reductions of of 5–10°C per 1000 m of elevation, windier upland conditions and longer upland winters combined with shorter summers) and inland/coastal differences (*e.g.* milder coastal winters but cooler coastal summers; Met Office 2011). In regard to the former point, relatively few European Lower Palaeolithic sites are associated with upland landscapes (defined here as elevations above 1000 m): the rich artefact landscapes of northeastern France, southeast England and northern Germany are predominantly below 200 m, although the Spanish meseta reaches *c.* 600–800 m (above present day sea-level; Fig. 2.2). As to inland/coastal contrasts, the distributions

Table 3.13: Comparison of site evidence with suggested winter refuge criteria

Criteria	Lower Palaeolithic sites			
	Arago *(Levels Q & R)*	*Beeches Pit* *(Beds 4 & 5)*	*Bilzingsleben*	*Gran Dolina (TD-6.2)*
Ameliorated climatic conditions?	✓ (cave; temperate woodland)	✓? (deciduous woodland)	✗? (Jan. temp: -0.5 – +3°C)	✓ (Jan. temp: +4.3°C)
Access to fresh water?	✓ (River Verdouble)	✓ (Springs)	✓ (Springs)	✓
Access to animal foods?[1]	✓	✓ (e.g. aurochs)	✓ (e.g. rhinoceros, red deer)	✓ (e.g. deer)
Access to plant foods?[2]	✓ (e.g. boreal & deciduous forest species)	✓ (e.g. deciduous woodland species)	✓ (e.g. wood/ wetland species)	✓ (e.g. *Celtis* seeds)
Shelter opportunities	✓ (cave)	✓? (woodland)	✓? (artificial shelters?)	✓ (cave)
Access to fuel supplies?	✓ (Boreal & deciduous forest)	✓ (deciduous woodland)	✓ (oak-mixed woods)	✓ (open woodland)
Access to diverse habitats?	✓ (steppe, grassland, boreal & deciduous forest)	✗? (closed woodland, tufa springs)	✓ (montane, riverine, springs)	✓ (Mediterranean open woodland, steppe, permanent water)

Sources: Mania and Mania (2003; 2005); Preece *et al.* (2006); Rodríguez *et al.* (2011); Lebreton *et al.* (2016). [1]Indicators of anthropogenically-exploited species (*e.g.* Gran Dolina TD-6.2) or species' remains present at site (*e.g.* Beeches Pit); [2]Indicators of specific species (*e.g.* Gran Dolina) or general plant habitats (*e.g.* Bilzingsleben)

and potential resources of selected early sites such as Pakefield and Happisburgh I & III are noteworthy (although taphonomic and sampling bias should also be acknowledged: Parfitt *et al.* 2005; 2010; Ashton *et al.* 2008b; 2014; Cohen *et al.* 2012). More generally wind chill and exposure, whether in uplands or lowlands, increases the risks of hypothermia and frostbite and was presumably avoided if possible. Local variations, *e.g.* in valley side or slope aspect, would also impact on the attractiveness of a particular habitat for a sustained occupation, with differing exposure to dominant winds and sunlight impacting on levels of warmth, dryness and illumination (Blundell 2020). Benefits and variations at similar scales were also highlighted during the Stage 3 Project. Davies *et al.* (2003, 210–211 & fig. 11.5) emphasised the presence of mixed mosaics of ecotones in the Dordogne, the Ardennes and the Middle Danube for example, with preferential MIS 3 settlement patterns in the side valleys of the Ardennes perhaps reflecting a need for greater shelter from prevailing winds. Those locales with favourable micro-climates are likely to have been preferred local 'winter landscapes', and the focus of late autumn or early winter residential moves. Such movements are described here as inter-seasonal relocations. The scale of such movements is more likely to be in the order of 10s, rather than 100s, of kilometres, since a strategy based on regular, larger-scale movements (*i.e.* multiple long-distance moves within each year) would be highly demanding in terms of habitual costs, and therefore an unlikely evolutionary strategy (see also Chap. 6). Thus, local variations in landscape settings and/or resource availability are likely to have structured hominin movements, creating habitual landscapes if not necessarily persistent places[26] (see also Ashton 2018; *cf.* Shaw *et al.* 2016; Shaw and Scott 2018).

Conclusion: a winter's tale

This chapter has sought to highlight the many practical challenges that would have been faced by Lower Palaeolithic hominins over the course of a mid-latitude European winter. The requirements of securing sufficient food and fuel and/or other raw materials for cultural insulation were by no means straightforward or insignificant. Yet the archaeological record (*e.g.* Roebroeks and van Kolfschoten 1995; Gamble 1999) demonstrates that these challenges were successfully met, initially in the south, ultimately across western Europe, and sometimes during the cooler phases of mid-latitude warm episodes (*e.g.* Ashton *et al.* 2008a; Parfitt *et al.* 2010). The challenges may also

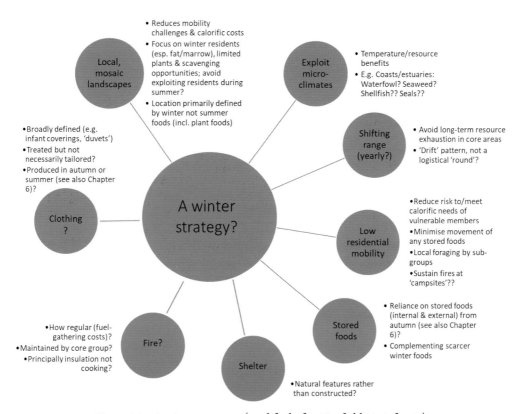

Figure 3.14: A winter strategy (modified after Hosfield 2016, fig. 10).

have become more marked during the Middle Pleistocene, as seasonality increased. Consideration of the available resources (*e.g.* fur and fat-bearing winter residents, winter plant foods) highlights the availability of the raw materials for successful survival through local residency, so long as we are willing to accept the need for, and possible presence of, selected 'survival behaviours' (*e.g.* rudimentary cultural, or perhaps biological, insulation, internal and perhaps external food storage, a blend of residential and logistical mobility) amongst the repertoire of Early (although they were rarely to be found north of the Alps) and Middle Pleistocene hominins (Fig. 3.14). These behaviours may very well have been at least as integral to their survival as the enduring stone tools that so dominate the archaeological record.

However, the composition, reliability and 'success' of these behavioural repertoires, set against the demands of survival, geographical variations in habitats and climatic regimes, and the limited evidence for key components in the archaeological record, may have been variable. The punctuated expansion in European hominin distribution from the Early to later Middle Pleistocene may suggest in particular that reliable winter food provision and/or cultural insulation strategies were rather different amongst *H. antecessor* and *H. heidelbergensis*, with significant change after *c.* 500–600 kya, at least to the west of the Rhine (*e.g.* Dennell and Roebroeks 1996; Hosfield and Cole 2018; see also Chap. 6).

From a 'year in the life' perspective however it is important to look beyond winter and consider the other seasonal demands: as the year turned to spring, other challenges lay in store for Europe's earliest inhabitants ...

Notes

1. The other result of challenging winters is of course extirpation (local extinctions) of hominin groups.
2. This tendency was revealed by sensitivity tests on the MCR method: there was often a discrepancy between MCR estimates derived from modern coleopteran faunas and the monthly temperatures measured at nearby meteorological stations, with MCR winter temperature estimates usually too warm (Pettitt and White 2012, 35).
3. Related to the ribs.
4. Allen's rule states that the limbs and other appendages of animals living in cold climates tend to be shorter than in animals of the same species living in warm climates. Bergmann's rule states that body mass increases in colder environments.
5. Minimal sustainable ambient temperature (MST) is 'the minimum temperature at which an animal can maintain normal body temperature by raising its basal metabolic rate to its maximum sustainable level, in humans usually about three times normal BMR' (White 2006, 568).
6. Lower critical temperature (LCT) is 'the lower limit of the thermoneutral zone within which a mammal can regulate its core temperature solely by controlling its thermal conductance ... as the ambient temperature falls below this level, homeostasis can only be maintained by progressively increasing internal heat production, and incurring the additional energetic costs associated with this increase in heat production' (Aiello and Wheeler 2003, 148).
7. 1 clo is approximately equivalent to the level of insulation provided by a western business suit (Aiello and Wheeler 2003, 150).
8. Comparison with species such as horse and red deer (Ryder 1977; Bennett and Hoffmann 1999) suggest that a winter coat would be a likely component of all-over hominin body hair.

9. A further chronological exception can be found on the geographical margins of Europe, at Gesher Benot Ya'aqov (Goren-Inbar *et al.* 2004; Alperson-Afil 2008).

10. Experienced, elder 'fire keepers' may also have played a key role in the production of foraging tools and other artefacts.

11. If fires were only a requirement in the hours of darkness (although in the shortest months this might extend to up to *c.* 15 hours in northern Europe; see Table 1.1) then during the daytime there would be no need for individuals to remain at the hearth sites, and overall fuel demands would also be reduced.

12. Normal body temperature in modern *H. sapiens* is *c.* 37°C.

13. The critical external temperature threshold for hypothermia is variable: MacDonald (2018) suggests a safe limit for humans accustomed to clothing of *c.* -1°C, while Gilligan (2017) suggests a safe naked limit for maximally biologically adapted modern humans of *c.* -5°C. The risk of hypothermia is also exacerbated by wind-chill (Yankielun 2007).

14. The thermoneutral zone is the environment in which body temperature is maintained with the lowest energy expenditure and oxygen consumption, and it is 5–7°C higher in human children than in adults (Mateos *et al.* 2014).

15. Unfitted, 'simple', clothes have been suggested to provide 1–2 clo of thermal protection in still-air conditions, but poor protection from windchill (Gilligan 2010, table 1).

16. Although fire has a wider range of other potential benefits (*e.g.* predator defence), and clothing 'replacement intervals' are difficult to estimate.

17. Some of these sites, with their long-term, if not continuous, occupations, are currently the strongest evidence for the persistent places concept in the European Lower Palaeolithic (*e.g.* Ravon 2018).

18. MacDonald (2018) suggests that hide shelter coverings might last 1–2 years after hair removal and heavy greasing with animal fats on both sides.

19. Palimpsest effects have been considered because of the apparent animal/vegetation discrepancies, e.g. evergreen taiga/montane forest and mammoth and woolly rhinoceros, and inter-species contrasts between climate model predictions and the modern climatic tolerances of small mammals such as lemming, Arctic fox and souslik (Stewart *et al.* 2003).

20. This is working on the assumption that Lower Palaeolithic hominin metabolism was not markedly different from that of modern humans (following Buck and Stringer 2014a).

21. Pemmican is a concentrated, energy-rich, mixture of dried lean meat and rendered fats, and sometimes also dried fruits (Speth and Spielmann 1983, 19). The name derives from the Amerindian *pimii* (≈ grease; Collins Dictionary 1989).

22. However, it is important to note that wolf howling also has a number of other purposes, such as reunion, social bonding, spacing and mating (Theberge and Falls 1967; Harrington *et al.* 2003), and thus might be an unreliable 'carcass cue'.

23. Based on band sizes of *c.* 40 (see also Chap. 4: Box K), such a group might have included *c.* 12–14 adults (after Kelly 1995, 213).

24. Similar seasonal patterns were documented amongst the !Kung of the Kalahari: local foraging distances increased to 10–15 miles (16–24 km) in the late dry season, when the desirable foods in the immediate vicinity of the waterholes had been consumed (Lee 1968).

25. Although resource exhaustion, and other factors such as the availability of water and bedding or the hygiene of the camp, would mean that residential moves, if and when they occurred, were probably relatively long-distance events (see also Kelly 1995, 126 & fig. 4–7).

26. Shaw *et al.* (2016, 1440) defined Palaeolithic persistent places as 'showing evidence for repeated and frequent use over long periods of time – both open and sheltered sites (e.g. caves and abris) over at least one interglacial phase'.

Chapter 4

Springtime – a land awakening ...

Spring renewal

The spring months in mid-latitude Europe would bring longer days[1] (Table 1.1), gradually improving temperatures, changing precipitation regimes, new plant growth and the gradual 'closing'-in of the landscape, and a dramatic change in the 'soundtrack' of life (Box J). But to what extent can we reconstruct spring conditions for the Early and Middle Pleistocene? While many palaeoclimatic measures emphasise maximum (July) and minimum (January) temperatures, specific studies do offer spring perspectives. Analysis of herpetofauna has provided valuable estimates of temperature and precipitation for a series of key Early and Middle Pleistocene sites in Spain (Table 4.1 & Figure 4.1; Blain *et al.* 2013; 2014; 2016). While conditions vary over time, latitude and local topography, the general trends from March to May of increasing temperature and reducing precipitation (with the exception of Gran Dolina TD-6.2) are familiar, as is the significant variability.

In terms of food resources, spring is likely to have been associated with ungulate migrations, the emergence of various species from hibernation, and, by late spring, a variety of animal births, covering both ungulates and birds. From a vegetation perspective, spring fresh growth occurs in a wide variety of plant species, supporting both primary and, indirectly, secondary consumers. For Lower Palaeolithic hominins, spring thus brought with it a range of new challenges and feeding opportunities.

Spring relocations?

I have argued elsewhere that inter-regional seasonal hominin migrations at the scale of hundreds of kilometres are an unlikely strategy (Hosfield 2016; see also Chap. 6). However, spring *is* likely to have stimulated shorter distance moves from winter residences, as hominins tracked and/or responded to animal movements, either to

Box J: Sounds in the Lower Palaeolithic world

The nature of academic writing and publishing, and perhaps also the nature of archaeologists themselves, has often resulted in descriptions of the human past which lack the colour and noise, the sheer vitality, of human life. This can be frustrating, not least because ethnography, anthropology and primatology repeatedly remind us of just how vibrant life really is. An interesting recent exploration of the sounds of the past was created for the British Mesolithic site of Star Carr (https://soundcloud.com/jonhughes409/star-carr-sonic-horizons-rough) and highlighted a range of the possible sounds likely to be familiar to the Lower Palaeolithic hominins of Europe: the dawn chorus, bird-song, owls hooting, the screech of raptors, mammalian cries (*e.g.* bellowing, shrieking, howling), thundering hooves, branches cracking under-hoof or paw (or hominin foot), animals or humans pushing through undergrowth, animals browsing on trees, rainfall, wind in the trees, running water, the slosh of waves on a lake or sea-shore, the crackle of fire, the crisp ring (or not so crisp, depending on skill!) of stone knapping, and hominins talking, shouting, crying, laughing, maybe even singing and making music (Mithen 2005), and perhaps even occasional moments of near silence ...

intercept animal migrants or to relocate within or close to preferred animal habitats (*e.g.* summer feeding grounds). Changing weather conditions, and plant availability, although the latter may have been more ubiquitous, are also likely to have been key factors. Spring mobility might also have been driven by over-exploitation of local animal and/or plant resources around winter residences over the preceding months, especially if levels of winter mobility were low (see Chap. 3).

Table 4.1: Spring temperature and precipitation data for selected Spanish Early and Middle Pleistocene sites (Blain et al. *2013; 2014; 2016)*

Site	Month	Temperature (°C)			Precipitation (mm)		
		Mean	SD	Range	Mean	SD	Range
		Early Pleistocene					
Fuente Nueva-3	March	12.8	2.7	6.0–15.0	66.0	22.0	40–130
	April	13.8	2.4	8.0–16.0	69.0	18.0	50–110
	May	17.2	2.0	13.0–19.0	54.0	20.0	40–110
Gran Dolina TD-6.2	March	8.3	1.7	–	63.8	7.4	–
	April	10.0	1.5	–	82.5	7.1	–
	May	14.1	0.8	–	93.8	11.6	–
		Middle Pleistocene					
Aridos I	March	11.5	2.8	6.0–14.0	57.6	15.0	40–100
	April	13.6	2.7	8.0–16.0	60.9	14.0	50–90
	May	16.4	1.9	12.0–18.0	47.6	18.0	30–80

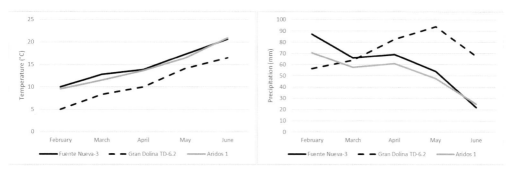

Figure 4.1: Spring temperature (left) and precipitation (right) trends for selected Spanish Early and Middle Pleistocene sites (Blain et al. 2013; 2014; 2016).

Table 4.2: Selected environmental characteristics for temperate grasslands, woodlands and boreal forests (Kelly 1983, table 3)

Biome	Primary biomass	PP/PB[1]	SB/PB[2]	Herbivore body size	Secondary biomass distribution
Temperate grassland	1600	0.38	4.3	Large	Gregarious
Temperate deciduous forest	30,000	0.04	0.5	Medium	Dispersed
Boreal forest	20,000	0.04	0.2	Large	Dispersed

[1]Primary productivity/primary biomass; [2]Secondary biomass/primary biomass

Kelly's (1983; 1995) review of hunter-gatherer mobility provided an important perspective on the potential roles and scales of residential and logistical mobility at mid-latitudes. While the environmental contexts of extant hunter-gatherer groups are not an ideal comparison (a common problem when exploring the warm stage forests of Lower Palaeolithic Europe), Kelly noted that hunter-gatherers tended to move frequently in high biomass areas with relatively inaccessible plants and animals[2] (*e.g.* temperate deciduous and boreal forests; Table 4.2): examples from boreal forest groups suggested 50–60 residential moves/year (Kelly 1983, table 6), while between 30–60 moves/year were recorded for five non-tropical groups (Kelly 1983, table 7).

A highly mobile Lower Palaeolithic lifestyle is likely since a reliance, at least in part, on fauna in low effective temperature environments impacts on hunter-gatherer mobility behaviour due to the significant search, pursuit and commuting costs associated with a mobile prey. While exact dietary models in the Lower Palaeolithic are uncertain and while costs in European warm stage forests may have been less than those in Kelly's (1983) boreal forest examples, a general model of regular residential moves,[3] initiated when commuting costs become excessive and local resources are inadequate, seems likely to have applied. As for the scale of those moves, residential move *distances* increase with lower Effective Temperature (ET; Kelly 1983; 1995): this reflects relatively long logistical trips (due to large animal territories and increasingly segregated resources, and because additional benefits, such as hides, reduce the 'cost' of these trips[4]) and the resulting exhaustion of resources over a relatively wide area during each residential occupation. Consequently, when residential moves take place they cover longer distances: the Montagnais and Nunamiut moved an average of 64 km and 69.5 km/residential move (Kelly 1983, table 1). Thus spring-time relocations in the Lower Palaeolithic may have been relatively long-distance affairs.

Animal behaviours and their impacts on springtime residential moves may have been relatively predictable however, either as a result of traditional behaviours or because of the ecological preferences and requirements of species. For example, modern red deer seek a combination of good cover and food supply in close proximity (Corbet and Harris 1991). Such conditions would be widely found in the mosaic landscapes of European Early and Middle Pleistocene warm stages, and comparable habitat preferences could have been a structuring factor in logistical prey searches, and perhaps also in residential moves. Moreover, the modern red deer populations in the BPF demonstrate considerable range fidelity (Kamler *et al.* 2008), which is at its strongest during the summer amongst females,[5] and is probably linked to traditional birthing areas (their ranges are also at their smallest in the summer: 4.6±0.2 km²). Similar behaviours in the past could potentially have structured hominin mobility. But are such parallels appropriate?

As noted in Chapters 2 and 3, such modern examples are useful but should not be applied too literally, not least because our appreciation of Pleistocene animal behaviours has shifted markedly in the last two decades thanks to methodological breakthroughs in dental wear and isotopic analysis. This has built on, and in some cases transformed, older models which essentially applied modern animal behavioural data to their Pleistocene equivalents (*e.g.* Gamble 1986). For example, the strontium signals from Isernia suggests ranging behaviour of up to *c.* 50 km for rhinoceros (contrast with the data in Table 3.8; see also Chap. 3), related to seasonal migrations (Lugli *et al.* 2017). Intra-regional movements across such distances, whether in whole or in part, would not be beyond the scope of hominins, and broadly chime with the scales of residential mobility reported by Kelly (1983). Migratory behaviour in other species has also been explored, although usually from a Late Pleistocene perspective (Britton *et al.* 2009; Julien *et al.* 2012).

The scales of hominin spring movements would thus have partly varied according to particular mammal species' behaviours. However, availability of other food patches (*e.g.* conifer trees with carbohydrate-rich inner-bark) is likely to have been at least as, if not more, important, especially since reliable resource provision was likely a key factor in hominins' reproductive success. Favourable micro-climates may also have been an important structuring factor impacting upon spring movements. Thus inter- (and intra-) seasonal residential movements would likely also have been influenced by different topographic settings: undulating terrain can provide greater variability in food resources, habitats and climatic conditions over shorter horizontal, although not necessarily vertical, distances. Such variations are evident for example in Greece during MIS 11, where the cooler conditions associated with Lake Ohrid are primarily attributed to its higher altitude (693 m), in comparison with the lower sites of Ioannina and Tenaghi Philippon (Kousis *et al.* 2018). Similarly, the differences between the winter temperatures at Gran Dolina (TD-6.2; December–February: 4.3–5.0°C) compared to Aridos (December–January: 8.8–9.5°C) may in part reflect altitudinal differences (*c.* 1080 m and *c.* 650 m above sea level, respectively) – although temporal differences between the phase(s) of warm stages represented at the two sites are likely to be a partial factor as well (Blain *et al.* 2013; 2014).

Any Pleistocene mobility, whether migrations or daily foraging trips, would be characterised by both obstacles and hazards. The latter are often thought of in terms of the larger carnivores that are commonly found in both open-air and cave deposits: lion, hyena, scimitar-toothed cats, bear, and others (*e.g.* Fig. 4.2), with significant variations between the Middle and Early Pleistocene (see also Chap. 2). These would be a clear predatory threat: Rodriguez-Gomez *et al.* (2017) highlighted the importance of prey in the 45–90 kg (*i.e.* hominin adult) size range for a wide range of Pleistocene predators (Canidae, Felidae, Hyaenidae, Ursidae). The dangers may also have been particularly marked for lone individuals and/or for groups of children (the latter being a smaller and possibly easier prey item, from a carnivore's perspective). The specific nature of threats would have varied between ambush predators (*e.g.* lions and other large felids) and cursorial predators (*e.g.* wolf and hyaena; Van Valkenburgh *et al.* 2016) and might therefore also have varied between relatively closed and open

Figure 4.2: Reconstruction of M. meridionalis (southern mammoth) carcass exploitation by hyaena, at Fuente Nueva-3 (right; modified from Espigares et al. 2013, fig. 7).

interglacial forest habitats. The threats from carnivores might also have been especially marked during the winter and early spring months, when carnivores' food
in general is scarcer and in poorer condition, but later springtime might also have
been a period of heightened danger as adults (*e.g.* wolf mothers) hunted to feed new
litters. But there would be significant other hazards as well, including venomous
snakes such as the European asp (*V. aspis*) and Lataste's viper (*V. latastei*),[6] but also
many of the larger herbivores present in these landscapes. During and after birthing
(and during rutting; Chap. 6) both female and male adult horse, deer and bison can
present a considerable threat to hominins. Finally, other hominins may also have
posed threats. Saladié *et al.* (2012) suggested that the cannibalism at TD-6.2 is inter-
group, stemming from territory defence, with attacks and kills focusing on immature
individuals. Since attacking such individuals reduces the risk, presumably they were
only part of small groups (perhaps foraging parties comprised of infants and a small
number of adults?) when confronted.

Regular residential mobility also introduces greater movement costs to all, as
opposed to only being incurred by the logistical task group(s) (although higher commuting distances for the latter would indirectly 'cost' the entire group through the
likely impact on resource returns). Particularly subject to additional costs would be
pregnant females and those carrying infants (Wall-Scheffler *et al.* 2007; Wall-Scheffler
and Myers 2013), but the impacts would also extend to those suffering from temporary
or permanent post-cranial trauma (*e.g.* the Sima de los Huesos pelvis 1 individual;
Bonmatí *et al.* 2010).

Alongside animal and hominin threats, the spring landscape itself could present
hazards. In a Palaeolithic context rivers are often discussed as routeways: linear
features that permitted navigation through a wooded landscape (*e.g.* Ashton *et al.*
2006). This was undoubtedly true on occasions, and rivers would also be a potential
source of other resources: freshwater, stone (in the form of gravel bars and beaches),
water-loving plants, terrestrial animals drinking and perhaps even aquatic animals.
All these resources can be found at rivers but do not necessitate their crossing. The
latter might have been a risky activity for hominins: it necessitates getting wet and,
to a varying degree through the year, cold, and also leaves hominins vulnerable during
the crossing. There might well have been a very different perspective of the near
bank as opposed to the far bank for a Lower Palaeolithic hominin and rivers might
therefore constrain, as well as funnel, hominin movements. Pleistocene sediment
sequences give us some sense of the scale of these rivers: meandering channels of the
Warta river, south of Poznan, vary from *c.* 50–200 m (channel widths estimated from
Vandenberghe 1995, fig. 6). The nature of river channels and floodplains would also
vary significantly, across climatic cycles (Gibbard and Lewin 2002; Fig. 4.3), along the
length of each river from head to mouth (Howard and Macklin 1999), and through
the year. Rivers would be likely to flood from autumn–spring (Fig. 4.1), and might be
especially prone to spring flooding in those landscapes with more significant winter
snowfall (*e.g.* central and eastern Europe and upland zones), although it is likely that

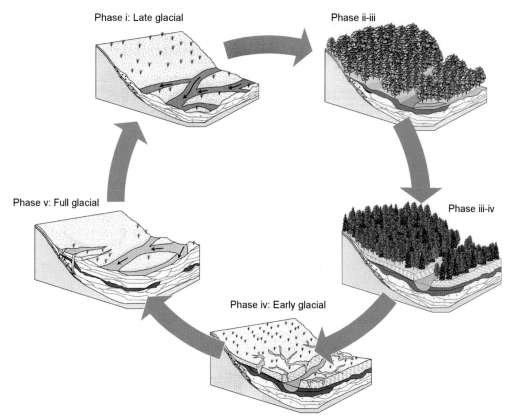

Figure 4.3: Schematic reconstruction of fluvial phases and landscapes across a glacial-interglacial climate cycle (modified from Gibbard and Lewin 2002, fig. 4).

interglacial forests would to some extent ameliorate flooding impacts in a way that our modern agricultural landscapes do not (*e.g.* Boardman *et al.* 1994; Thomas and Nisbet 2007). Nonetheless, the challenges of rivers might be a further constraining factor in limiting the scale of spring movements, and perhaps also structured the direction of movements. It is also important to consider mobility and activity across the entire landscape. The nature of the Lower Palaeolithic record has meant that behaviour is sometimes discussed through an overly-fluvial lens (Pope *et al.* 2016; Blundell 2020; Chap. 2: Box F). Yet the archaeological record makes it clear that not everything happens on the floodplains: *e.g.* the chalk landscape solution doline archaeology of Caddington (Sampson 1978), the lake-side settings of Schöningen and Bilzingsleben (Mania and Mania 2005; Urban and Bigga 2015) and the interfluve plateaux archaeology of Madrid (del Cueto *et al.* 2016). These settings highlight the interesting possibility that riverbanks and floodplains might have been little visited, or even avoided, at certain times of the year, *e.g.* during the most active spring floods.

Spring resources: plant and animal foods

Reduced plant food availability and the condition of other animals during the winter months (Chap. 3) are likely to have left hominins in a relatively poor condition come early spring. However, spring in Early and Middle Pleistocene Europe would be likely to have brought significant animal movements and, particularly in late spring, birthing of calves, foals, fawns and other young. Movements probably occurred at both regional and local scales: *e.g.* ungulate herds and birds migrating to summer feeding and/or breeding grounds (although attempts to demonstrate such seasonal movements by Pleistocene animals have had mixed success, and been focused on the Late Pleistocene; *e.g.* Pellegrini *et al.* 2008; Britton *et al.* 2011; Birch *et al.* 2016), or the association of horse with fresh spring grass (Sturdy and Webley 1988). Alongside this, the spring would bring growth and development of a diverse range of plant foods, from northern Europe (*e.g.* Mabey 2012) to the Mediterranean (*e.g.* Leonti *et al.* 2006), with a particular emphasis on the shoots and stems of leafy wild greens. Finally, in late spring and early summer, birds' eggs, nestlings and other small protein sources (*e.g.* grubs, lizards, reptiles, amphibians; Graves-Brown 1996) would also be available. This wide range of foods would be important, to offset the protein excess that would likely otherwise arise from over-reliance on poor condition spring ungulates (Speth and Spielmann 1983). But what would the specific nutritional sources available to Lower Palaeolithic hominins be, how would they be acquired, and who would acquire them?

Assessing the specific plant foods that were potentially available to Early and Middle Pleistocene hominins is difficult: plant identifications from pollen remains typically vary between species, genus and family level (Lowe and Walker 1997). The exact timings of individual species' growth and development through the year would also be impacted by specific climatic and weather conditions. Nonetheless, the outstanding macro-botanical evidence from Schöningen (Bigga 2014; Bigga *et al.* 2015; Urban and Bigga 2015) offers valuable insight: over 200 species of carpological remains (fruits and seeds) have been recovered, while two-thirds of the species from site 13 II-4 (the 'spear horizon') have documented dietary use in the ethnobotanical literature. Table 4.3 highlights potential spring-time dietary plants (all of which can apparently be eaten raw[7]) based on the Schöningen evidence, albeit inevitably biased towards aquatic and lakeshore vegetation. Bigga *et al.* (2015) also suggested catkins and the young leaves of willow, alder and birch as an early spring (and later winter) food, alongside the inner bark of a number of tree species, including birch and pine, as a source of sap and sugar. Interestingly, the double-pointed wooden artefact at Schöningen has been suggested as a possible bark peeler (after cutting the bark with a chopping tool or other sharp edge), although it is slightly longer than the ethnographic examples, mostly 40–50 cm in length, summarised by Sandgathe and Hayden (2003; Fig. 4.4). They note that large animal ribs or antler sections might also serve this purpose, as at the later Neanderthal site of Salzgitter-Lebenstedt (Gaudzinski 1999), while elephantidae tusk tips are another potential tool (Saccà 2012). Such tools would facilitate access to carbohydrates and sugar,

Table 4.3: Dietary potential, and seasonality data, for selected plant species available for spring collection and recorded at Schöningen

Botanical name	Common name	Edible parts	Nutritional & other features	Seasonal availability	Ecology
A. tripolium	Sea aster	Leaves, stems	–	April–October	Moist soils and saltmarshes (tolerates saline environments)
A. prostrata	Spear-leaved orache	Leaves, seeds	–	July–October (flowers)	Well-drained, dry or moist soils
C. demersum	Hornwort	Leaves	–	Young plant growth begins in spring; blooms in summer	Eutrophic, warm lakes
C. album	Fat hen	Flowers, leaves, seeds	Vitamins A & C; contains oxalate	June–October (flowers)	Pioneer plant; incl. lakeshores
C. palustre	Marsh thistle	Shoots & leaves (young)	–	June–September (flowers)	Damp and moist meadow & woodland soils
H. vulgaris	Common marestail	Leaves, shoots (incl. over-wintered spring stems)	–	May–August (flowers) Autumn–Spring (ideal harvesting)	Aquatic conditions (shallow standing or slowly flowing water)
M. spicatum	Water milfoil	Roots	–	July–September (flowers)	Eutrophic lakes (plant is submerged)
M. verticillatum	Whorled water milfoil	Leaves	–	Summer (flowers & fruits)	Eutrophic lakes (plant is submerged)
N. lutea	Yellow water lily	Leaves, roots, seeds	Fresh roots are toxic	Fruit & seeds follow flowering (June–September)	Shallow water & wetland (aquatic plant)
P. australis	Common reed	Leaves, roots, seeds, shoots, stems	Starch & sugar (roots)	–	Moist soils, bogs or fen wood (max. water depth: 1m)
P. hiercioides	Hawkweed ox-tongue	Leaves (young)	–	–	–
P. aviculare	Knotweed	Leaves, seeds	Rich in zinc	May–June (young leaves)	Pioneer plant
P. natans	Broad-leaved pondweed	Roots, shoots (young)	Starch-rich roots	May–August (flowers)	Mesotrophic, standing waters

Table 4.3: (Continued)

Botanical name	Common name	Edible parts	Nutritional & other features	Seasonal availability	Ecology
P. pectinatus	Fennel pondweed	Leaves, stems, roots	Bark must be removed from roots	–	Shallow standing or slowly flowing alkaline water (plant is submerged)
R. cf. caesius	Dewberry	Fruits, leaves	–	Late spring–early summer (fruits)	Fen woods and shores
R. maritimus	Golden dock	Leaves, seeds	Contains oxalic acid (reduced by cooking)	Spring–summer (leaves)	Mud-weed habitats along rivers and lake shores
S. sagittifolia	Arrowhead	Leaves, roots	Starch-rich tubers	June–August (flowers)	Still or slow flowing water and river banks
S. lacustris	Bulrush	Rhizomes, shoots, seeds	Starch & sugar-rich (fresh spring roots & autumn)	Spring & autumn	Slow flowing water, river banks and lakeshores
U. dioica	Nettle	Leaves, stems, shoots, roots	Vitamins A & C, iron & protein	February–June (young leaves)	Woods and river valleys; widespread

Sources: Lippert and Podlech (2001); Mabey (2012); Bigga (2014); Bigga et al. (2015). Bigga et al. (2015) list a variety of foods whose flowers and/or fruits/ seeds are available in the summer (or later), but whose leaves, shoots and/or stems are edible (e.g. C. palustre [Marsh thistle]) earlier in the year: in such cases the species are only edible when cooked (e.g. R. aquatilis [Common water crowfoot] or R. repens [Creeping buttercup]) they are not listed above. None of the above information should be used as a 'safety guide' with regards to the picking and consumption of wild plant foods. Any readers wishing to do so are strongly recommended to consult an appropriate, dedicated guidebook, such as Mabey (2012).

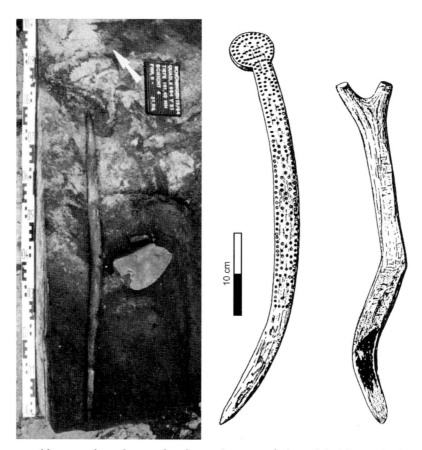

Figure 4.4: Double-pointed wooden artefact from Schöningen (left; modified from Schoch et al. 2015, fig. 7) and ethnographic bark peelers (right; Sandgathe and Hayden 2003, fig. 2).

and to a critical vitamin C food source in the case of pine and birch, during the first half of the year before fruits were available, and would also enhance access to larger, older trees with greater yields (Sandgathe and Hayden 2003; Bigga *et al.* 2015; Schoch *et al.* 2015).

Mabey (2012) suggested a range of potential raw plant foods currently available during spring (*e.g.* the leaves of Jack-by-the-Hedge [garlic mustard], sea beet [wild or sea spinach], dandelion, and ramsons [wild garlic] and the young leaves of beech), although some of these species have not been recovered from Lower Palaeolithic sites (Table 4.4). Mabey also highlighted the potential of coastal resources in late spring and summer, both seaweeds (*e.g.* carragheen and sea belt [kelp]) and other plants (*e.g.* sea beet). Thus, while the specifics of a plant diets in Lower Palaeolithic Europe are partially speculative, it is clear that there was considerable potential for collection and consumption of fresh stems and leaves during the spring, alongside a smaller number of mushroom species.

Table 4.4: Pleistocene records of potential spring plant foods (Mabey 2012), after Godwin (1975)

Species	Common name	Pleistocene?[1]	Earliest record[2]	Comments
A. petiolata	Jack-by-the-Hedge	No	Roman	–
A. ursinum	Ramsons	ND	ND	–
B. vulgaris	Sea beet	Yes	Ipswichian	–
B. nigra	Black mustard	Yes	Hoxnian	Young leaves
F. sylvatica	Beech	Yes	Cromerian	Young leaves
H. lupulus	Hop	No	Mesolithic	Young shoots & leaves
T. officinale	Dandelion	Yes	Hoxnian	Leaves & flowers; Identified to *Taraxacum* genus

[1]ND: No data available in Godwin (1975); [2]In light of Godwin's (1975, table 1) climate stage model (including the following sequence: Beestonian > Cromerian > Anglian > Hoxnian > Wolstonian > Ipswichian > Weichselian), 'Hoxnian' is cautiously interpreted as MIS 11 or MIS 9, and 'Cromerian' as spanning the early Middle Pleistocene. None of the above information should be used as a 'safety guide' with regards to the picking and consumption of wild plant foods. Any readers wishing to do so are strongly recommended to consult an appropriate, dedicated guidebook, such as Mabey (2012)

Despite the need for a balanced diet, animal foods are still likely to have been a significant component in European Lower Palaeolithic diets. Such resources might be especially critical in spring after the shortages of winter (although the animals themselves would also be in relatively poor condition in early spring), enabling critical rebuilding of body fat reserves and meeting the needs of pregnant females. Exploring the nature of animal exploitation in Lower Palaeolithic Europe is therefore a key step in understanding the seasonal lives of hominins. While later Palaeolithic periods offer unambiguous and frequent evidence for specialist, logistical hunting (*e.g.* Straus 1987; Bratlund 1996; Gaudzinski and Roebroeks 2000; Patou-Mathis 2000; Mellars 2004; White *et al.* 2016), much of the Lower Palaeolithic record is more in keeping with Bailey's (1981, 5) opportunistic exploitation: 'as the need for food arises, regardless of the effect on future supplies'. In part this is indicated directly by the variety of species represented by anthropogenically modified remains on key sites from both the Early (*e.g.* Huguet *et al.* 2013) and Middle Pleistocene (*e.g.* Parfitt and Roberts 1999), and perhaps also by the emphasis on adult animals in selected assemblages (*e.g.* 41.2% of the deer, horse and bison [based on MNI] in the Gran Dolina TD-6.2 assemblage; Saladié *et al.* 2012, table 4). Such age profiles might suggest a lack of concern with controlled exploitation (defined by Bailey as careful monitoring of the relationship between rates of exploitation and available food supplies). However opportunistic exploitation is also likely to reflect the dynamics of the food webs of which hominins were a part. Rodríguez-Gómez *et al.* (2012) proposed that prey in the 45–90 kg (*e.g.* wild boar, ibex) and 90–180 kg categories (*e.g.* red deer, Clacton fallow deer) supported the highest intensity of predation pressure, and competition between predators, in the Pleistocene. While it is clear that hominins were part of this pressure (*e.g.* the medium-sized deer focus at Gran Dolina 10.1; Rodríguez-Hidalgo *et al.* 2015) Rodríguez-Gómez *et al.* (2012) also highlighted a potential niche amongst heavier prey in the 360–1000 kg (*e.g.* bison, horse) categories: the butchery evidence from Gran Dolina TD-10.2 and Schöningen looks intriguing in this regard. Based on the reconstructed food web for Galería (*c.* 500 kya; Rodríguez-Gómez *et al.* 2012), and using the hominin-generated assemblages from Gran Dolina TD-10, *H. heidelbergensis* in the Atapuerca landscape would likely have been directly competing with lion (*P. leo*) for juvenile bison and horse (360–1000 kg), and red deer (90–180 kg), wolf (*C. lupus*) for red deer and possibly bison, and possibly *C. alpinus* [dhole] for red deer. It is also noteworthy that the heaviest prey (1000 kg+; *e.g.* rhinoceros) were also predated by just one or two species, and at the lowest intensity, in the reconstructed food web.

The variety of prey in hominin-generated assemblages may also reflect a range of other factors likely to impact on prey selection, including site occupation lengths and the age/sex/mobility of the resource catchers/collectors (Blasco *et al.* 2013). Bailey (1981, 5) suggested that opportunistic exploitation 'may persist as a long-term strategy in relation to staple resources where there is a considerable barrier to over-exploitation ... because of some other environmental limiting factor holding down human population levels', and argued that the dense forests of prehistoric Europe may have resulted in limited access to resources. While pollen sequences highlight that warm stage landscapes were not consistently covered by dense forest, either chronologically or spatially (Chap. 2), this is a reminder that prey would have had to be actively sought, especially during those times of the year when animals were disaggregated and/or visibility was reduced by vegetation. This argument has been proposed by Rodríguez-Hidalgo *et al.* (2015) as an explanation for the low-level exploitation of roe deer in the Gran Dolina TD-10.1 'bone bed', highlighting the pursuit costs, and therefore lower return yields, associated with its elusive and closed forest ecology. Bailey also suggested that, when resources were accessible and exploitation levels higher, periodic decreases in specific food items would result in a temporary reliance on other resources and/or habitats. While this is difficult to detect in Lower Palaeolithic contexts, the range of site habitats in the archaeological record (*e.g.* lacustrine, fluvial, plateau/interfluve, coastal) may reflect such flexibility.

While early spring animals would be in relatively poor condition with low fat levels (Speth and Spielmann 1983), particular elements could still be targeted to meet specific macro-nutrient needs. An emphasis on specific animal parts is suggested at Gran Dolina TD-10.2 (the 'bison bone bed'), where consumption of the tongue is indicated by slicing marks on the mandibles and the hyoids (Rodriguez-Hidalgo *et al.* 2017). The presence of cut-marks on the ribs and vertebrae in the same assemblage may indicate a focus on the fat and meat in the bison hump. There is evidence for such focused butchery and consumption from Early Pleistocene sites as well: deer tongue removal is suggested at Gran Dolina TD-6.2, as is hominin brain consumption (Saladié *et al.* 2012). Chewing on bones is also clearly demonstrated at Gran Dolina TD-10 and TD-6, with reference to experimental data (Saladié *et al.* 2013).

A potentially important source of food for hominins in the late spring and early summer months may have been ungulate newborns, although their relative helplessness would be offset to some extent by aggressive parental defence in certain species (*e.g.* horse and bison; Bennett and Hoffmann 1999; Daleszczyk 2004). The appearance of such resources would undoubtedly vary in timing, both by species and environmental conditions. In modern seasonal environments such as Europe, vertebrate births typically occur within a narrow time window, which is matched against peak forage availability and/or quality (Plard *et al.* 2014). There is also a tendency towards earlier rather than later births: early born young typically survive favourably compared to later born young as they benefit from a longer period of growth prior to winter, although very early births must survive harsh early spring conditions. Specific comparisons of modern 'birthing window' data to Pleistocene environments are therefore obviously problematic and are included here (Table 4.5) only as an approximate guide. The suggested birthing seasons for European bison, roe deer and red deer are based on broadly comparable climatic and habitat conditions however (the BPF, a deciduous woodland from northeast France with continental climate, and a woody/Mediterranean scrub landscape in Spain). It is also notable that the breeding and birthing periods for red deer and roe deer do not appear to vary according to latitudinal variations (data for modern British populations are comparable to that in Table 4.5; Corbet and Harris 1991). However, the data for wild horse are clearly much more problematic, reflecting the modern distribution of the species. In European captive herds of Przewalski's horse females typically gave birth between April and June, with breeding ending by late summer (Bennett and Hoffmann 1999). Nonetheless the overall patterns (late spring–early summer birthing and summer–early autumn breeding) are broadly consistent across all four species, and a comparable annual pattern in the warm stages of the Early and Middle Pleistocene seems likely, given the generally equivalent climatic conditions.

Table 4.5: Modern reproduction data for key ungulate species

Species	Mating season	Birthing season	Study area	Conditions	Reference
European bison (B. bonasus)	Aug–Sept	May–July	Białowieża Primeval Forest (Poland & Belorussia)	Coniferous & mixed forest (Jan./July temp.: -4.2/+18.6°C)	(Mysterud *et al.* 2007)
Horse (E. przewalskii)	May–Sept	May–June	Kalamaili Ungulate Protected Area (China)	Desert grassland (min./max. temp.: -38/+50°C)	(Chen *et al.* 2008)
Red deer (C. elaphus)	Sept–Nov	May–June	Sierra Morena (Spain)	Woody & Mediterranean scrub vegetation	(Carranza and de Reyna 1987)
Roe deer (C. capreolus)	July–Aug	Mid-April–mid-June	Trois Fontaines forest (France)	Deciduous forest (Jan./July temp.: -2/+18.5°C)	(Plard *et al.* 2014)

There are many attractions for carnivores in targeting young animals. In the Białowieża Primeval Forest wolf kills of wild boar are at their highest in spring and summer, when piglets are at their most abundant and vulnerable (Jędrzejewski *et al.* 2002) – this is in marked contrast to the adult boars, whose body structure, active defence behaviour and large groups (maintained all year round) make them a difficult and dangerous prey for both wolf and lynx (Okarma *et al.* 1995). This season of piglet availability also overlaps with the birthing of the wolves' own cubs, and similar synchronicities might also have been characteristic of hominin birthing cycles. Other animal behaviours might also have enabled late spring predation by hominins. Modern British red and roe deer newborns are left alone and isolated from their con-specifics for significant periods between feeds (until they start to accompany their mothers after 7–10 days [red deer] and 6–8 weeks [roe deer]; Corbet and Harris 1991), potentially providing opportunities for hunting. It is difficult to assess whether such behaviour would also have occurred in the context of Pleistocene predators, although this hiding strategy of predator defence is commonly used by modern species in closed, forested habitats (Daleszczyk 2004).

A further characteristic, and benefit, of spring and early summer hunting may well have concerned the spatial and numerical distributions of prey of all ages. While it is clearly inappropriate to extend specific animal group sizes and social behaviours from the present day into a Pleistocene ecosystem (see also Chap. 3: Box H), modern studies offer interesting perspectives on changing group size, structure and location across the year. Investigations into lynx hunting in the BPF have shown that the adults' prey, typically roe deer and red deer fawns, is spatially-concentrated in the spring–summer, reflecting both seasonal abundance and the low mobility of juvenile deer (Okarma *et al.* 1997). Specific tracking of red deer in the BPF similarly highlighted small spring–summer ranges for both males (13.6 ± 1.2 km^2) and, especially, females (4.6 ± 0.2 km^2; Kamler *et al.* 2008), while the daily ranges of BPF wolf packs were at their smallest in May (average: 9.3 km^2), with movements concentrated around breeding dens and rendezvous sites (Jędrzejewski *et al.* 2001). Comparable Pleistocene prey concentrations might therefore have permitted relatively small hominin home ranges in the first part of the year, a potential benefit if their own birthing was concentrated in this period, as suggested by Mussi (2007).

However mixed-species assemblages sometimes show limited evidence for a focus on newborns, perhaps because of the limited returns (in terms of body size): only 11.8% of the deer, horse and bison [based on MNI] in the Gran Dolina TD-6.2 assemblage were juveniles (Saladié *et al.* 2012, table 4). This may in part reflect the hiding strategy of predator defence (*e.g.* used by deer in forested habitats), which would present obvious challenges (finding the prey) as well as opportunities (Daleszczyk 2004). The other main method of predator defence, following, would also present both challenges and opportunities to hominins. Although hiding is more typical in forested environments, European bison in the BPF rely on a following strategy, and mothers of very young calves (especially in their first week of life) are aggressive towards intruders, with threats such as shaking horns, charges and infrequent fights (cows' attacks on humans in the BPF, while rare, are most common during calving and winter; Daleszczyk 2004; Haidt *et al.* 2018). Speculatively, the potential risks associated with such responses might have impacted on who was actively involved in the kill during spring (*e.g.* the involvement or not of adolescents), although valuable contributions to other tasks, such as locating prey, could potentially be made by young and old. In the case of the first few days after birth the BPF bison calves stay close to their mothers, with other animals seldom recorded in the vicinity: such small dyads might have been difficult for hominins to locate if comparable reproductive behaviours were followed by Pleistocene bison, although the habit of the BPF bison cows to use vocal calls when calves vanish from sight might provide auditory cues (Daleszczyk 2004).

Yet while some assemblages show limited evidence for the targeting of new or recently born young in the late spring/early summer, this strategy is evident at Gran Dolina TD-10.2 (MIS 11–9), where seasonality and age profile data from dental eruptions, use-wear and microwear indicates significant bison hunting at this point in the year (and in autumn/early winter as well [Chap. 6]; Rodríguez-Hidalgo *et al.* 2016). Moreover, there is clear evidence at TD-10.2 for a mono-specific focus: bison remains unsurprisingly dominate the 'bison bone bed' (98.4% NISP; MNI=60).

The general spring behaviours of ungulates would also present likely challenges to hominins. In modern populations spring sees aggressive red deer behaviour amongst the stags: belligerent acts associated with antler casting in March/April have been documented in the herds from the Žehušice game reserve, Czech Republic (Bartoš 1985). While Bartoš does not indicate whether such aggression would likely be extended into predator defence, it does at least suggest that the animals would be in heightened states during this period, while more generally Geist (1998, 182) has noted that red deer will both strike with their forelegs and kick with their hind legs. However hunting during such periods might still have been attractive to hominins, not only because it would provide significant animal food shortly after the end of winter but also because after antler casting the Žehušice bachelor groups tended to disintegrate, thus potentially reducing the size of an adult male red deer resource available at any particular place and time, until the late year re-aggregation associated with the autumn rut (Bartoš 1985).

A rather different type of spring-time (and early summer) resource centres on birds. Their exploitation in the Palaeolithic has received renewed attention recently (*e.g.* Blasco *et al.* 2014; Negro *et al.* 2016), most notably with regards to Neanderthal use of raptor and corvid feathers (Finlayson *et al.* 2012) and other items, such as eagle talons (Radovčić *et al.* 2015). But birds also have a range of characteristics favouring their use as a food: edible, easy to catch at the egg and nestling stages (as long as the species' behaviour is known), widespread in most landscapes, nests that are often conspicuous and with contents that are seasonally predictable and, with the exception of adult owls and raptors, they do not pose significant dangers to humans, unlikely many other species of mammals, snakes and lizards (Negro *et al.* 2016). Based on average values for grouse/partridge and duck, Hockett and Haws (2003) suggest protein and fat values of 21.7 g and 14.1 g/100 g respectively. As noted elsewhere (Chap. 3), migratory waterbirds overwintering in different parts of Europe were therefore a potentially significant source of winter fats and their presence has been recorded in the zooarchaeological assemblages from a number of Lower Palaeolithic sites, although with little direct evidence for cut-marks (*e.g.* Boxgrove; Harrison and Stewart 1999). Yet many other bird species could also provide reliable sources of accessible food at other times of the year, with chicks, nestlings and eggs being of particular importance during the late spring and early summer. The Boxgrove avifauna for example spans 19 taxa, including waterfowl, seabirds, songbirds and gamebirds, while 16 taxa including waterbirds, gamebirds, pipits and wagtails and corvids have been recorded at the Sima del Elefante (Harrison and Stewart 1999; Núñez-Lahuerta *et al.* 2016). As Negro *et al.* (2016) have emphasised, the distinctive characteristic of birds to reach their adult mass, or even exceed it, while still flightless, favours a human strategy of targeting the flightless nestlings just as they reach their maximum weight. Eggs meanwhile provide a valuable source of fat, protein, albumin and essential carotenoids (Negro *et al.* 2016). While birds are a small resource (the mean body size for the bird class is 37 g, with few species above 1 kg; Negro *et al.* 2016), they would nonetheless help to meet some of the regular and high-quality food requirements of weaned children. These foods would likely be within the foraging abilities of younger children (although certain species might require climbing to access them): of the Boxgrove and Sima del Elefante species, for example, the present day representatives of grey partridge (sedentary), teal (partial migrant) and common quail (migratory) all nest on the ground, amongst tall, dense vegetation (grey partridge), in heath and scrub (teal), and in open country (common quail; Harrison and Stewart 1999; Núñez-Lahuerta *et al.* 2016). However, since such relatively 'easy' access might also mean that eggs and/or nestlings were eaten on the spot, the archaeological visibility of such behaviour would likely be low – which indeed it is.

A further question, albeit relevant throughout the year, concerns whether fledged birds could have been caught? Krech III (2005) has observed that the flesh (and eggs) of a diverse range of species are consumed by Eskimo groups using a range of hunting technologies. Many of the simpler technologies and techniques outlined by Krech would seem to be within the compass of Lower Palaeolithic hominins: throwing sticks and stones as birds flew past in narrow valleys; catching moulting waterfowl; catching fledgling birds and letting their cries attract their parents. Thus, fledged bird hunting may have been a stable component of hominin foraging and nestling and/or egg gathering in particular may have been a core focus of child foraging groups, and perhaps adults too.

However the Lower Palaeolithic evidence for bird exploitation is modest: a cut-marked radius of a bird at the Sima del Elefante (level TE9a: *c.* 1.2 mya; Huguet *et al.* 2013); three striations on the radius of a medium-sized bird at Gran Dolina TD-10.2 (Rodriguez-Hidalgo *et al.* 2017); and multiple incisions, probably cutmarks, on the distal metatarsus of a large-sized bird at Dursunlu, Turkey, where there is a large and species-rich bird assemblage (*c.* 0.9 Ma; Güleç *et al.* 2009). As a rule the frequency of evidence greatly increases from the Middle Palaeolithic onwards (Negro *et al.* 2016). While this scant Lower Palaeolithic record might partly reflect bird consumption without tool-use (*i.e.* no cut-marks), it is notable that at Gran Dolina TD-10.1 there is no evidence of any sort for a hominin role in the accumulation of the bird (and rabbit) remains (Rodríguez-Hidalgo *et al.* 2015): butchering marks, intentional break-age or human tooth marks are all unknown. It therefore seems likely that any bird exploitation was focused towards egg and perhaps also chick/nestling collection, rather than the hunting of fledged birds.

Animals and plants in the diet ... but how?

And so to the critical questions: how significant were mammal foods within the spring diet, which species were favoured and what were the relative roles of hunting and aggressive (or passive) scavenging in their acquisition? While the evidence base is limited, what is often notable in those Lower Palaeolithic assemblages with cut-marks is the emphasis on medium-, and occasionally large-, sized mammals: varying combinations of horse, red deer, bison and, less frequently, rhinoceros and straight-tusked elephant, are present at sites such as Boxgrove, Schöningen, Soucy, Aridos and Gran Dolina TD-6.2 and TD-10 (Parfitt and Roberts 1999; Lhomme 2007; Yravedra *et al.* 2010; Saladié *et al.* 2011; van Kolfschoten *et al.* 2015a; Rodriguez-Hidalgo *et al.* 2015; 2017). The primary access evidence (*e.g.* cut-mark frequency and distribution; butch-ery processes; relationships to carnivore traces) and evidence for probable hunting weapon technologies (spears) is suggestive of hunting over confrontational scavenging for medium-sized game in both the Middle and late Early Pleistocene (*e.g.* Parfitt and Roberts 1999; Saladié *et al.* 2011; Schoch *et al.* 2015; Rodriguez-Hidalgo *et al.* 2015; 2017), although statements vary in their level of certainty:

> The evidence provided by the cut-marks and the presence of carnivore gnawing overlying cut marks is evidence for very early access to the carcass by hominids, although the evidence available to date, with the possible exception of a puncture wound in the scapula from GTP 17, does not allow us to distinguish definitively between hunting or confrontational scaveng-ing as the main method of carcass procurement [at Boxgrove]. The circumstantial evidence, however, favours the former method. (Parfitt and Roberts 1999, 414–415)

By contrast, at Gran Dolina TD-10.2 a variety of lines of evidence strongly suggest primary access through communal hunting (and the use of the cave itself as a kill-butchery site): the presence of usually rare elements (*e.g.* hyoid bones), the strong representation of early stage butchery tasks (*e.g.* tongue removal, evisceration and skinning), the systematic and intensive nature of the overall butchery process (other tasks included disarticulation, dismemberment and bone breakage), the general fre-quency and distribution of cut-marks and the catastrophic and seasonal mortality profile (Rodriguez-Hidalgo *et al.* 2017).

The mode of acquisition for elephantidae (predominantly straight-tusked elephant; *P. antiquus*) remains uncertain at all of the key European sites (*e.g.* Villa 1990; Wenban-Smith 2013; Konidaris *et al.* 2018), principally because of the limited butchery evidence (although this may be due to the thickness of the animals' flesh). However the number of sites with associations between stone tools and elephant remains, including Aridos, Ambrona, Barranc de la Boella, Castel di Guido, Ficoncella, Fuente Nueva-3, Marathousa 1, Notarchirico, La Polledrara, Southfleet Road and Torralba (Villa 1990; Villa and Lenoir 2009; Piperno and Tagliacozzo 2001; Yravedra *et al.* 2010; Saccà 2012; Espigares *et al.* 2013; Wenban-Smith 2013; Aureli *et al.* 2015; Mosquera *et al.* 2015; Santucci *et al.* 2016; Konidaris *et al.* 2018) seems to suggest that exploitation, if not killing, of mired elephants at water sources was a regular, if not habitual, component of Lower Palaeolithic survival. It is difficult to assess to what extent such places were actively visited/monitored in anticipation of such opportunities, or whether single-animal sites represent one-off encounters, in part because 'failed' visits will be archaeologically invisible. While some sites represent single animals (*e.g.* Southfleet, England and Marathousa, Greece) and are perhaps most parsimoniously interpreted as unique events, others (*e.g.* Torralba and Ambrona, Castel di Guido) clearly suggest repeated visits and exploitation of carcasses on multiple occasions (Villa 1990; Villa *et al.* 2005; Villa and Lenoir 2009; Saccà 2012). The latter seems unsurprising in light of the possible importance of elephant as a Palaeolithic food source in terms of its fat and meat composition (Reshef and Barkai 2015; Solodenko *et al.* 2015; Agam and Barkai 2016): alongside the brain, Konidaris *et al.* (2018) have emphasised that the fat content, and other soft tissues, in the hind limb and foot cushion may also have been an attractive target on elephant carcasses.

The evidence for rhinoceros butchery (*e.g.* at Boxgrove and Gran Dolina TD-6) is perhaps even more surprising, given the potential aggression of the species and the absence of natural predators of adult animals (although it has been suggested that *P. leo* could have preyed upon juveniles; Rodríguez-Gómez *et al.* 2012). In light of this it is perhaps possible that rhinoceros were occasionally hunted by hominins, although the zooarchaeological evidence strongly suggests the infrequency of this strategy.

Compared to elephants and rhinoceros medium and large-sized mammals such as red deer, horse and bison are much more frequent in Lower Palaeolithic assemblages and would still provide substantial quantities of food, especially if hunted rather than scavenged. Despite the potential risks of hunting these size classes of animals (90–180 kg [red deer] and 360–1000 kg [horse and bison]), this emphasis on larger ungulates is supported by spear throwing experiments by Milks *et al.* (2019): their demonstration of improved hit-rates associated with larger targets is unsurprising but is also an interesting challenge to the common view that larger game would be more challenging to hunt. Species' ecology and behaviour may be further factors in the size patterns of hunted/butchered carcasses, such as the relatively limited evidence for roe and fallow deer butchery, as opposed to larger deer species, at Boxgrove. This is despite the presence of both roe and fallow deer in the palaeosol deposits, with the former abundant (Parfitt 1999a; Parfitt and Roberts 1999). A potentially interesting perspective on this pattern is found in the BPF, where wolf predominantly hunts red deer. Jędrzejewski *et al.* (2000) have argued that roe deer's very small groups and secretive and elusive style of life in the forests may make them a difficult prey for wolves – as has also been suggested to explain the paucity of roe deer in the Gran Dolina TD-10.1 assemblage.

Although there is limited direct seasonality evidence available for European Lower Palaeolithic sites generally, a late spring/early summer-time occupation has been suggested for locality 5 (level 1) at Soucy, where remains of red deer, horse and bovids are associated with handaxes and retouched flake tools. Spring exploitation of animals (bison) is especially strongly documented at Gran Dolina TD-10.2 (Rodríguez-Hidalgo *et al.* 2016). The paucity of calves within the butchered Soucy fauna (juveniles and young adults are listed by Lhomme 2007) is suggestive of limited targeting of spring newborns in that instance, perhaps because of aggressive parental behaviours or the difficulty of locating mothers and calves? However, there is evidence for very young calf kills amongst the Gran Dolina assemblage (21 of the 60 bison are less than 2 years old, and 35% of the individuals are linked to late spring/early summer deaths; Rodriguez-Hidalgo *et al.* 2017).

But to what extent were these larger mammals dominating the animal food diet? Ethnographic studies have frequently highlighted the uncertainty of hunting (*e.g.* 2–3 large antelope killed per hunter per year amongst southern African hunter-gatherers; Liebenberg 2008), although the environmental contrasts and the marginal territories of some modern hunter-gatherer groups make direct parallels of limited value (but see also Marlowe 2005). However, comparisons of biomass accessibility (secondary biomass/primary biomass) for tropical savannah (3.8), temperate deciduous (0.5) and evergreen (0.2) forest, and temperate grassland (4.3) suggest that prey densities are unlikely to have been substantially greater in the higher latitudes, and may have been significantly lower than densities in low latitude savannahs (Kelly 1983, table 3). It also seems unlikely that Lower Palaeolithic hominins were significantly *more* successful than modern hunter-gatherers. Using a different approach, the modelling comparison of Pleistocene and modern environments presented by Rodríguez-Gómez *et al.* (2012) also suggested markedly lower prey biomass (kg/km^2) for three European Pleistocene assemblages (Venta Micena, Atapuerca-Galería IIa/IIb and Amalda V) in comparison to the Serengeti National Park, Tanzania. It seems likely that a reliance on large game hunting would be highly risky. But would hominins have been able to access a wide range of small mammals? This seems unlikely for certain species, such as hare or rabbit, given their distinctive attributes (*e.g.* small, fast) and the probable need for either textile technologies or other 'tools' (*e.g.* warren-based mass harvesting using nets, fences, water or fire; Cochard *et al.* 2012), and indeed the archaeological record offers little evidence of such exploitations, in contrast to the Middle and Upper Palaeolithic (Fiore *et al.* 2004; Cochard *et al.* 2012; Fa *et al.* 2013). However there are occasional Lower Palaeolithic glimpses of a wider animal exploitation strategy, with evidence of beaver exploitation at Arago Cave (Lebreton *et al.* 2017; Chap. 3), occasional modifications of lagomorphs (*e.g.* at Sima del Elefante and Terra Amata; Huguet *et al.* 2013; Morin *et al.* 2019), and the evidence for cut-marks on tortoise remains, in two different levels, also at Sima del Elefante (Blasco *et al.* 2011). Frustratingly however there is little or no seasonal dimension to these records. While it is tempting to suggest that tortoises were exploited in the spring as they emerged from hibernation, given their slow-moving nature it seems more likely that they could have been exploited throughout much of the period of the year during which they were active.

Can we say anything more about likely hunting strategies? The settings for hunting are difficult to know beyond occasional well-preserved sites (*e.g.* the Schöningen lake-shore and the Boxgrove palaeosol; see also Chap. 6), and much of the available palaeoenvironmental evidence is strongly suggestive of closed and open woodlands (Chap. 2), although early spring landscapes would be more open prior to the completion of annual vegetation regrowth. The difficulties of hunting in woodland were highlighted by Liebenberg (2008), who noted that tracking is much harder in areas of woodland and thicker vegetation and may require the more cognitively demanding speculative tracking (*i.e.* interpreting signs and building hypotheses of animal behaviour). Wetland habitats and other 'fixed' points (*e.g.* horse 'shades' or ruts; see also Chaps 4 & 5) may therefore have been a regular and more reliable focus for hunting and/or confrontational scavenging.

In discussing trapping among recent hunter-gatherers Holliday (1998) described an edge-of-the-woods strategy in the boreal zone, characterised by seasonal fishing and trapping in the forests and hunting aggregated herds at the tundra/forest boundary. This was driven specifically by the low biomass in the boreal forests and involved strategies and technologies unlikely to have been present in the Lower Palaeolithic but it may have interesting implications for earlier strategies too given the Pleistocene evidence for mosaic landscapes. An edge-of-the-woods approach would offer relative security, woodland food (*e.g.* nestlings and birds' eggs, inner bark, spring fungi, elusive ungulates such as roe deer or mother–fawn dyads) and fuel sources (if needed), but also access to open grasslands and larger-bodied ungulates, potentially in greater numbers (*e.g.* migrating spring herds). The role of vegetation cover in possible hominin hunting strategies was highlighted by Geist (1998, 210), who emphasised that red deer, while not vulnerable to wolf packs unless exhausted or sick, can be susceptible to ambushes by bear in thickets. This would fit with many views of Lower Palaeolithic-type spears as short-range, thrusting weapons (Churchill 1993), although Milks *et al.* (2019) have highlighted their throwing potential.

Wetlands also appear to be a common element in the mosaic landscapes suggested by the evidence from a number of key sites (see also Chap. 2). The productivity and value of wetland habitats has been emphasised by Nicholas (1998) with reference to their general accessibility and stability – reflecting the year-round presence of water and the diversity of the available resources (*e.g.* plant foods such as cattail, animals drawn to drink and wetland dwelling animals). What is particularly important is the sheer diversity of wetland habitats – swamps, bogs, fens, marshes, mires and estuaries all meet Nicholas' (1998, 721) definition: '(a) is seasonally or periodically water-covered or saturated, (b) supports hydrophytic vegetation [plants that grow wholly or partly submerged in water], and/or (c) has hydric soils [permanently or seasonally saturated by water, resulting in anaerobic conditions]', as do the shallow-water margins of lakes, ponds, river margins, springs and waterholes. The creation and/or extension of wetlands by beavers can also lead to the concentration of other animal species, especially ungulates and other herbivores, in these habitats, as a result of their water dependency (Turner 1999; Liarsou 2013). Nicholas (1998) rightly notes that wetlands can be both attractive (resources and ambushing opportunities) and dangerous (other predators, insects), and it seems likely that hominins would utilise wetlands for both plant and animal exploitation and to meet their own freshwater needs, but not as residential settings.

The availability of animals around wetlands would also vary depending on the demands of individual species (*e.g.* horses' water requirements), which would be accentuated during drier seasons, for obvious reasons (see also Chap. 5): present-day ponies in the UK's New Forest increase their feeding in bogs and wetter heathlands during such periods (Corbet and Harris 1991). The European Lower Palaeolithic record certainly provides us with plentiful examples of wetland exploitation – from the Boxgrove waterhole (Roberts and Pope 2009) and the Schöningen and Bilzingsleben lake-shores (Mania and Mania 2005; Stahlschmidt *et al.* 2015a) to the elephant remains associated with wetland or near-wetland settings in sites across the length and breadth of Europe. Such settings might be more open (in part as a result of animal activity), potentially emphasising the importance of distance weapons within medium/large-mammal hunting strategies. The potential effective ranges of 15–20 m for Schöningen-type spears proposed by Milks *et al.* (2019) looks especially intriguing in this regard.

From day-to-day the spatial focus of hunting may also have shifted, in order to avoid or reduce, at least as far as ungulates were concerned, the behavioural depression of prey availability. This involves the lessening of certain indicators (*e.g.* smell or sound) that might otherwise trigger high alertness in ungulate prey. It may be evident amongst the wolves of the BPF in the rotational use of their territories: they return to the same parts every 6 days on average in the autumn–winter (this rotational behaviour may also relate to the patrolling and defence of territories; Jędrzejewski *et al.* 2001). It seems possible, if not likely, that Pleistocene ungulates would also be cued into the indications of hominin hunters and, thus, hominin hunting mobility strategies might have been structured accordingly.

While the direct evidence for plant food consumption is limited, the rich palaeoenvironmental evidence from Schöningen has nonetheless highlighted the potential for spring plant foods (Bigga *et al.* 2015; Table 4.3): catkins and young leaves of willow, alder and birch are emphasised and wetland habitats would be a key source of USOs. Evidence for plant food consumption has also been found indirectly through dental striations in the Sima de los Huesos samples, which suggest the probability of an abrasive diet involving hard, poorly processed foods such as roots, stems and seeds (Pérez-Pérez *et al.* 1999), although there is no seasonal dimension to this evidence.

Finally, spring foraging may also have been about more than just immediate food returns. The improving climate and lengthening days of the spring might well have offered opportunities for hominins to identify animal territories and key plant resources (*e.g.* fruiting trees in blossom), perhaps with a view not only towards short-term gathering and hunting, but also looking ahead to summer and autumn opportunities (see also Chap. 5). Implicit in this is the assumption that Lower Palaeolithic hominins were capable of linking present observations with future events (*e.g.* blossom and fruit, or birds' building nests and eggs), anticipating their future needs and gathering knowledge for the future (see also Chap. 2: Box D). This is inevitably somewhat speculative, but it would seem to be an integral component to survival in a changing, highly seasonal landscape such as Europe.

European *Homo*: always a hunter?

The question of whether habitual hunting was part of the behavioural repertoire of the very earliest Europeans is a long-standing one, in part because of the clear evidence for significant predatory competition in the Early Pleistocene (Rodriguez-Gomez *et al.* 2012; 2017; Rodríguez *et al.* 2012; see also Chap. 2) and the suggestion that hominins may have successfully exploited the remaining flesh and all the bone nutrients on carcasses which the machairodonts (sabre and dirk-toothed cats; Fig. 4.5) were unable to fully consume.[8] It has thus been suggested that Early Pleistocene hominins followed a fundamentally scavenging lifestyle (Turner 1992; Arribas and Palmqvist 1999). A key early question however was whether hominins were (Arribas and Palmqvist 1999) or were not (Turner 1992) able to successfully compete alongside the giant, bone-cracking hyaena *P. brevirostris* (*e.g.* Turner and Antón 1996, figs 1–3), prior to the latter's disappearance around *c.* 0.5 mya[9] and the general faunal shift that saw the sabre-tooth cats and giant hyaena replaced by modern African species (*e.g.* lion, leopard and spotted hyena). To some extent this debate has been overtaken by the clear demonstration of an Early Pleistocene presence in southern Europe, but the feasibility of a scavenging mode at this time is clearly critical to the question of when European *Homo* became habitually reliant upon hunting. This is highlighted by the diversity of predators in Early Pleistocene landscapes (Rodríguez *et al.* 2012), including social hunters (*e.g.* wild dogs & other canids, *C. crocuta*, *Homotherium*), solitary ambush hunters (*e.g.* jaguar & *Megantereon*), and dedicated scavengers (*e.g.* *Pachycrocuta*).

Much of the recent discussion of scavenging and hunting has focused on earlier hominins and East Africa, with methodological and taphonomy-driven debates exploring four main models of carcass access: hominin hunting; hominin access to fleshed carcasses through confrontational scavenging; hominin scavenging of carcasses that had not been disturbed by other carnivores; hominins passively scavenging defleshed carcasses at other carnivores' kills (Domínguez-Rodrigo 2002). In particular,

Smilodon   *Homotherium*

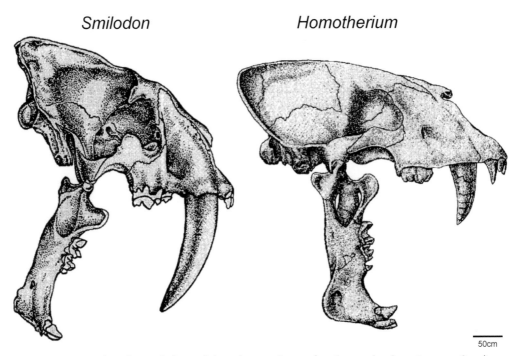

50cm

Figure 4.5: Craniodental morphology of the sabre-tooth cats (Arribas and Palmqvist 1999, fig. 2).

the respective cases for access to marrow on defleshed carcasses (*e.g.* Blumenschine 1991; 1995) and primary access to fleshed carcasses (*e.g.* Domínguez-Rodrigo 2002) have been vigorously debated. From a Lower Palaeolithic dietary perspective what would seem to be critical in Europe's seasonal landscapes is access to fats as well as flesh, with marrow and brains being particularly important (Blumenschine 1991), although there would also be potential for accessing fat in the viscera and around muscle tissues, *if* the carcass was intact (Domínguez-Rodrigo 2002). However various counter-arguments regarding the yield from scavenged carcasses have also been made, highlighting the very limited opportunities at hyaenid and canid kills, the lack of potential for such small yields to be widely shared, and particularly questioning the potential of passive, as opposed to confrontational, scavenging as a strategy (*e.g.* Domínguez-Rodrigo 2002; Fig. 4.6).

Within an Early Pleistocene European landscape the efficiency, or otherwise, of the machairodonts as carcass consumers, and their inferred role as carcass generators but not destroyers, is thus a critical issue (*e.g.* Turner 1992; Arribas and Palmqvist 1992). But the potential effectiveness of *Homo* as a confrontational scavenger is also key, give the possible limitations of passive scavenging. There are obvious risks (Treves and Naughton-Treves 1999), but Bunn and Ezzo (1993) have also emphasised the potential abilities of earlier *Homo*, noting the tendency of larger African carnivores to displace smaller ones at a kill in one-to-one or otherwise equal confrontations, the

Figure 4.6: Animal carcass in Garden Route Game Lodge, South Africa (Source: Zenith4237 [Wikipedia Commons]; details in Fig. acknowledgements).

use of vocalisations and branch-waving by chimpanzees and the stature and potential group sizes of the hominins (further accentuated in the European species). The levels of danger around European Pleistocene carcasses are difficult to assess, although the diverse range of documented carnivores, particularly in the Early Pleistocene (*e.g.* Rodríguez-Gómez *et al.* 2012; Rodríguez *et al.* 2012), suggests a relatively high level of risk, and probably more rapid carcass consumption, than in the BPF in the present (*e.g.* Selva *et al.* 2003; see also Chap. 3). This may have been particularly true in relatively open habitats (see also Blumenschine and Cavallo 1992), although targeting abandoned kills and/or seeking refuge through tree-climbing might offer means of reducing risk (Blumenschine 1991). While thrusting spears may not have been a component of Early Pleistocene technology, I suspect that a means of keeping other predators at rather more than arm's length would still have been key, in both hunting and scavenging encounters. Stones, and a throwing physiology (*e.g.* Roach *et al.* 2013), seem to be the most obvious solution.

A European hominin scavenging strategy might have been further facilitated by the structure of the environment: Palombo (2010) has suggested that open, fragmented landscapes between *c.* 1.5 and 1.0 mya would have produced partitioned resources and potential niches which a flexible, opportunistic hominin could have exploited. The scale of that resource is ambiguous however. In the East African record the accumulation of the hominin-modified remains of large numbers of individual animals at specific locations has been interpreted not only as carcass transport away from the kill, to avoid carnivore competition, but also as carcass transport for the purposes of feeding others, rather than purely as 'refuge seeking' (Domínguez-Rodrigo 2002). Unfortunately, there are few comparable sites in the earliest European records, although the evidence at Fuente Neuva-3 is suggestive of primary access (to the elephant carcass), perhaps by confrontational scavenging, and exploitation of both flesh (cut-marks) and marrow (percussion marks, with the bones broken when fresh; Espigares *et al.* 2013). Given the nutritional and energetic demands of early European *Homo*, and the seasonal fluctuations in animal condition and other food availability, passive scavenging does seem an unlikely candidate as a major strategy.

However, all modes of carcass access were probably used at some point in the Lower Palaeolithic, since it is likely that specific scenarios of hominin (and other carnivore) access to carcasses varied from event to event, reflecting both evolutionary factors (*e.g.* hominin cognition) and local ecology (*e.g.* numbers of individuals involved, the speed with which the carcass was reached). While the balance of carcass access scenarios might have changed over the course of the European Early and Middle Pleistocene, a fluid blend of strategies was probably retained (as has increasingly been recognised in species traditionally 'pigeon-holed' as obligate predators or scavengers, *e.g.* hyaena).

Nonetheless, Espigares *et al.* (2013) query the ability of Early Pleistocene *Homo* to *generally* out-compete predators the size of *P. brevirostris* (*c.* 110 kg), although they also note that stone-throwing against individual animals might enable successful confrontational scavenging. Madurell-Malapeira *et al.* (2017) have similarly stressed the significant competition and limited evidence for primary hominin access to carcasses at the Orce sites and suggested that these levels of competition may have persisted to the very end of the Early Pleistocene (*e.g.* at Vallparadís). The range of cursorial predators in Early Pleistocene landscapes (*e.g.* canids and hyaenids) would also have presented challenges to *Homo* in terms of reaching carcasses quickly enough. A predominantly scavenging strategy in landscapes occupied by *P. brevirostris* also further complicates the demands of accessing fats as well as protein. Moreover Rodríguez *et al.* (2012) have suggested that Early Pleistocene carnivore population densities may have been relatively low (although carnivore diversity was high), due to low prey biomass, which is suggested by the low species diversity and comparisons with net primary productivity in present day East and South African ecosystems. All this would seem to place further pressure on any sort of scavenging-dominated lifestyle.

The evidence for hunting and, if present its timing and character, in the Early Pleistocene is therefore critical. Rodríguez-Gómez *et al.* (2016) have questioned the role of hunting at the earliest Orce sites, stressing the very high levels of competition and suggesting that *if* hunting was used then *Homo* population densities must have been low. The earliest assemblages from Atapuerca are therefore key here, particularly Sima del Elefante (*c.* 1.1–1.2 mya) and Gran Dolina TDW4 (Huguet *et al.* 2013), which pre-date the clear evidence for primary access hunting at Gran Dolina TD-6.2 (Saladié *et al.* 2011). While there are few remains with anthropogenic modifications at Sima de Elefante (levels TE9–TE14; n=18, 0.6% of the total remains), those that are present represent a diverse range of activities: skinning, dismembering, defleshing and bone breakage. The frequency of modifications is higher at TDW4 (n=34, 3.7% of the total remains; for comparison, 13% of the remains at TD-6.2 are modified), and again suggest a diverse range of activities, including evisceration (Huguet *et al.* 2013). Carnivore activity is present in both sites, in greater proportions than hominin modifications (respectively 2.5% and 9.2% of remains, compared to 4.8% at TD-6.2). Both assemblages therefore suggest opportunistic exploitation (the modified fauna are predominantly large and medium-sized ungulates; *e.g.* bovid, red deer, rhinoceros), with TDW4 functioning as a natural trap into which ungulates accumulated. However, Huguet *et al.* (2013) concluded that hominins had primary access (due to the range of butchery processes, the type of carnivore modifications, and the order of tooth marks and cut-marks), probably through a mixture of hunting and confrontational scavenging. A further interesting aspect of TDW4 is the possibility that hominins regularly monitored the location for carcass opportunities, suggesting landscape memory and the role of revisited places.

A blend of hunting and confrontational scavenging in the Early Pleistocene is perhaps also supported by the longer-term trends in European predator competition. While high levels of predation pressure and predator competition (*i.e.* low prey biomass per predator species) have been suggested to occur throughout the Pleistocene, they may have been especially high in the Early Pleistocene (Rodríguez-Gómez *et al.* 2012). This has been argued on the basis of reconstructed food webs for Venta Micena (Early Pleistocene), Galería IIa/IIb (Middle Pleistocene) and Amalda V (Late Pleistocene), all of which excluded *Homo*. These high competition levels may suggest that carnivore population densities were lower than modelled, but they also highlight the difficulties faced by *Homo*, whose inclusion in the food webs would further enhance the competition. It both supports the notion that high levels of competition in the Villafranchian phase of the Early Pleistocene (*c.* 2.6–1.2 mya) delayed or hindered the expansion of *Homo* into and/or through Europe,[10] but also raises question marks over the hunting behaviours of Early Pleistocene *Homo*, with reference to both their inherent abilities and the levels of competition. There is certainly a marked contrast with the successful behavioural adaptations suggested for *Homo* during the later Middle Pleistocene, based on both reconstructed food webs and site-specific evidence.

Overall, the majority of the earliest sites are ambiguous in terms of hunting evidence, although primary access to flesh and/or marrow is suggested at selected sites, both at Atapuerca (Sima de Elefante and Gran Dolina TDW4) and elsewhere (*e.g.* with regard to the *M. meridionalis* carcass at Fuente Nueva-3; Espigares *et al.* 2013). However clear evidence for primary carcass access through hunting does emerge at the very end of the Early Pleistocene period. This is most convincingly demonstrated at Gran Dolina TD-6.2 (Rodríguez-Gómez *et al.* 2013; Garriga *et al.* 2017), and Rodríguez *et al.* (2012) have suggested that *Homo* entered the predatory guild, with a focus on prey in the 90–360 kg range, in the Early Galerian (*c.* 1.2–0.78 mya).

Food, and other stresses?

The timing of the appearance of a widespread reliance on hunting, the possible difficulties of scavenging, the uncertainty of individual hunting episodes and the relative importance of animal foods in the seasonal north all raise the question of dietary reliability in the European Lower Palaeolithic. This issue would have been especially relevant in late winter and early spring, when food availability was at its worst. Although the sample is very small, hominin 1 from Gran Dolina (the holotype of *H. antecessor*) suffered from a stress episode in early childhood, resulting in a disturbance in the formation of the dental tissues (Bermúdez de Castro *et al.* 1999). However, the later *H. heidelbergensis* fossil sample from the Sima de los Huesos suggests a mild pattern of dental trauma, with a frequency of enamel hypoplasia by individual of no more than 40%. The majority occurred between birth and 7 years (Bermúdez de Castro and Pérez 1995), presumably linked to difficulties in meeting the specific dietary demands of early weaned infants (Chap. 2), and/or the health of lactating females. These patterns were repeated in the larger Sima teeth sample examined by Cunha *et al.* (2004), who argued that the relatively low prevalence of linear enamel hypoplasias and plane-form defects was best explained by low levels of developmental stress, although the third year of life was, relatively, the most stressful. Notably, there is a significant reduction in the prevalence and severity of enamel hypoplasia in the Sima sample in comparison to the larger Neanderthal sample (Bermúdez de Castro and Pérez 1995). The relatively low European latitude of Atapuerca and its possible general role as a refugia, and/or the specific climate(s) associated with the Sima population, may be significant here, as might relatively low populations – perhaps enabling a less stressed hominin life than that experienced across Europe as a whole by Neanderthals. However dental hypoplasia data do not, necessarily, mean that the Sima population was heathier than Neanderthals, since individuals may die from acute stresses either before their developing teeth can register the stress or after their teeth have finishing growing (Cunha *et al.* 2004). This caveat is important given the available age at death data for Lower Palaeolithic hominins (Bermúdez de Castro *et al.* 2003a; Kennedy 2003; see Table 4.6).

This evidence for at least occasional development stress highlights the potential issue of seasonal dietary shortages and the question of fallback foods. These can be thought of as secondary, low ranking foods that are often relied upon in times of seasonal stress (Leonard *et al.* 2010), although it is important not to ignore the flexible roles of specific food items: inner bark for example has been described ethnographically as a staple, a supplement, starvation food, and a 'treat' (Sandgathe and Hayden 2003). In the Lower Palaeolithic seasonal stress would be most likely to occur, although not exclusively, in the winter and early spring, when both plant foods were less widely available and animals were in relatively poor physical condition. Potential fallback foods might therefore include underground storage organs (Hardy 2010; see also Chap. 3), fungi (although their energy contributions are limited; Kalač 2009), scavenged carcasses (Gamble 1987; see also Chap. 3) and perhaps also cannibalism. Alongside their obvious short-term impacts, seasonal fluctuations in food supply might also impact on multi-generational health, if the principles of the Developmental Origins of Health and Disease (DOHaD) hypothesis (*e.g.* Barker 2012; Gowland 2015)[11] also applied to Early and Middle Pleistocene hominins.

Table 4.6: Lifespan estimates for Eurasian Early and Middle Pleistocene hominin species (after Bermúdez de Castro et al. 2004; 2017; Kennedy 2003)

Site	Child (1–<5 yrs)		Juvenile (5–<10 yrs)		Adolescent (10–<20 yrs)		Young adult (20–35 yrs)[1]		Old adult (35+ yrs)[1]		n
	n	%	n	%	n	%	n	%	n	%	
H. antecessor											
Gran Dolina TD6	2	25.0	1	12.5	5	62.5	0	0.0	0	0.0	8
H. heidelbergensis											
Sima de los Huesos	1	3.6	0	0.0	18	64.3	6	21.4	3	10.7	28
European Middle Pleistocene[2]											
Various sites[2]	2	7.7	2	7.7	10	38.5	10	38.5	2	7.7	26
H. erectus											
Zhoukoudian	0	0.0	0	0.0	15	68.2	6	27.3	1	4.5	22

[1]Division between 'young adult' and 'old adult' after Bermúdez de Castro *et al.* (2004); [2]Pooled data from Abri Suard, Arago, Atapuerca (TG site), Coupe Gorge, La Chaise, Lazaret, Mollet, Montmaurin, Steinheim, Petralona, Pontnewydd, and Vergranne

During times of shortage might cannibalism have been a regular strategy? The evidence from Gran Dolina certainly makes it very likely that cannibalism was conducted: *H. antecessor* remains, along with bison and red deer, were the most commonly modified remains in the TD-6.2 assemblage. However, the antecessor remains are associated with multiple occupation events (identified through micro-stratigraphic evidence; Carbonell *et al.* 2010), and were mixed in with herbivore consumption, leading Saladié *et al.* (2012) to conclude that this cannibalism was not a response to dietary stress, but might instead have been a repetitive social practice (albeit with a dietary benefit).

Considering health more broadly, evidence for trauma and pathologies is, unsurprisingly, clearest in the large sample from the Sima de los Huesos. Traumas include bilateral temporo-mandibular arthropathy (*i.e.* jaw pain, including 'locking'), ear hyperostosis (possibly leading to deafness), cranial erosions and lesions and maxillary osteitis (bone inflammation) associated with a dental apical abscess (Pérez *et al.* 1997). In some cases there is evidence for a period of post-trauma survival and this may indicate a degree of intra-group care. Examples include the rounded edges, indicating bone healing, associated with the left browridge lesion on the frontal bone fragment AT-764 (Pérez *et al.* 1997) and the anomalous growth in the left maxilla of Sima cranium 5, which suggests a fall or impact months before death, followed by infection, facial deformation and probably pain and functional maxillary disability (Gracia-Téllez *et al.* 2013). The postural and locomotive impairments on the Sima pelvis 1 individual, who was probably in their 5th or 6th decade of life, suggest that they were able to perform a limited range of activities (Bonmatí *et al.* 2010) and may also be indicative of social care at this time. However, there is also evidence for perimortem trauma, of which at least some appears to most likely reflect inter-personal violence (*e.g.* cranium 17; Sala *et al.* 2015; 2016).

Hunting may also have been a significant source of skeletal trauma, although a wider range of causes than just hunting has been highlighted by Trinkaus (2012). Assessing responses to humans in modern animal populations with a view to extending them back into the Pleistocene is fraught with difficulty, since the predator–prey relationships of the past have fundamentally changed. However a study of European bison responses to human disturbance in the BPF offers three potentially useful observations: first, aggression was rare (bison typically flee from human encounters); secondly, males were more aggressive than females, with associated differences in the timing of attacks (♂: mainly during the rut; ♀: mainly during winter and calving); and finally, when attacks did occur they were provoked by humans approaching too closely (the average distance was *c.* 21 m; Haidt *et al.* 2018). Mech *et al.* (2015) recorded that bison have also been known to kill wolves in self-defence, while horse aggression (kicking, biting) is well known and, like the bison 'aggression trigger range' in the BPF, would be a significant concern for hominins reliant on seemingly short-range spears (Milks *et al.* 2019). Red deer can also 'box' with their forefeet if other warnings are ignored (Corbet and Harris 1991), although flight, rather than confrontation, might be a more likely response to hominin hunters, given the reliance of hinds and fawns on 'hiding' rather than 'following' strategies. Scavenging, especially confrontational scavenging (Bunn and Ezzo 1993), may also have been a source of trauma (Treves and

Figure 4.7: Confrontational scavenging (© Chris Crump; http://www.chriscrumpartist.com/).

Naughton-Treves 1999), while the rich array of large, predatory carnivores in Europe's Early and Middle Pleistocene landscapes is well documented (Fig. 4.7).

Seasonal deteriorations in food availability, particularly in late winter and early spring, and their impacts on health, raise the possibility of self-medication through plant use. While Hardy *et al.* (2013) discussed self-medication in the context of Neanderthals at El Sidrón, their rationale (the widespread examples of self-medication amongst higher primates and other animals; and that hominin gut reduction from *c.* 1.8 mya would have resulted in the need for increased external plant processing and enhanced knowledge of plant properties) would be just as applicable to European Lower Palaeolithic hominins. In that regard it is noteworthy that 32 of the plant species identified at Schöningen have medicinal aspects (Bigga *et al.* 2015; Hardy 2018). A further indicator of medical care is the evidence for tooth-pick grooves in the teeth from the Sima de los Huesos (Gracia-Téllez *et al.* 2013), although there are also notable examples of dental trauma at the site (*e.g.* Pérez *et al.* 1997).

The potential for habitual Palaeolithic healthcare has been emphasised recently by Spikins *et al.* (2019). While the majority of pre-*H. sapiens* evidence is associated with Neanderthals, Spikins *et al.* also highlighted Lower Palaeolithic examples, including specimens from the Sima de los Huesos. Moreover it is likely that much healthcare would be archaeologically invisible: medical data associated with modern wilderness activities such as hiking indicate that as little as 2–4% of common injuries and other ailments would be archaeologically visible, and that the vast majority would require minor, low cost care such as wound cleaning, and the provision of food, water and/or warmth. The importance and likely deep origins of modest healthcare is evident in the presence of such behaviours in primates (*e.g.* wound treatment and assisted childbirth), social carnivores, and the frequency of ailments in contemporary hunter-gatherers that, *e.g.*, prevent hunting participation. The latter highlights that low-cost healthcare would have significant evolutionary benefits, by maintaining the health of both individuals and, through the sharing of food and other resources acquired through foraging, the group. While not all specific arguments put forward by Spikins *et al.* (2019, table 2) for the adaptive role of healthcare in Neanderthal populations are transferable, their point that small groups would be especially vulnerable to the loss of individuals with experience, skill and knowledge is certainly applicable. Moreover, while the conditions of Lower Palaeolithic occupations were generally comparable or favourable in comparison to those of the Middle Palaeolithic, mortality risks from cold, seasonal food variations, high mobility, and hunting would still be present in Lower Palaeolithic Europe.

Who were the foragers ...?

In light of the strong arguments for a mixed diet and the availability of a diverse range of foodstuffs from late spring onwards, would these various different foods have been habitually acquired by different members of the local group or band? Labour division,

Box K: Fission and fusion in the Pleistocene

While Dunbar's (1998) social brain hypothesis predicts large communities by the time of the European Lower Palaeolithic (Chap. 2), labour division implies smaller day-to-day groups. Why may these have been necessary? Large communities present obvious ecological challenges (*e.g.* ensuring adequate food provision) but are critical for the maintenance of genetically healthy populations. Temporary shifts between larger and smaller groups (social fission–fusion) may well have been the solution to these tensions. The fission–fusion model of Grove *et al.* (2012) for evolving hominin social systems emphasises increasingly complex, multi-level systems, in response to the foraging constraints imposed by higher latitudes (*i.e.* reduced diversity and density of resources, larger foraging areas, lower population densities) and larger group sizes (implied by cranial increases). Temporary fission into small foraging sub-groups avoids the problems of foraging in increasingly large groups (areas become so large and travel times so high that foraging returns are inevitably in deficit), while fusion provides the benefits of large communities (*e.g.* reduced predation risk and access to an increased pool of mates).

To some extent the European Lower Palaeolithic seems to fall between discussions of the earliest central places (*e.g.* food sharing, stone caches and/or routed foraging in Oldowan landscapes; Isaac 1981; Potts 1994; 1988; 1991) and the regional-scale social landscapes and 'release from proximity' (Gamble 1998b) that characterise the later stages of hominin evolution (Grove and Dunbar 2015). Yet while the material transfers of Europe's Lower Palaeolithic lack the evidence for the significant 'scaling up' that is seen in the Middle Palaeolithic/Middle Stone Age (MSA), the temporal and spatial patchiness of foraging resources would seemingly require fissioning sub-groups, temporary loss of co-presence, and therefore pre-determined aggregation locations (Grove and Dunbar 2015). Middle Pleistocene encephalisation may therefore reflect changes in group structure, as well as size. On a more practical level, Grove and Dunbar highlight the critical importance of central places to fission–fusion behaviours, noting that shelter, water, tool-making raw materials and carcass locations could all have marked out places as localities for aggregations (*i.e.* residential camp-sites; Box L). However, they also note that such central places need not have been imbued with 'profound social meaning', at least at the start. While the degree of profundity is undefined (as is the timing of the 'start' of central place-focused behaviour), I agree and suggest that places may have had greater instinctive meaning (*e.g.* security through aggregation, and through the trust that such aggregations would have helped to create, in part through their being associated with food and other resources) than social.

But when would such fusion occur in a European Lower Palaeolithic context, and what would it look like? If the 'rule of three'[1] is applied to the suggested community number of 128 (Gamble *et al.* 2014, table 6.1) for *H. heidelbergensis*, then possible numbers for the sub-community social groupings can be estimated (rounded-up/down) as 43 (overnight camping group or band), 14 (foraging group) and 5 (intimate group). It seems likely that foraging groups would be the main focus during day-to-day foraging, with camping groups or bands re-aggregating at the end of the day (what I have chosen to term daily fission–fusion or dF–F). But would the 'rule of three' extend upwards for Lower Palaeolithic hominins, towards the mega-band or endogamous band (384 individuals using the above numbers) and even the tribe (n=1152)? Such social sub-division, or more, is certainly predicted by Grove *et al.* (2012, fig. 3), with 4–5 sub-groups predicted for *H. heidelbergensis* (and probably 4 for *H. antecessor*). Given Mussi's (2007) argument for seasonal conception, it seems likely that any fusion would most likely occur during the late summer/early autumn (labelled here as seasonal fission–fusion or sF–F), and this would chime with the relatively abundant food resources at that time of year (Chaps 5–6). It might even include opportunities for multi-band hunting, collaboratively or otherwise, of large ungulate aggregations.

What is noteworthy however is that the archaeological record offers very little material evidence for such multi-level social fusions. That does not mean they did not exist, but if they did then such fusions apparently did not result in either the architecture of aggregations (in contrast for example with the Upper Palaeolithic mammoth bone settlements of central Europe) or the exchange of materials from remote sources (*e.g.* exotic lithics). The former should not surprise us, as overnight or short-term camps such as Foxhall Road[2] are characterised by a lack of architectural investment during the Lower Palaeolithic. However, the absence of the latter is perhaps more noteworthy. The 'mega-band' may also have been reliant on the appearance of at least rudimentary language, as a key element of regional hominin interactions (Grove and Dunbar 2015), which might limit such structures to the later Lower Palaeolithic hominins (see also Chap. 2). Finally, it is also noteworthy that notions of larger 'breeding' groups in the earlier Palaeolithic of mid-latitude regions may to some extent be challenged by genetic and palaeoanthropological analysis: *e.g.* Ríos *et al.* (2019) have suggested that the El Sidrón Neanderthals suffered from in-breeding (see also Chap. 5: Box O).

Moreover, the spatial scale of such communities is also worth considering, not least because it is well known that annual range sizes increase with a dependence on hunting (Kelly 1995). Grove *et al.* (2012, fig. 1) suggest population densities of 0.1–0.2/km^2 for hunter-gatherers at 40–50° latitude (based on their best fit trend line). This would imply an area for the endogamous band (n=384) of 1920–3840 km^2

(a hypothetical circular territory with a diameter of 49.4–69.9 km). Such distances would certainly not be insurmountable obstacles to seasonal aggregations, although none of this discussion establishes where such aggregations occurred or indeed how they were organised. Comparably calculated estimates for overnight bands (n=43) indicate areas of just 215–430 km^2 (a hypothetical circular territory with a diameter of 16.5–23.4 km). These would be within the suggested daily foraging ranges of modern hunter-gatherers (c. 15–25 km), and the ranges of individual foraging groups would be smaller again (although these might also vary in size depending on the resources being acquired; *e.g.* static plants or mobile animals). However, those densities are based on modern hunter-gatherers (from Binford 2001) and should therefore be interpreted cautiously: an alternative density estimate of 0.002/km^2, based on lithic transfer data and a core area/home range distinction (see Chap. 5: Box O), presents more challenging area and distance estimates (*e.g.* 192,000 km^2 [a diameter of *c.* 495 km] for the endogamous band [n=384] and 64,000 km^2 [a diameter of *c.* 285 km] for the community [n=128]).

Finally, as Grove and Dunbar (2015) emphasise, fission–fusion behaviour at all scales can be linked to four significant, and cognitively demanding, hominin abilities: displacement (the use of language to refer to subjects removed in time and/or place from the setting of the communication), object permanence (knowing that objects or agents continue to exist when they are not present), mental time travel (modelling future encounters with individuals who are not continually present) and inhibition (suppressing instinctive reactions while mentally simulating future events). Such abilities have been documented in a range of fission–fusion species (Grove and Dunbar 2015 and references therein), and while the precise sophistication of these abilities amongst Lower Palaeolithic hominins may be debated, their presence seems undeniable given the apparent needs for fission–fusion foraging and mobility set by the environmental characteristics of mid-latitude Europe.

---

[1] The 'rule of three' is the observation that each social grouping (*e.g.* communities, mega-bands and tribes, in the case of hunter-gatherers) is approximately three times larger than the grouping below it (*e.g.* communities of *c.* 150, mega-bands of *c.* 500, and tribes of *c.* 1500; Gamble *et al.* 2014, table 2.1).
[2] At this site the cluster A assemblage from the grey clay was interpreted as imported artefacts, used and discarded by no more than five individuals. These hominins were sitting around a campfire, or another feature such as a tree or a carcass, protected from the wind by the slope (White and Plunkett 2004).

whether defined by age, sex or other criteria, also connects to broader concepts of group organisation, and in particular to the respective roles of logistical and residential mobility (Chap. 3: Box I) and the concepts of group fission and fusion (Box K). While many of the arguments for both logistic and residential mobility are indirectly based (see also Chap. 3), just occasionally we see direct glimpses. The recently uncovered Happisburgh III footprints (Ashton *et al.* 2014) appear to include both adults and children, which is suggestive of an entire foraging group or band in that particular instance, rather than a logistic, task-specific, adult-only group (although a predominantly adult male group has been suggested by the footprint sites at Ileret, Kenya; Hatala *et al.* 2016).

Ethnographic studies have repeatedly documented the differences between male and female foraging, including the differing characteristics of female and male hunting (Bliege Bird and Bird 2008; Gilby *et al.* 2017). Typically, women hunt less, hunt different types of prey (*e.g.* small, relatively immobile species) and hunt in different ways to men, who tend to hunt larger and, in the context of hunting failure, riskier prey (*e.g.* Bliege Bird *et al.* 2009; Marlowe 2007; Gilby *et al.* 2017). Thus, female foraging tends to ensure regular provisioning, with more irregular contributions from males (whether through pair bonding/food exchange or through group level sharing/ status signalling; Gilby *et al.* 2017). These patterns are frequent, and repeated across different ecological contexts, and the evolutionary reasons behind them have been fiercely debated (*e.g.* Gurven and Hill 2009; Gurven and Hill 2010; Hawkes *et al.* 2010). Of particular relevance to the European Lower Palaeolithic may be Marlowe's (2007) suggestion that there is reduced division of labour in less seasonal, more productive habitats, and that labour division likely emerged after pair bonds. It is therefore possible, at least on environmental and/or life history grounds, that some separation of foraging tasks may have been present in the European Palaeolithic.

But are such models and principles really applicable to Lower Palaeolithic hominins? It is all too easy to fall into modern comparisons and unsupported statements about the roles of females and males, of all ages, in the Pleistocene, not least because the evidence is limited. Trauma evidence from later in the Palaeolithic certainly highlights the potential dangers of hunting (Berger and Trinkaus 1995; Trinkaus 2012), and in the case of Neanderthals the evidence from trauma and mortality profiles is suggestive of shared exposure to day-to-day tasks and their attendant risks across males, females and children (Pettitt 2000; Kuhn and Stiner 2006). Unfortunately, the available fossil evidence from the European Lower Palaeolithic is insufficient to robustly test whether a similar pattern exists, and moreover danger and/or potential trauma would also be present in various other walks of life such as non-hunting foraging (*e.g.* from other carnivores), day-to-day living (accidents) and even intra- or inter-group violence. Nonetheless the risks, and inefficiencies, of involving the whole group on a hunting trip are also highlighted by Kelly (1983), in the context of residential and logistical mobility, although the exact nature of those risks are left unspecified (*e.g.* missing the game? vulnerability of, and potentially other costs to, individuals?).

These various dangers, combined with other threats and risks such as disease, starvation and childbirth (Kennedy 2003), are reflected in the apparently small populations throughout the Palaeolithic (despite the seeming reduction in birth-spacing in

Homo), which is increasingly evident in the latest palaeogenetic data for Neanderthals (Kennedy 2003; Sánchez-Quinto and Lalueza-Fox 2015). It is possible that if danger was recognised, then it was seen in the overall context of life, rather than with regards to one activity or another. Moreover, day-to-day group sizes may well have been small – based on modern day hunter-gatherer data (Kelly 1995; Gamble *et al.* 2014), the predictions of the social brain hypothesis (Chap. 2 & Box K) and the evidence from *in situ* Lower Palaeolithic 'camp sites' such as Bilzingsleben (Mania and Mania 2005). Might participation have been dictated by the matching of an individual's skills to the task, a fluid approach to the composition of foraging parties (Bliege Bird and Bird 2008), and sometimes simply 'all hands to the pump', rather than any pre-defined notions of what females and males *should* do? One potential example of this might be the participation of at least older children in communal hunting parties, since a need for multiple individuals is suggested both by the demands of carcass transport (Saladié *et al.* 2011) and possibly by the specific hunting strategies used. The latter are harder to establish, but the suggested repeated hunting of individual horse 'families' at Schöningen (Voormolen 2008) and the 'bison bone bed' evidence from Gran Dolina TD-10.2 (Rodriguez-Hidalgo *et al.* 2017) would seem to be suggestive of at least small-scale cooperative hunting on occasions. Moreover, a relatively early initiation to the involvement of children in aspects of hunting, with a gradual intensification in the extent and nature of their participation as they aged, would also be beneficial given the evidence for hunting returns peaking later in life (*e.g.* Kaplan *et al.* 2007).

In spring, the renewed availability of food, and the need for all in the group to build up their energy stores after the winter, might well have favoured the involvement of all, male and female, young and old, in foraging tasks, although perhaps focusing on different food sources. Child foraging, in particular, might have been especially critical in spring (when eggs, chicks and other low-risk foods are available) and perhaps also in late summer/autumn (focusing on fruits and nuts). The late spring and early summer periods, especially, would offer longer days during which foraging returns could be maximised. While child involvement in foraging in modern hunter-gatherer societies clearly varies, Hawkes *et al.* (1995) documented the significant involvement of Hadza children in foraging activities both in/near camp and away from camp. Children from age 3 upwards were involved in in/near camp foraging, while from age 6 upwards girls and boys took part in long berry-collecting trips (up to 10 hours), in the company of adults. While the children's high foraging return rates are clearly not transferable, the study highlights the cognitive and physical potential of young children to participate in food-getting, feed themselves, and potentially provision others. Moreover while 'children's foods' might be less available at higher latitudes, there would seem to be a sufficient range of plant foods (Tables 4.3, 5.6 & 6.3) for sub-adult foraging to be a significant component of Lower Palaeolithic strategies. In spring (Table 4.3) such foods could include new shoots and leaves.

While different foraging groups might increase levels of potential vulnerability for some/all of the hominins, this could to some extent be counter-balanced by prey species contracting into relatively small home ranges during their own birthing seasons, thus reducing the separation between different foraging groups. This might be especially important for late stage-pregnancy hominin females or new mothers, for whom access to animal foods, amongst a wider mixed diet, is likely to have been critical to successful pregnancy and births (*i.e.* the production of healthy offspring).

So far however much of this discussion has ignored the changing stages of life, in particular sexual maturity, and, as Graves-Brown (1996) has noted, biology clearly plays a role in the formation of gender. The demands of childcare, both in terms of tasks (*e.g.* feeding, carrying, nurturing and protecting) and energy budgets (to fuel gestation and/or lactation), might argue against a major engagement of late stage pregnancy/breastfeeding females in ungulate hunting, for a number of reasons. Hunting of mobile prey loads additional energy demands onto individuals, particularly on those carrying helpless offspring, although this could be any older individual. Prey are sensitive to the presence of hunters, at the very least in visual, auditory, and olfactory terms, and the crying of infants, with its sometimes unpredictable nature as any parent will know, would likely be a potential source of game taking flight. Hunting also has a temporal dimension, one which might not always coordinate with the breastfeeding 'timetables' of infants, although amongst the !Kung San infants are fed much more frequently than in most western societies (Lummaa *et al.* 1998).[12] Thus hunting and pregnancy/breastfeeding may not have overlapped, due to the physical demands of the later stages of pregnancy, and possible incompatibility between young infants and the quiet of the hunt. However, the later childcare stages can at least in part involve alloparenting (*i.e.* individuals other than the parents acting in the parental role which might include older siblings; although short inter-birth intervals would result in the frequent presence of pre-weaned infants within small groups), and therefore there appears to be no reason why females who were not in the later stages of pregnancy/breastfeeding could not have hunted. This is certainly supported by specific ethnographic examples: Bliege Bird and Bird (2008) report the involvement of women in the hunting of mobile prey in various groups, including the Kubo (capturing pigs, cassowaries and bandicoots) and the Netsilik and Copper Eskimo (caribou drives and seal hunts), although there are other groups in which women only acquired plant foods or immobile prey. Moreover, it is important to recall that 'hunting' is a multi-stage process, encompassing searching (and potentially pursuing), killing, butchering, and transporting: any adult individual, and perhaps also adolescents, could be valuably involved in the hunting of mobile prey without necessarily actively partaking in all of the above tasks.

What of the pre-reproductive adolescent females? They could surely be just as effective an 'apprentice' hunter as adolescent boys – while there seem to have been small reductions in *H. heidelbergensis* female stature compared to males (Chap. 2), there could equally be contrasts in hand/eye coordination, observational skills, memory (of animal behaviour and landscape topography) that might well favour the involvement of specific individuals, whether girls or boys, in the hunting of mobile prey. But from a selection perspective there are potentially problems with such a strategy, since an investment in young females (*e.g.* 'training' in hunting techniques, gained experience) has a reduced pay-off compared to males, as the adult life of the former is likely to include periodic intervals of pregnancy and lactation. A more limited participation in mobile prey hunting by adult females has been suggested ethnographically: many of the percentage estimates for the proportion of foraging women's subsistence efforts allocated to the pursuit of mobile prey are below 20%, although these reflect a combination of factors (Bliege Bird and Bird 2008). Thus, a dedicated investment in young males, rather than males and females, might have evolutionary benefits for an individual group, especially given the lengthy learning and practice periods suggested for skills such as spear throwing (Milks *et al.* 2019). Moreover, the exposure of young, and reproductive adult, females to hunting risks clearly negatively impacts on group survival as a whole. I am certainly not proposing the explicit social expression and/or enforcement of Palaeolithic gender roles: the 'you can't come hunting, you're a girl ...'-type view. But it is possible, given their potential reproductive importance later in life, that there was increased survival potential, *i.e.* fitness, in those groups where young females avoided, by choice or instinct, certain tasks. Their value would also be accentuated by the likely small size of groups, relatively short life expectancies (Table 4.6), the hazards of childbirth, and the overall costs and investments involved in the models of reproduction, growth and development followed by large-brained and large-bodied hominins (Geist 2003).

If there were differences in the foraging focus of males and females, what other types of food-getting might females, especially sexually mature females, have engaged in? By contrast with medium- and large-sized mammal hunting, the foraging of static foods (*e.g.* plants, eggs and nestlings) and the hunting of smaller game removes at least some of the former activity's attendant risks, while specific residential locations and moves (perhaps minimising foraging distances to plant food patches?) would also reduce travel times and the energetic demands of this type of foraging. There are a range of modern ethnographic examples of female hunting, typically of smaller game, with prey including armadillo (Ache, Paraguay) and lizards (Martu, Australia; Bird 1999; Bliege Bird and Bird 2008). Moreover, such foods are also unlikely to be ignored by mobile prey foragers, especially if a hunt was unsuccessful, although they may have lacked the skills and/or knowledge to effectively forage these 'other' foods. Foraging of static foods or low mobility prey might also have been undertaken by individuals, potentially of any age or sex, with temporarily or permanently impaired mobility.

It is thus possible that the foraging skills and experiences gained by females through childhood, and/or reinforced during later gestation and lactation, were likely to relate more to immobile plants (*e.g.* fruits, seeds, herbs, roots, fungus), static animals (*e.g.* eggs and nestlings), animal-created foods (*e.g.* nectar/honey), and low-mobility animal foods (*e.g.* frogs, reptiles and small mammals; Bliege Bird and Bird 2008). By contrast reproductive females may have had less (but not necessarily no) appropriate knowledge and practical skills (and fewer opportunities to practice and maintain them) relating to larger, mobile prey, hunting, although it seems unlikely that the two 'sets' of expertise were mutually exclusive.

Finally, who would have been involved in foraging for more reliable resources? Bliege Bird and Bird (2008) observed that, amongst the Australian Martu, younger women often remain at camp to care for smaller children but are more likely to join foraging groups when targeting resources that children can acquire on their own. The degree of such foraging might also be related to the likely productivity, and be reduced during seasons when pests and other dangers could only be avoided through vigilance (Bliege Bird *et al.* 2009). Hawkes (1996) similarly observed that females must balance foraging against childcare (*i.e.* keeping offspring safe) and that in different habitats the balance between these two concerns will shift (*e.g.* more emphasis on foraging in a safe locality; although the integration of children into foraging will shift the balance towards this activity). While I would expand 'females' into a broader group of 'child-carers' (as this could be older siblings, pregnant/breastfeeding females, older males/females, and other demographic 'groups'), this is a notable consideration from both a landscape and a seasonal perspective. What is a safe locality in the Lower Palaeolithic world? Firesides would be one possibility, although as noted elsewhere (Chap. 3), hearths were probably an irregular occurrence, at least at certain times of year, and unless fire was a mobile technology (*e.g.* flaming brands, smouldering polypores) would be of little use in foraging. Cave mouths (unless already containing other predators) would also offer a degree of defence but the numbers of such sites are limited for this period, albeit possibly due to taphonomic factors (Chap. 3) and would again be of limited use during foraging. Simple group numbers might also offer protection against carnivores, especially lone ambush predators, although numbers would inevitably be temporarily reduced if different, complementary foraging strategies were used. Threats would also vary between seasons, depending on the behaviour of carnivores and herbivores (*e.g.* elevated aggression related to the protection of newborns in late spring/early summer; Daleszczyk 2004) or desperation brought on by winter starvation), or the levels of insect activity, especially in the summer.

Estimating the size of these groups is inevitably speculative, and would no doubt vary by season, food type and patchiness, and other variables such as predation risk or proximity to 'border' areas and potential inter-group encounters (as also documented amongst chimpanzees; Pusey and Schroepfer-Walker 2013). Starting from the social brain's theoretical band size prediction of 43 for *H. heidelbergensis*, foraging groups of 14 individuals can be suggested on the basis of the 'rule of three' (Gamble *et al.* 2014; figures for *H. antecessor* would be slightly smaller). Since not all band members are adults (with perhaps 30–50% of the group made up of adolescents or younger; Kelly 1995), that would suggest that each foraging group might contain 7–10 adults, *if* adults and children were equally divided, and assuming that all adults were healthy and able to forage. If not (and there would be reasons for a skewed division between, say, mobile prey hunting groups and other foraging groups), then the latter groups might easily contain fewer than half a dozen adults and 10–20 children, of various ages and degrees of independence.

Food sharing ... or, where is everyone else?

The issue of overlapping or complementary food-getting roles and activities, whether divided on male/female lines and/or otherwise (*e.g.* different stages of the life-cycle), is critical given the need for adults, and perhaps also older children, to support unproductive young in the human life history model (Chap. 2), and in spring, to meet the wider nutritional needs of a group denuded by winter. Trust in others, in this case to deliver and share particular foods, has been highlighted more broadly by Spikins (2012) with respect to handaxes and notions of goodwill, trustworthiness and reciprocal altruism in Acheulean societies. An interesting complement to trust in food providers might also be found in the issue of residential fires. Twomey (2013) highlighted the danger of free-riders where fire is concerned, *e.g.* with respect to individuals' (non) contributions to fuel gathering while benefitting from a communal fire [the concept of which is generally supported by the scale of the known fire traces], and argues that free-riding is a significant constraint on the evolution of human cooperation. There is thus a need to monitor the fire-tending and fuel gathering intentions and actions of fellow group members, and to discourage any free-riding (*e.g.* stealing cooked food) – all of which Twomey related to future-directed group-level cooperation. As Twomey has also suggested, this monitoring may be a practical example of effective Theory of Mind (ToM; see also Chap. 2) during this period: *e.g.* 'I *believe* that you *think* that I *intend* to keep the fire burning in order that you will share your hunting kill with me' (modified from Dunbar 2007) would appear to be an example of third-order intentionality. Similarly, although not at all unique to the European Lower Palaeolithic, Nowell (2010) has highlighted the increased importance of cooperation and trust in association with ground, as opposed to tree, dwelling (*e.g.* who acted as 'sentry' and delayed their own gratification [sleep]?). The likely fission–fusion lifestyle of European Lower Palaeolithic groups (Box K), reflecting resource distribution and seasonality, the provisioning demands of large-brained, altricial hominins, and other issues such as fire management and ground dwelling, thus does seem likely to have required significant cooperation and trust (Spikins 2012).

However, if the hunting of mobile, medium and large mammals, with its attendant unpredictability and other costs, was typically undertaken by specific foraging subgroups, how was the meat made available to the other members of the group (and trust rewarded)? In short, how was food shared? In modern hunter-gatherers the answer seems clear: meat is returned to the campsite for sharing and consumption, although there is also evidence for consumption of elements of the carcass at the kill site (*e.g.* Buck *et al.* 2016). But is this equally likely for Lower Palaeolithic hominins? The question is worth asking, not least because there is so little evidence for open-air campsites. This may in part reflect the likelihood of significant residential mobility (Chap. 3: Box I) and the association amongst recent hunter-gatherers of high residential mobility with ephemeral shelters (Kelly 1995, fig. 4.6): *i.e.* campsites were present but are hard to find in the archaeological record (Chap. 3). Residential mobility might also have been used to minimise the distances between favoured hunting areas and dependents, and the locations and durations of each new residential site may even have been partly dictated by hunting strategies (although other location factors shouldn't be ignored: *e.g.* safety, natural shelter, and access to other key resources such as other food sources and raw materials). The fragmentary nature of the Lower Palaeolithic record makes it difficult to resolve this issue (Box L), but it is notable that high resolution open-air sites such as Boxgrove and Schöningen include no examples of 'campsite' or sleeping site-type evidence (*e.g.* fire traces).

Yet while Bilzingsleben's interpretation as an open-air camp site remains debated (Chap. 3), there are other, much more convincing, candidates. Saladié *et al.* (2011) have proposed Gran Dolina TD-6.2 as a home base, in light of both the technological behaviours (*e.g.* the diverse range of introduced raw materials and the full knapping sequences; Carbonell *et al.* 1999) and the comprehensive butchery tasks (skinning, dismembering and/or disarticulation, defleshing, evisceration, periosteum removal and possible tendon removal) undertaken at the site. The intensive carcass processing at TD-6.2 indicates a safe, controlled environment, to which carcasses of various sizes were transported, sometimes whole, to reduce risks of theft, injury or death (Saladié *et al.* 2014). Moreover, the lack of carnivore modifications on the cannibalised hominin remains has been interpreted as evidence that hominins could and did control the cave as a home base, during a longer occupation period within which the cannibalism occurred. The later Gran Dolina occupation in level TD-10.1 has also been interpreted as a residential base camp, in light of the systematic and comprehensive butchery, including marrow extraction (with the level of bone breakage in marked contrast to, *e.g.*, hunting camps), fragmented remains, and a paucity of carnivore evidence (Rodríguez-Hidalgo *et al.* 2015). Finally, there is evidence that such sites were used flexibly. Four differing types of occupation have been proposed at Arago cave: long duration home base; temporary seasonal habitat (secondary campsite); hunting stopover; and bivouac (de Lumley *et al.* 2004). The key point in this context however is that all these sites sometimes indicate carcass transport *into them*, at least sometimes of complete animals, and therefore delayed consumption – this implies the presence of inhibition in the cognitive make-up of these hominins (Nowell 2010). But what are the practical demands of moving animal foods?

The rich insights into Lower Palaeolithic butchery techniques which Boxgrove (Parfitt and Roberts 1999), Schöningen (Voormolen 2008; van Kolfschoten *et al.* 2015a) and Atapuerca (Saladié *et al.* 2011; Huguet *et al.* 2013; Rodríguez-Hidalgo *et al.* 2015) have provided highlight significant opening-up and breaking apart of the carcass: *e.g.* skinning, dismemberment/disarticulation, filleting and marrow bone breakage. But where did such processes occur? The above sites suggest that they occurred

Box L: Lower Palaeolithic campsites?

The evidence for campsites or home bases in the European Lower Palaeolithic record is very mixed. In part this is a product of preservation and taphonomy (see also Chap. 2: Box F), but it may also reflect archaeologists' uncertainty as to what we should be looking for. A key difficulty of discussing campsites in the archaeological past concerns the terminology that we commonly use: a 'home base' is defined by the OED as 'a place from which operations or activities are carried out', while 'campsite' is, slightly unhelpfully, 'a place used for camping'. These terms, and the widespread interpretation and/or use of them in westernised societies, can potentially lead to a check-list type approach (*e.g.* family or household-based shelters, fires, middens, 'domestic' activity areas), with the obvious danger of seeking 'our' own behaviour in the deep past. Such an approach can be likened to Mania and Mania's interpretation of Bilzingsleben (2005, 114) as a 'socio-cultural environment with living structures, the use of fire and special activity areas'. Yet as Gamble (1999, 169) has observed: 'If our goal is to find shelters in order to use them as a summary of social life in the early Palaeolithic, then I have no doubt that we have the analytical ingenuity and imagination to find them'.

Insights from the ethnographic and ethnoarchaeological literature can also be problematic. While there have been significant ethnoarchaeological studies of hunter-gatherer campsites (*e.g.* Binford 1978; Gamble and Boismier 1991), many of these studies are either based in very high or low latitude habitats (*e.g.* Binford 1991; Nicholson and Cane 1991), and/or relate to modes of life (*e.g.* sedentism and houses/huts; Boismier 1991; Fisher and Strickland 1991) for which we have no clear evidence in the Lower Palaeolithic record.

But does a seasonal, ecological framework offer an alternative approach for seeking out Lower Palaeolithic campsites? Before tackling this, what might be the nature and duration of a campsite's use (while acknowledging that any useful definition will most likely encompass considerable variability)? Duration is perhaps easier to address: a minimum of an overnight occupation is suggested, although this, in itself, intrinsically links the concept of a campsite with sleeping. However, residential occupations could be longer, perhaps significantly so, as is indirectly indicated by the residential move frequencies given in Table 5.11. The nature of these sites is more uncertain, since many of the activities known to occur at proposed campsites (*e.g.* sustained butchery at Gran Dolina TD-6.2) also occur at proposed kill-butchery sites (*e.g.* Boxgrove and Schöningen, which are logistical foraging or activity sites in the terminology of Binford 1980 and Kelly 1983). How are they different? Repeated visits do not seem to be a distinguishing criterion (all of the above sites show evidence for this), and while sustained occupation might be, the resolution of the record currently makes this difficult or impossible to test for as a rule. However, the

Caune de l'Arago is a rare exception to this, and different styles of habitation have been defined across its long sequence on the basis of variations in lithic technology, raw material exploitation, the hunted fauna, seasonality indicators and the thickness of the archaeological layers. Four types of occupation were identified: long duration; temporary seasonal; hunting stopover; and bivouac. Activities were more diverse during the longer occupations, with fewer and/or very specialised activities during short stays (de Lumley *et al.* 2004). The presence of 'domestic' activities might also be a useful distinction (*e.g.* at La Noira: B. Hardy *et al.* 2018), but common candidates for 'domestic' features (*e.g.* hide processing or controlled fire) are rarely detectable (fire) and/or occur (skinning) at both 'campsites' and 'kill-butchery' sites.

Given some of the challenges outlined elsewhere, what attributes might a Lower Palaeolithic campsite require from a seasonal perspective? Natural or anthropogenic shelter, warm/dry ground conditions (if possible) and safety from predators (enabling activities such as food sharing and tool-making/modification, and, perhaps at least as importantly, a sense of security) would seem to be core requirements. For this reason the term campsite is used here in a broad sense, encompassing concepts such as home bases, sleeping sites and residential sites. This overall definition may seem limited, but it is grounded in the key survival requirements of Lower Palaeolithic Europe, while emphasising the role of such sites in 'social security' as well as in provisioning, manufacturing and sleeping (Gamble 1991). It is place-focused, but similarly to Gamble's (1999) 'gatherings' re-interpretation of Bilzingsleben it puts little emphasis on formal structures (*i.e. évidentes* structures) or material investment in the place. Any investment, particularly at longer-occupied sites, is suggested to be emotion-based instead.

In light of this, what might the archaeological record of a Lower Palaeolithic campsite look like? In short, modest. Artefact and food residues could reasonably be expected (*i.e. latentes* features[1]), but not necessarily in a manner that would distinguish the 'campsite' from an 'activity' site – nor would greater or lesser quantities of materials provide an easy distinction, given the complexities of temporal resolution. Campsites might also be small, to emphasise the ethics of sharing – which might also occur through the positioning of hearths, if present (Gamble 1991). It is perhaps for these reasons, rather than a more fundamental 'site-less' existence, that campsites in the Lower Palaeolithic are so difficult to find in the absence of fire traces (*e.g.* Beeches Pit, Foxhall Road?) or permanent natural shelters (*e.g.* Gran Dolina TD-6.2). Nonetheless there are still occasional indicators of the re-use of campsites (*e.g.* at TD-6.2; Carbonell *et al.* 2010), and by extension the importance of specific places within Lower Palaeolithic landscapes, despite the general absence of architectural investment: *e.g.* two of the three hearths at Beeches Pit (area AH) intersect, obviously indicating discrete burning events (Preece *et al.* 2006), but might their overlapping 'footprints' perhaps also suggest repetitive use and organisation of this residential space?

---

[1] *Latentes* features are arrangements of material categories such as flaked stone and bone refuse, in contrast to *évidentes* structures, such as hearths and other well-defined features (Gamble 1991, 11).

both in-the-field (Boxgrove, Schöningen) and at home bases (Atapuerca). In the latter case, and since fat and marrow utility were a key factor in carcass transport decisions (*e.g.* Rodríguez-Hidalgo *et al.* 2015), it is likely that marrow-rich elements were preferentially transported, although accessible and/or non-portable elements might have been consumed at the kill (*e.g.* blood, tongue, eyes?). Such 'snacking' behaviour is perhaps suggested by bison tongue removal at Gran Dolina TD-10.2 and also by the tooth mark evidence on ribs (Rodriguez-Hidalgo *et al.* 2017). The transport of valuable carcass parts from kill-butchery sites is also suggested at TD-10.2 by the relative paucity of bison long bones: this does not appear to be a taphonomic bias and indicates the further movement of meat and marrow to other locations, presumably residential sites.

Such transport evidence is of wider cognitive significance because, while rates of hunting and scavenging success are difficult to estimate, studies of both extant hunter-gatherers (*e.g.* Lee 1968)[13] and, to a lesser extent, other mammalian social predators indicate frequent hunting failure (*e.g.* 43–47% hunting success for Scandinavian wolves preying upon moose and roe deer, *c.* 1/3 success for spotted hyaena in the Masai Mara; Holekamp *et al.* 1997; Wikenros *et al.* 2009), although modern hunter-gatherer perspectives are complicated by occupations of marginal landscapes (Domínguez-Rodrigo 2002). If the Lower Palaeolithic experience was similar and access to fresh 'meat' not an everyday experience, then it seems likely that kill-butchery sites were highly emotive and socially charged occasions (the presence or threat of presence of other carnivores would presumably further heighten the hominins' state of alertness). The evidence for carcass transport is therefore significant, as it suggests that hominins were able to inhibit, at least in part, the urge to immediately consume the animal bounty. Lurking scavengers were no doubt sometimes a key factor in the decision to transport all/part of the carcass (although on-site butchery lasting several hours still appears to have sometimes occurred; Pope *et al.* in press), but the demands of wider group provisioning are likely to be a further factor.[14] However, while successful access to meats, fats, blood and other resources through co-operative action may have helped to build and maintain social cohesion and trust (see also Spikins *et al.* 2014), it may also have exposed or highlighted intra-group tensions (*e.g.* in terms of access to particular parts of the carcass). The participation of multiple butchers, generating relatively 'haphazard' cut-marks, at Qesem Cave may indicate one approach to the issues of food access and/or sharing (Stiner *et al.* 2011).

However, there is also likely to have been considerable variability in these behaviours. Saladié *et al.* (2011) highlighted the range of factors which can impact on carcass transport decisions, including the distance between the kill/butchering site and the home base, the number of animals to be processed, the weight of the carcass (and the so-called 'schlepp effect'; Klein 1976), the number of participants in the hunting/scavenging expeditions, the location and time of day of carcass acquisition and the risk of predation by other carnivores during the initial processing (and perhaps also during the transporting phase).

How would marrow (and meat) yields be moved? One option might be to skin the carcass and then use the hide as a rudimentary carrying bag, drawn together by the hooves (*e.g.* Klein 1976). Internal organs such as stomachs could perhaps be used in a similar fashion, depending on the degree of in-the-field butchery. Carcass-laden hominins moving through interglacial forests or grasslands would presumably have been vulnerable to being tracked, trailed and/or attacked by other scavenging predators – there is extensive zooarchaeological evidence for them in the vicinity of key open-air sites (*e.g.* Table 2.2). But would that vulnerability be significantly greater than that encountered while processing the carcass for several hours, as Pope *et al.* (in press) have recently argued for the GTP-17 locality at Boxgrove, at a kill-butchery site? It seems unlikely, although carcass exploitation in closed woodland habitats may have been less risky given that competition generally increases in open-air habitats, where carnivores benefit from the greater visibility (Saladié *et al.* 2014). What all this does favour is the importance of a means of potentially repelling scavengers that came too close, potentially while both processing at the kill/acquisition site and then moving the carcass. As well as disadvantaging and ambushing weapons (Churchill and Rhodes 2009), and seemingly both thrusting and short-range throwing weapons (Milks *et al.* 2016; 2019), long wooden artefacts such as the Schöningen spears may have played an important role in carcass defence (and perhaps also in confrontational scavenging). They would seem to fit the bill much better than hand-held stone artefacts (due to the insufficient distance between you and the target), and may have been used in combination with thrown stones and other objects (while the latter are obviously useful you need to keep replenishing your supply). Overall however there is clear evidence for carcass transportation at Gran Dolina and other sites (*e.g.* selective movement of bison cacass portions at Isernia la Pineta; Hohenstein *et al.* 2009), and given its costs, risks and involvement of multiple individuals would seem to be a clear indicator of social cooperation, delayed consumption and food sharing (Saladié *et al.* 2011), and presumably provisioning of anyone not involved in that particular foraging party. Moreover, its occurrence at both TD-6.2 and TD-10 indicates the presence of this behaviour amongst both *H. heidelbergensis* and *H. antecessor*. Cognitively, the demands of food sharing behaviour (*e.g.* carcass transport) can be linked to the construction of mental maps and staying on-task (*i.e.* interference control), and thus to enhanced working memory (Nowell 2010).

So what of those instances of comprehensive butchery in open-air settings (*e.g.* Boxgrove and Schöningen). Did this involve moving the rest of the group to the meal? Well first, what do I mean by moving to the meal? I do not see it as feasible that vulnerable infants and children, in particular, were spending significant time at the kill/scavenging site itself, simply due to their likely exposure to other carnivores,[15] although from the comprehensive butchery evidence at Schöningen it does appear likely that the hunters were able to defend their kills at this open-air site (Starkovich and Conard 2015). However, the remainder of the group might 'trail' the hunters and thus move to, or remain at, a location nearby: perhaps defendable, certainly less littered with the flotsam and jetsam of butchery, but facilitating rapid access to the butchered carcass. This would maximise access for all to some of the key, vitamin C-yielding, organs and other internal parts of mammals (*e.g.* brain, liver, spleen and testicles; Speth 2017). Buck *et al.* (2016) observed that much gastrophagy occurs out-of-camp amongst the Hadza, and that chyme[16] might be a valuable weaning and old age food, as it is easy to process and digest (this would presumably also be true of rotten meat and fish; Speth 2017). This strategy would remove the need to move the kill (although it would not remove potential carnivore threats), which might be significant depending on the general state of the carcass as a whole. This last issue might be further magnified in the case of a scavenged carcass, where the likely state of such a resource would presumably limit the potential (or value) to move it to another location.

To some extent this choice of moving the meal (carcass transport) or moving the group might be defined by energetic pay-offs, which are difficult to model with any confidence. For example, would it be more costly (in terms of energy expenditure and subsequent food requirements, and/or risk) to move late stage-pregnancy females, lactating females and newborns and young children to the kill, as opposed to the smaller hunting group moving the meal? The latter option seems particularly preferable in the spring if Mussi (2007) is correct and peak conception was most likely to occur in late summer, when food availability (= enhanced energy intake and body fat) and light conditions would favour female fertility. This would schedule the significant food requirements of the 2nd and 3rd trimesters against late winter and early spring, seasons characterised by more limited food availability, although birth itself would occur close to the beginning of the next resource-rich period (Taylor 1997). In such a situation it seems especially likely that complete, or disarticulated parts of, carcasses were returned to the non-hunting group during cold or cool seasons, to minimise the latter's energetic expenditure and exposure to the general risks of mobility.[17]

The wolves and lynxes of the BPF offer an interesting perspective here – the activity areas of wolf mothers during late spring/early summer are markedly smaller than usual, due to the particular challenges of the pups needing near-constant care and attention (Jędrzejewski *et al.* 2001), while the distances between lynx mothers' kills are shortest when small, vulnerable kittens have to be brought to each kill (Okarma *et al.* 1997). There might have been a similar seasonal variation amongst hominin strategies, although the hominin experience is substantially complicated by the greatly extended periods of helpless infancy and then childhood (Chap. 2). A 'food to group' strategy would probably be further favoured if cooking was involved (Twomey 2013), although how habitual cooking was remains a point of debate (see Chap. 3). However, a warm spring or summer day might favour the 'group to carcass' strategy, as might the need for groupwide access to carcasses in springtime, after the harshness of winter.

The issue of differential access would also apply to other foods. Sandgathe and Hayden (2003) note that inner bark, initially soft and moist, will rapidly dry out if not eaten or processed. If it was targeted as a low risk food item by adult–child foraging groups then it seems likely that it was also rapidly, and perhaps predominantly, consumed by them, although it might also have served as a valuable on-the-go 'snack' food during hunting trips. Amongst modern foragers, despite them having more 'container' options, there is extensive evidence for consumption while collecting (Hawkes *et al.* 1995), and Woodburn (1968) observed that male Hadza hunters sometimes satisfied their hunger on hunting trips by eating berries. In the context of possible foraging sub-groups in the Lower Palaeolithic there is evidence that diets were not always equivalent. Analysis of the enamel surfaces from 190 teeth at the Sima de los Huesos documented a significant sex-related difference in dietary striation patterns, interpreted as the result of males and females consuming foods with different consistencies (Pérez-Pérez *et al.* 1999). While the striation evidence as a whole suggested an abrasive, plant food diet, this particular study is nonetheless supportive of the notion of different male and female foraging groups (although consumption of different foods might happen for other reasons) *combined* with only limited or partial sharing.

These discussions of food-sharing also highlight the time dimension – whether through carrying food to the reminder of the group or bringing the group to the meal. When this is added to the time requirements of hunting (searching, killing [potentially involving pursuit times and or waiting in ambush] and butchery) and foraging (searching and collecting) it seems likely that access to the food would more often than not occur towards the end of the day. Why is this significant? Wrangham (2009) has emphasised the greater time demands of raw food consumption, noting for example that human chewing times are just *c*. 10–20% of the equivalent great ape values – and a major factor in this difference, which in part also reflects dietary preferences, is the process of cooking. The acquisition of animal food, it seems, might have been just half the battle. The cooking of animal foods would improve their value, not necessarily because cooking enhances their nutrient benefits, but rather because the body spends less energy in processing them. This also enables them to be consumed in the relatively short daylight hours, particularly in spring, autumn and winter, available at the end of the hunting day, or even after dark in the glow of the fireside. However, this cooking hypothesis does not, as noted in Chapter 3, tally with the currently available evidence for fire. While this may be a consequence of taphonomy, Henry *et al.* (2018) have questioned the value of cooking from a costs/ benefits perspective, and Speth (2017) has explored the potential of rotten food as an alternative to cooking. Hominin survival in environments with short foraging days is not therefore clear supporting evidence for fire and cooking, although from a cultural insulation perspective fire may still be a significant technology in the Lower Palaeolithic occupation of seasonal Europe (*cf.* Stahlschmidt *et al.* 2015b).

Springtime babies?

The evidence from a variety of mammals in mid-latitude environments (Table 4.5) suggests a concentration of births in late spring and early summer. Would this also apply to Lower Palaeolithic hominins? Mussi (2007) has argued for seasonal patterns in conception and birth amongst the earliest Europeans. This is partly based in the well-known pattern that under-nourished women conceive more rarely, as well as having more pregnancy difficulties and delivering children who face more health difficulties (Mussi 2007). Specifically, female fertility is highly dependent upon energy intake, especially fat, body mass, and light (*i.e.* photoperiod). This becomes especially significant when seasonal variations in food supply are considered, potentially resulting in poor diet, stress and/or low body weight: winter amenorrhea, for example, is reported among late 19th century Inuit. Mussi (2007) favoured the late summer/early autumn, when animals were fatter, as a key period for peak conception – babies would then be born in late spring/early summer, when resources were again becoming abundant (see also Taylor 1997), although the energy-demanding 2nd and 3rd trimesters would occur in the least productive periods of winter and early spring. Mussi estimated an additional requirement of 340–450 kcal/day, with 70 g of protein/day, during those periods.

Late summer peaks in conceptions would therefore place significant additional seasonal pressures on spring and, to a lesser extent winter, food provision, especially of high-quality foods. These pressures could potentially have been met through an emphasis on medium and large-sized mammal exploitation, despite the inherent risks of hunting (time, potential failure and physical dangers). Seasonal spring births might also reduce hominin group mobility at this time of year, as is seen in a variety of herbivore and carnivore species in the BPF (Jędrzejewski *et al.* 2001; Kamler *et al.* 2008), making larger social aggregations unlikely.

A reproductive consequence of Early and, especially, Middle Pleistocene encephalisation and skeletal change concerns the mechanics of birthing. Trevathan and Rosenberg (2000; Rosenberg and Trevathan 2002) have emphasised that both cranial size and broad, rigid shoulders would impact upon hominin birthing, with a need for rotation at least of the shoulders, although they concluded that it was unclear whether rotation of the cranium would have been required prior to the late Middle Pleistocene (see also Ruff 1995). Thus, while Franciscus (2009) concluded that both the Neanderthal and modern human lineages would have had difficult births and obligate midwifery, the key question from a Lower Palaeolithic perspective is whether obligate midwifery would have extended to Neanderthals' European ancestors. In the Sima de los Huesos pelvis 1 the largest dimension of the pelvic inlet is transverse, compared to sagittal (*i.e.* anterior–posterior) in the midplane – indicating that the Sima hominin's birthing process probably involved rotating of the foetus and complicated deliveries (although the Sima pelvis is male, both sexes show twisted birth canals in modern humans; Arsuaga *et al.* 1999; but see also Weaver and Hublin 2009). This might imply habitual cooperation from adults (and adolescents?) during birth.

Complex birthing procedures have also been emphasised by Kennedy (2003), with reference to two particular aspects of Palaeolithic demography: why are so many individuals dying young and how did societies survive when so many individuals were not fulfilling their reproductive potential? While some of the young adult deaths which dominate the fossil profiles (Table 4.6) may reflect hunting traumas, and while the Sima data may be unrepresentative of the living population (Bermúdez de Castro *et al.* 2004), Kennedy particularly emphasised potential childbirth-related deaths. The age profile tallies with ages at sexual maturity, estimated as 12.5 years for African *H. erectus*, 13–16 for Asian *H. erectus*, and 13.5–16.5 for modern humans, with Neanderthals argued to be at the lower end of the modern range. These deaths might partly reflect, and lead to, an absence of experienced older midwives. Kennedy (2003) concluded by noting that the small size of human populations until the end of the Palaeolithic, argued on the basis of low genetic variability and various archaeological data (*e.g.* Bocquet-Appel *et al.* 2005; Bocquet-Appel and Degioanni 2013; Schmidt and Zimmermann 2019), might perhaps have been caused by frequent deaths of reproductive individuals, leading to bottle-necking and loss of local lineages. This would certainly seem to chime with the European evidence, in particular the highly varied character of *H. heidelbergensis* fossils and the source–sink nature of occupations (Dennell *et al.* 2011).

Childbirth and demanding infants: energetics and social costs

Aiello and Key (2002) have previously demonstrated the significant reproductive costs for large-bodied *Homo* (see also Box M), and the likelihood that hominins from *H. erectus* onwards adopted an essentially modern human model (*i.e.* reduced lactation periods and shorter inter-birth intervals) rather than an australopithecine/great ape-type model (*i.e.* longer lactation periods and wider inter-birth intervals). The first set of costs is therefore the energy demands of pregnant and lactating females and the need to make animal kills and other high energy foods available to them. Mussi (2007) highlighted that the odd human habit of hunting much larger animals than itself, in marked contrast to primates, is best explained by food sharing practices that resulted in group-wide benefits, given the lack of evidence for substantial storage. Provisioning, both immediate and possibly delayed (*e.g.* through rotten meat storage; Speth 2017), thus seems likely. This would seem to support a model of multiple small foraging groups and daily fission–fusion, as a means of maximising the dietary breadth of any provisioning, and reducing the risk associated with specific foraging failures. Mussi also argued in favour of provisioning males – as it would enable females to access additional high-quality foods and allow them to externalise some of their high repro-ductive costs, as Aiello and Key (2002) and others have also argued.

But there are further issues to be addressed. A modern human model of repro-duction (short inter-birth intervals, early weaning, long childhood, and rapid brain growth) would have required access to high-quality food resources (Kennedy 2003). Bogin and Smith (1996) characterised the dietary requirements of childhood as low in volume, but dense in energy, lipids and protein. Aiello and Key (2002) emphasised meat and tubers with respect to *H. erectus*, while Kennedy (2003) specifically stressed marrow and brain tissue: animal foods would be critical in European latitudes but underground storage organs (USOs) and other foods would also be significant (Hardy 2010). Methods of food preparation would also be important: a 'rotten meat strategy' (Speth 2017) could limit the amount of processing time required to prepare the food for weaned infants, thus partially reducing a key component of care costs (Kramer and Otárola-Castillo 2015).

Thus, early weaned infants and pregnant/lactating large-bodied females would have had specific dietary requirements, which extended throughout the seasons and probably over 2–3 years for each breast-fed infant and over 3–4 years for each weaned infant. It is also probable that the balance of breastfeeding and weaned foods during the transition to the latter fluctuated by the seasons, reflecting variations in food availability (see also Joannes-Boyau *et al.* 2019). How were these dietary needs met? The most obvious solution for Aiello and Key (2002) was extrasomatic (external to the body) nursing from mothers, in the form of food provision, carry-ing and protecting. However, while it is difficult to estimate the exact costs of such nursing, the addition of more frequent offspring results in higher daily costs, due to extrasomatic nursing overlapping with gestation and lactation. It results in a burden of significant, possibly unfeasible, energetic costs on the mothers, and thus food provision, carrying and protecting must instead be provided through nursing by others. But who provides this nursing?

This question of food provision in human evolution has often been linked, in recent times, to the grandmothering hypothesis: food-gathering by post-menopausal women to support their daughters and nieces (Hawkes *et al.* 1998), targeting those foods (as mothers also do) which children cannot acquire for themselves. This provisioning would enhance the nutritional welfare of weaned children at and immediately after the arrival of their new siblings, when their mothers were able to forage less frequently and/or effectively, improve their survival chances, and thus the grandmothers' greater longevity was selectively passed onto subsequent generations. The model appears to explain various distinctive aspects of human life history: *H. sapiens*' long lifespans after menopause (the menopause occurs at approximately 50 in modern day humans, with a range of 45–54, with little evidence for cross-cultural variation[18]); late age at sexual maturity; early weaning; and high fertility (*i.e.* short inter-birth spacing; Kennedy 2003; Robson and Wood 2008). It is noteworthy that it also occurs amongst hyenas, another highly successful social mammal (Bogin and Smith 1996). Hawkes (2016) has subsequently noted that the grandmothering contribution might be especially important in more seasonal environments, such as Europe, where the availability of foods that just-weaned children could handle was reduced.

But is this model applicable in a Lower Palaeolithic context? Our knowledge of hominin lifespan at this time is limited due to the fossil record, although the ages that are available for both *H. antecessor* (Gran Dolina TD-6) and *H. heidelbergensis* (various sites; Table 2.8) are not clearly indicative of long lives (Kennedy 2003). To what extent are these patterns robust? The demographic profile from the Sima de los Huesos does not appear to be stable, with insufficient numbers of individuals reaching 40, when compared against living hunter-gatherers (and chimpanzees; Bermúdez de Castro *et al.* 2004). It is thus likely that the Sima accumulation is anthropogenically-formed, with intentional and preferential selection of younger individuals 10–20 years old. There is also no evidence for Lower Palaeolithic burial traditions, and in a hominin society without burial, and with little clear evidence for settlement sites, the visibility and detection of older individuals is likely to be extremely low. It is also true that the ages of adults have sometimes been systematically mis-estimated, and the bones of older individuals are less likely to be preserved (Hawkes and O'Connell 2005).

Nonetheless, the more reliably aged sub-adults do dominate ancient samples, and the systematic removal of older adults from the fossil record due to bone mineral depletion is unlikely due to the distinctive bone biology of Early and Middle Pleistocene humans (Kennedy 2003). Kennedy's review of various samples from the Middle and Late Pleistocene are especially revealing: at the Sima de los Huesos (MNI=34) 56% of individuals had died before age 20, and 88% by 30, with only one female living past 30 (and one individual of unknown sex living past 40); at Zhoukoudian (n=22) there were twice as many juveniles (0–14 years) than adults (15+), and only one individual older than 50 (Table 4.6). It also seems likely that life expectancy at the Sima was typically in the 40s, a view tentatively supported by anterior dental wear rates in the sample (Bermúdez de Castro *et al.* 2003a). Neanderthal samples showed a similar picture, and Kennedy argued that the patterns were also consistent with recent, pre-contact hunter-gatherer groups:

> Long-lived individuals have, no doubt, always been present in human societies but their numbers remained too low to provide the significant and active role demanded by the GMH [Grandmothering Hypothesis]. (Kennedy 2003, 561)

However, other life history estimates *are* suggestive of early weaning, and a relatively late age at maturity: Nowell and White (2010) suggested sexual maturity at around 13 years old, and first birth at around 15–16.5 years old. This is significant because longevity and delayed maturity are strongly correlated (Aiello and Key 2002). Hawkes (2016) has also presented genetic evidence in favour of the grandmothering hypothesis: the CD33 allele, which is protective against LOAD (late-age Alzheimer's dementia), is derived in humans and occurs at levels four-fold higher than in chimpanzees. *If* protection against LOAD is the allele's main phenotypic effect then it would only have been favoured if there were fitness benefits arising from cognitive competence, such as that required by foraging or the retention of accumulated 'cultural' knowledge, in later life. The distribution and other characteristics of the CD33, and other related, alleles indicate that they evolved before the emergence of modern humans in Africa at *c*. 150–200 kya: the question from a Lower Palaeolithic perspective is how much longer before?

In light of this mixed picture, Kennedy (2003) stressed alloparenting more generally (Aiello and Key 2002), and highlighted the diverse roles of parenting, many of which do not always overlap: for example feeding, protection, nurturing and socialisation. While typically referred to as the 'grandmothering' hypothesis, it is also important to note that lifespan advantages of human females over males are relatively small: there is also scope for males to be involved in childcare in this model (Aiello and Key 2002). Contemporary variations on the grandmothering hypothesis have been explored by Bogin and Smith (1996) with reference to the Hadza alongside the extended family groups and shared childcare of the Philippine Agta and the childcare provided by fathers amongst both the Agta and the Aka of central Africa.

Significant alloparenting has also been suggested by modelling of accumulated care costs for Pleistocene mothers (Kramer and Otárola-Castillo 2015). This is an important consideration since a suggested reproductive period of *c*. 15–40 (derived from upper age estimates and growth and development data) and a suggested weaning age of 2.5–4 years old (based on enamel hypoplasia patterns; Cunha *et al.* 2004) means that a Middle Pleistocene female might have had seven children over her lifespan, although not all of these would probably survive to reproductive age, given the available demographic data (Kennedy 2003; Bermúdez de Castro *et al.* 2004). By varying inter-birth intervals, age of (foraging) independence and age of dispersal, Kramer and Otárola-Castillo's (2015) model suggested that care costs are at their greatest when intervals are short (3 years in the model) and independence is late (20 years). In such circumstances the provision of additional care (*i.e.* cooperative breeding) from other adults is likely to be needed, although costs can be partially reduced depending on levels of infant survival. An inter-birth interval of around 3 years in the European Lower Palaeolithic has been suggested by dental/weaning evidence (Cunha *et al.* 2004), although age of independence may be slightly shorter, depending on the duration of adolescence (*e.g.* Schwartz 2012). Kramer and Otárola-Castillo's model also highlighted the potential role of juveniles in provisioning, noting in particular that they can focus on and share easy-to-acquire resources (*e.g.* fruits) and close-to-camp tasks (as they tend to be less efficient long-distance bipeds than adults), thus 'freeing up' adults to focus on contrasting resources (*i.e.*

distant, hard-to-acquire items such as mobile animal prey). The model thus has interesting implications for the timing of the appearance of task specialisation, age division of labour and inter-generational cooperation between mothers and juveniles (although *cf.* Kuhn and Stiner 2006 with respect to Neanderthals). Similarly, Kramer and Russell (2015) have suggested that cooperative breeding (*e.g.* through juvenile helpers) may have occurred before breeding system changes such as a shift to monogamy/pair bonding.

This potential combination of alloparenting and extended childhood again highlights the importance of 'safe spaces' in Lower Palaeolithic habitats – the equivalent of wolf dens, for example, in which many of the associated activities (*e.g.* feeding, protection, nurturing and socialisation) could occur. While there are occasional clear 'indoor' examples (*e.g.* Arago Cave), the total number of such sites is limited, with very few outdoor examples (Chap. 3). While this might be taken as an argument against the alloparenting model, I think it more likely that such sites, utilising natural features such as large tree-throw pits, hollow trees and small rock outcrops, are missing due to their light archaeological 'footprints'. Such sites might be especially critical during a late spring/early summer birthing season.

An alternative to the grandmothering hypothesis concerns the potential role of pair-bonding and cooperative parenting (*i.e.* male provisioning of female partners and offspring). This has been highlighted as a means of enabling extended childhoods and alleviating the impacts of the markedly higher female energy costs, due to reproduction, that emerged with early *Homo*. Power and Watts (1996) highlight higher levels of male investment as a key factor in *Homo's* successful evolutionary shift to big-brained and burdensome offspring. The potential presence of pair-bonding in the European Early and Middle Pleistocene is supported by the marked reduction in sexual dimorphism in later *Homo* (McHenry 1996; Aiello and Key 2002; Arsuaga *et al.* 2015), with dimorphism in the Sima de los Huesos sample argued to be comparable to modern populations (Arsuaga *et al.* 1997a). However, Nelson *et al.* (2010) have suggested that Middle and Late Pleistocene social systems of *H. neanderthalensis* and early *H. sapiens* may have been more promiscuous than contemporary populations.

Pair-bonding and provisioning models are grounded in reliable food returns. While much older literature has strongly emphasised the importance of male-hunted animal foods in the diets of hunter-gatherer societies, these notions have been strongly criticised (Bird 1999; Hawkes *et al.* 2010), and resupported (Gurven and Hill 2009; 2010), in recent years. The ethnographic observations of Hawkes *et al.* (1998) stressed that male hunting is a relatively unreliable means of provisioning mates and offspring and proposed that males have more to gain from investing their energies into mating opportunities rather than into provisioning. Such models have therefore emphasised the role of female hunting in hunter-gatherer diets and Graves-Brown (1996) suggests a complementary contrast between stable/reliable moderate yield resources (female) and unreliable, high yield resources (male). This questions the likelihood of pair-bonded males primarily meeting the provisioning needs of partners and offspring. The recent re-evaluations of the role and importance of plant foods in early Palaeolithic diets are also significant (Bigga *et al.* 2015; Hardy *et al.* 2015; Hardy 2018), since they highlight the possibility that hunting, or at least high yield/high risk hunting, despite its archaeological visibility (*e.g.* Saladié *et al.* 2011; van Kolfschoten *et al.* 2015a; Rodriguez-Hidalgo *et al.* 2017), may have been a less prominent component of daily behaviour.

Moreover, what are the benefits to males stemming from the greater cooperation inherent in pair bonding? One possibility is increased personal fitness,[19] but this would only be the case if paternity could be certain – and this would seem unlikely if life in seasonal Europe involved regular group fission and fusion stemming, for example, from the existence of multiple task-specific foraging groups (*i.e.* dF–F). Thus, as Aiello (2007) noted, male provisioning may well have been at the group level rather than at the individual level (*i.e.* it is not primarily focusing on mates and offspring). Two further benefits of group-level provisioning have also been proposed. The first is the reduction in the inter-birth intervals (as male cooperation and provisioning of weaned young help to reduce female daily energy expenditure [DEE] costs), thus increasing the number of females available for mating at any one time. The second, related, benefit is increased mating opportunities for individual males: ethnographic studies have suggested that family/household shares of male hunting returns are actually low, with group-wide sharing occurring to signal the hunter's status (Hawkes *et al.* 2010). In this regard it is also notable that where males have more mating opportunities they tend to spend more time engaged in foraging activities that yield shareable foods (*i.e.* hunting; Hawkes 1996). Finally, Coxworth *et al.* (2015) emphasised mate-guarding in the context of changing life history: selection for longevity results in more long-lived individuals, both male and female, but produces a bias in sex ratios as older males enter the competitor pool for still-fertile females. There is thus greater competition amongst males, both old and young, for females, despite the reduced birth intervals in the modern human model. As males' average success rate declines, defending or guarding a current mate, despite its inherent costs, becomes a more viable mating strategy than seeking a succession of new mates. However, as with the paternity certainty issue above, is such mate-guarding really feasible in the seasonal environments of Europe?

Relevant to these discussions of group *vs* individual provisioning and 'proximity strategies' such as mate-guarding is the question of whether we can see evidence for social units such as 'families' or 'households' in the fine-grained sites of the Lower Palaeolithic record. The evidence is extremely limited: the strongest case has been made at Bilzingsleben, on the basis of the three semi-circular 'hut' structures (Fig. 3.10). However, there has been much debate around the taphonomy and spatial integrity of that site (see also Chap. 3). A potential challenge to the concepts of 'families' and 'households' can be found in the butchery evidence at Qesem Cave, which has been described by Stiner *et al.* (2011, 230) as 'redundant, abundant and

Box M: Being born and growing up in the Lower Palaeolithic

Why should we consider life history? Hopkinson *et al.* (2013) emphasise that life history properties impact on hominin ranging behaviour, locomotion, diet, energetic requirements, subsistence strategies, childbirth, ontogenetic development, demography, communication and technology. In short, the question should really be, is there any reason why we should not discuss life history!

Modern humans have a highly distinctive collection of life history traits and stages (see also Chap. 2): infancy (from birth to weaning), childhood (from weaning to the eruption of the first permanent molar [M1] tooth – which occurs between 5.5 and 6.5 years on average), juvenile (from the M1 eruption to the onset of puberty; with brain growth complete by age 7 on average), adolescence (marked by puberty and then the adolescent growth-spurt) and adulthood (with early adulthood marked by the end of the growth spurt, the attainment of adult stature, and the achievement of full reproductive maturity) (Bogin and Smith 1996). After age 7 a modern human child becomes more capable of processing an adult-type diet, while their nutritional demands for brain and body growth diminish. This might have interesting implications for the participation of Palaeolithic children in more energetic activities such as animal foraging. By comparison, living primate life histories contain just three stages: infant, juvenile and adult, a model which has also been inferred for australopithecines (Bogin and Smith 1996): childhood is conspicuous by its absence. Yet as Hopkinson *et al.* (2013) have emphasised, the wider benefits of childhood are considerable: reduced inter-birth spacing (as a result of early weaning); additional years of slow growth (but *cf.* Robson and Wood 2008), enabling expanded behavioural experiences that enhance developmental plasticity and offer opportunities to practice technology, social organisation and other aspects of culture; and childhood has been demonstrated, amongst mammals with a juvenile phase, to favour survival. So when does childhood first appear?

Thankfully there is a rich body of evidence and data, from theoretical, skeletal and living biological sources, that enables us to discuss models of life history in extinct hominins. Of particular importance is the evidence for strong correlations between a range of life history traits (LHVs and LHRVs; see also Chap. 2): brain size, body size, age at sexual maturity, age at first birth, gestational length, lifespan and dental development (Kennedy 2003). This last trait, from a palaeoanthropological perspective, is particularly significant (Dean *et al.* 2001). Comparisons with apes highlight two basic models in brain growth. Ape brains grow rapidly before birth and relatively slowly thereafter (reflecting their relatively small adult size). Human brains, in contrast, grow rapidly both before and after birth, combined with a relatively large neonatal brain size. When considering the type of model appropriate to extinct hominins, Bogin and Smith (1996) have argued that an adult brain size of up to 850 cc is achievable simply by lengthening the fetal stage of growth, while maintaining slow rates of postnatal brain growth: *i.e.* an ape model. By contrast adult brain sizes above 850 cc can only be achieved by adopting rapid postnatal brain growth and slow body growth: *i.e.* a human model. If slow postnatal brain growth was maintained by larger brained hominins, and all the emphasis was placed on fetal growth, then the baby's head would simply be too large to be born. It was therefore proposed by Bogin and Smith (1996) that a new, 'human', model brings a childhood stage into hominin life history for the first time. This would be a time of rapid brain growth, with the associated provision of high-quality foods, while there would be an associated decline in the length of the infancy stage. They rejected the idea of an adolescence stage for early *H. erectus*, based on the advanced physical development, for his age, of the Nariokotome Boy. However, they do propose such a stage for later *H. erectus*, because of the species' complex behaviours – adolescence is clearly seen as a key period of learning in their model.

So what model of reproduction is likely to have existed in the European Lower Palaeolithic? Returning to life history traits, the correlation of longevity and delayed maturity reflects fundamental selection pressures: a delayed age of first reproduction and a long period of growth, resulting in a large adult body, is only justifiable if adult mortality is very low and if life spans are very long (Aiello and Key 2002). However, these large body sizes bring both costs and benefits. Larger mothers can deliver more energy to their offspring, but those larger bodies also require more energy to maintain basic functions. Aiello and Key (2002) have estimated that the changes in the body mass of *H. erectus* females led to increases of 30–40% in daily energy expenditure (DEE) compared to an average australopithecine value. Using their methodology and the body mass estimates from the Sima de los Huesos (Chap. 2), DEE increases of *c.* 58% are suggested for *H. heidelbergensis*. Moreover, DEE is further increased by gestation (typically by 20–30% in mammals) and lactation (Aiello and Key (2002) suggested a value of 37% for *H. erectus*, although this may be conservative). The DEE requirements for *H. erectus* females are therefore significantly higher than those for the australopithecines, and would increase further in later *Homo* species – and if an ape-like model of reproduction continued to be used, then the *Homo* females would be making a considerably higher investment in each child. Great apes wean their young at between *c.* 6–8 years, when the infant is *c.* 6.7 × their birth weight on average, and close to self-sufficiency (Kennedy 2003). An alternative is a modern human model of reproduction: a reduced lactation period (36 months on average in pre-industrialised societies, when infants are *c.* 2.7 × their birth weight on average; Bogin and Smith 1996; Kennedy 2003), and shorter inter-birth intervals. If this model is used, there are two advantages: an increase in the number of offspring per fixed time period and reduced energetic costs per offspring (Aiello and Key 2002). In short, humans have the potential for greater lifetime fertility than any ape: 'the "bottom line", in a biological sense, is that the evolution of human childhood frees the mother from the demands of nursing and the inhibition of ovulation related to continuous nursing. This, in turn, decreases the inter-birth interval and increases reproductive fitness' (Bogin and Smith 1996, 709). However, Hublin *et al.* (2015) have highlighted the mosaic nature of life history evolution, and the dangers of simple dichotomies between 'ape' and 'human' models, demonstrated particularly strongly with reference to *H. erectus*.

Dental data has provided a key perspective on changing life history, as in Dean *et al.* (2001), which focused on the australopithecines and early *Homo* (*i.e. H. habilis*), *H. erectus* (*sensu lato*), the Neanderthals and *H. sapiens*. Their conclusions were that enamel growth rates in the australopithecines and early *Homo* resembled apes (both fossil and living), that *H. erectus* (*s. lato*) patterns were also shorter than modern humans and that modern dental development might be a late development, perhaps associated with the Neanderthals. Similarly, Schwartz (2012) has argued that the M1 emergence age estimates for the Nariokotome Boy (KNM-WT 15000) fall between those for chimpanzees (3.0–3.5 years) and humans (4.7–7.0 years), but that the suggested age at death (8 years old) and large brain size suggest rapid maturation and a more ape-like model of growth and development. Set against this background, is there any evidence for childhood and adolescence, and a truly human model of growth and development, in the European Lower Palaeolithic? Of critical importance in beginning to resolve this question has been the *H. heidelbergensis* and *H. antecessor* fossil material from the Sima de los Huesos and Gran Dolina localities at Atapuerca (Carbonell *et al.* 1995; Arsuaga *et al.* 1997b; Bermúdez de Castro *et al.* 2017). Analysis of the dental material from TD-6 hominins 1 & 3 and Sima hominin 18 suggests that both European species are similar to modern humans in their dental development, implying both prolonged maturation and new life history stages (childhood and adolescence; Bermúdez de Castro *et al.* 2003b), although the patterns are not as derived in some respects and these new stages may not be quite as long as in modern humans (*e.g.* delayed M3 calcification; Bermúdez de Castro *et al.* 1999). By extension that would also suggest relatively short birth intervals, high rates of postnatal brain growth, extended offspring dependency, marked adolescent growth spurt, and delayed reproductive cycles.

Overall, these various models and data strongly suggest that European Lower Palaeolithic hominins, whether *H. antecessor* or *H. heidelbergensis*, are likely to have been characterised by a broadly human model of reproduction and growth and development, characterised by increasing brain size (Chap. 2) and the birth of increasingly helpless (secondarily altricial) infants. It is therefore clear that early life was significantly demanding for Palaeolithic *Homo*, for adult females, infants, and perhaps adult males and older siblings too.

heavy-handed', and as possibly indicative of the involvement of multiple individuals and more individually-focused approaches to feeding on shared resources. Thus, while pair-bonding may have existed, the limited evidence for living structures and/ or spatially differentiated living spaces makes it harder to envisage a clear distinction between the 'family' and the group. In such a circumstance, it seems likely that provisioning and sharing operated at the group level.

Either way, whether through male investment or grandmothering/alloparenting, such provisioning enables the gradual learning of skills and knowledge combined with physical development, resulting in productive foraging returns across the relatively long adult lifespan, although this was probably shorter than for contemporary *H. sapiens* (Kaplan *et al.* 2007; Table 4.6; Chap. 5). A further potential consequence of food sharing by adults is that it is likely to encourage sharing in juveniles (Kramer and Otárola-Castillo 2015), potentially generating another source of alloparenting. However, it is perhaps also important not to mistakenly give an impression of an 'alloparenting and provisioning idyll': this would make Lower Palaeolithic hominins unlike both our closest extant cousins and the vast majority of 'humanity', both past and present. I have no doubt that there was both inter- and intra-group conflict on occasions, and the Sima de los Huesos material provides examples of this (Sala *et al.* 2015; 2016), as does the TD-6.2 cannibalism (Carbonell *et al.* 2010; Saladié *et al.* 2012).

Daily living: local spring lives?

Residential moves in early spring were probably driven by food resource mobility and availability, and hominin choices, not least because different animals are likely to have been found in different topographic settings. However, a possible spring tension might have revolved around the need to monitor animal movements (*e.g.* Nunamiut men travel widely in spring to search out moving caribou; Kelly 1995) against the need, at least at a group level if not at a pair-bonded level, to adequately provision pregnant (3rd trimester)/lactating females. Moreover, since female reproductive success requires continuous and reliable nutrition (*e.g.* Graves-Brown 1996), might females have strongly influenced the timing, distances and 'destinations' of spring-time residential moves in particular? Residential groups were presumably also small in spring, due to both the mobility costs of aggregation, and the need for high quality food provision (there would thus be a clear disadvantage in large numbers of individuals exploiting a local area).

Other than larger-scale movements associated with monitoring/tracking game, spring life may well have been a relatively localised affair. An interesting perspective comes from the behaviour of breeding female wolves in the BPF, observed in 1994–96. Their ranges reduced during the 2 months prior to birth and, during birth and early nursing in May–June, they confined their activities to an average of 17 km^2, just 10% or less of their mean home range (Okarma *et al.* 1998). While hominin births may not have been so seasonally constrained (Mussi 2007), and while foraging group (≈ wolf pack) sizes, suckling/breastfeeding durations and child carrying options were clearly different, the wolf data nonetheless offer an interesting perspective on the scale of the reduction of a social carnivore's range, as a result of the infants' high suckling demands, combined with their low mobility and vulnerability to other predators.

Wolves also offer an interesting alternative perspective on the use of, and mobility within, the home range. As Jędrzejewski *et al.* (2001) note, wolves move for three fundamental reasons: to search for and kill prey; to mark their territories; and to join other pack members (after a temporary separation) at dens, kills or other places. Both the first and third of these reasons would certainly be priorities for Lower Palaeolithic hominins too. In the spring–summer, wolf movements in the BPF were concentrated around their breeding den and rendezvous sites and the areas used on consecutive days overlapped extensively: in other words, there were slow increases in the cumulative areas covered by the wolves' movements over successive days. In the case of the wolves it is unsurprising that this strategy coincides with the peak seasonal abundance of red deer and wild boar young. Might such constrained home ranges have characterised the spring and early summer lives of Lower Palaeolithic hominins too? If this was the case, then spring residential sites, and their core foraging areas, would likely have been carefully selected, to try and minimise the danger of local resources becoming depleted too rapidly.

Two further complicating factors associated with residential mobility would be the movement and maintenance of fire and the portability of any shelters (in whole or in part). While there are various effective materials which can be used to move glowing embers (*e.g.* the polypore fungi *F. fomentarius*[20]), the mean distances collated for subarctic and continental mid-latitude forest hunter-gatherers (North America and Asia) by Binford imply travel times of at least 1–2 days (see Table 5.11). Regular residential mobility would also argue against significant short-term fuel stockpiling (*cf.* Twomey 2013), as the unburned fuel would either have to be transported or abandoned (although 'sites' could perhaps be returned to over longer intervals, as has been argued for the Middle Palaeolithic using the concept of persistent places; *e.g.* Shaw *et al.* 2016; Pope *et al.* 2018). This may be a further argument in favour of a sporadic, rather than habitual, reliance on fire, although the condition of fuel would generally improve as spring progressed. With regards to shelters, MacDonald (2018) highlights the energetic costs of transporting any covering hides or structural elements (*e.g.* branches), although the latter could presumably be replaced by local vegetation.

What all the above highlight is that mobility, whether in the form of residential moves or logistical foraging, was never a leisurely walk in the woods. As well as monitoring the signs of resources, animal or plant, keeping watch for predators, assessing the weather, individuals must surely have been logging 'signposts' in the landscape (*e.g.* springs, stream confluences, distinctive trees or rocks, particular viewpoints) – the means by which they could re-find other members of the group at the end of the day and 'close' the daily fission–fusion circle. These are, as MacDonald (2007) has argued specifically for hunting, cognitively demanding tasks.

Given the relatively shorter days of early spring (see Table 1.1), it is likely that much of the available time was dedicated to meeting dietary needs. However other activities would still have occurred. One of the most significant may well have been antler collection. This material, known as a tool resource from artefacts such as the red deer antler soft hammers at Boxgrove (Pitts and Roberts 1997; Wenban-Smith 1999), would have been most easily acquired when shed. In the case of red deer stags this typically occurs at the end of winter, suggesting that late winter and early spring, perhaps February–April, was a key period for antler collection. Such activities have been suggested at Soucy 3 (level P), where c. 40 shed antlers of red and fallow deer were apparently collected (Lhomme 2007). This is a task that could potentially be assigned to older children, with antlers averaging around 1 kg in weight, although the largest modern red deer stags produce antlers up to 15–20 kg (Geist 1998). The exploitation of antler has been explored for later periods, and Tejero *et al.* (2012) have demonstrated that antler segments can be obtained through bifacial cutting and percussion, with handheld cores and large flakes used as percussion tools. Such tools were certainly available in the Lower Palaeolithic. Perhaps more uncertain is the question of when such materials would have been processed. Any processing would have been time-consuming and presumably required a secure space in which to work. Tejero *et al.* took 15–25 minutes to cut and remove each element (a tine or point) from medium-sized red deer antlers (with beam widths ranging between 30–48 mm). Perhaps such activities were delayed until the longer and warmer days of late spring or even early summer, although it seems likely that processing would be prioritised prior to residential moves, in order to minimise the weight of transported antler. This presumes that antler soft hammers, and perhaps other artefacts, were curated, an argument supported both by the time investment in their production and by the wear evidence of sustained use documented at Boxgrove (Pitts and Roberts 1997).

Spring: breathing new life ...

Spring highlights many of the challenges faced by Lower Palaeolithic hominins in Europe, and the tensions in possible survival strategies. New foods (migrating animals, newborns, fresh plant growth) would become available, but their exploitation would need to both meet the high-quality food requirements of a relatively highly encephalised hominin species and its slow growing altricial young, while limiting the residential mobility costs experienced by late-pregnancy females (Fig. 4.8). Cognitively-demanding fission–fusion, provisioning, and some form of labour division, would seem to be necessary parts of any solution, and the effectiveness of those strategies in extracting resources from highly seasonal landscapes may have been a key factor behind the changing distributions of *Homo* between the Early and Middle Pleistocene.

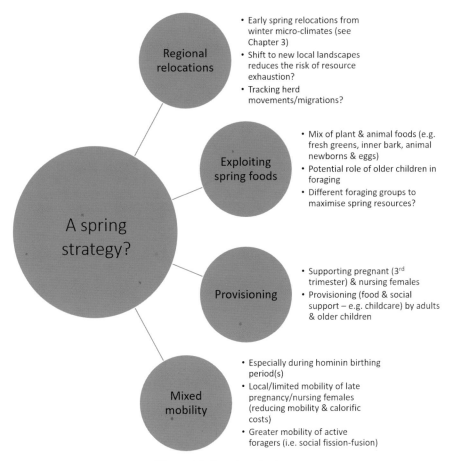

Figure 4.8: A spring strategy.

Notes

1. Current March and May days are respectively *c.* 2–3 and *c.* 5–7 hours longer than January days, with larger differences at more northern latitudes (*e.g.* London: +3 hr [March] and +7 hr [May]; compared to Madrid: +2 hr [March] and +5 hr [May]) (data source: https://www.gaisma.com/en/).
2. Primary biomass is inversely correlated to edible plant foods because humans eat the reproductive parts of plants (*e.g.* nuts and seeds) or stored carbohydrates (*e.g.* tubers): where primary biomass is high plants invest more energy in structural maintenance and sunlight capture, whereas where primary biomass is low, plants invest in reproductive tissue (Kelly 1995).
3. Residential moves could also have served as foraging trips at the same time (Kelly 1995).
4. This would be significant if, as argued in Chapter 3, clothing and other forms of hide use (*e.g.* as shelter coverings) were a component of Lower Palaeolithic behaviour in Europe.
5. In this study summer was defined as April–July, and the red deer data is therefore discussed in Chapter 4 in the context of birthing (Kamler *et al.* 2008).
6. These species are typically active for 7–8 months, approximately between early spring and late November, and are known from various sites including Gran Dolina (Blain *et al.* 2009).

7. None of these discussions should be used as a 'safety guide' with regards to the picking and consumption of wild plants. Any readers wishing to do so are strongly recommended to consult an appropriate, dedicated guidebook, such as Mabey (2012).
8. The argument that the machairodonts were unable to fully consume the flesh on carcasses (*e.g.* Turner 1992) is also significant because it further complicates the interpretation of cut-mark data with regards to questions of primary or secondary access.
9. This extinction was probably due to being out-competed by the incoming modern African carnivores, including the spotted hyaena *C. crocuta* (Turner 1992; Kurtén 1998, 238–242; Arribas and Palmqvist 1999).
10. It is also possible that early hominin populations in Europe were very small and widely dispersed, although this has further implications for how viability was maintained.
11. The DOHaD hypothesis argues that good maternal health is critical to disease prevention during the later life of their offspring, since malnutrition during the fetal and early years stages impacts gene expression and developmental plasticity. Good maternal health spans conception (as the developing foetus relies on the mother's stores of protein and fat), pregnancy (as a female is born with all the eggs she will ever release), and breastfeeding (since malnutrition impacts on the infant's growth), and the consequences of poor nutrition are therefore transferred across the generations (Lewis 2018).
12. !Kung children's crying is also less prolonged than amongst western children, perhaps because of !Kung society's continuous or near-continuous carrying habits (Lummaa *et al.* 1998).
13. Woodburn (1968, 54) observed that 'as many as half of the adult [Hadza] men fail to kill even one large animal a year', although it is unclear what proportion of these men never attempted to hunt large animals.
14. Domínguez-Rodrigo (2002) has suggested that food transported to communal sites must have been high yielding, to avoid social stress – an interesting, if indirect, measure of foraging efficiency at sites with evidence of carcass transport such as Gran Dolina TD-6.2 (Saladié *et al.* 2011).
15. For this reason I am unsurprised by the recent critique of the claimed hearths on the Schöningen lake-shore (Stahlschmidt *et al.* 2015b): a kill-butchery habitat, with its associated attractions to scavengers, seems an odd location for fire-making and cooking.
16. Chyme is the pulpy, semi-fluid mass of partly-digested food which passes from the stomach to the small intestine.
17. If the non-hunting group was located at a woodland residential site there would also be additional potential benefits of warmth and security (see also Chap. 3).
18. This contrasts markedly with living primates: for example, where menopause is reported in the chimpanzee it occurs at roughly the same age as in modern women, but is associated with a lifespan of *c.* 45 years (Kennedy 2003; Robson and Wood 2008).
19. Personal fitness refers to the number of offspring that an individual begets, regardless of who rears and/or supports them.
20. Jonny Crockett (Survival School; https://www.survivalschool.co.uk/jonny-crockett/) has observed that 'cold' charcoal is an excellent fire-lighting material and an easily portable resource (Crockett pers. comm.).

Chapter 5

Summertime ... was the living easy?

Hot days ... and dry days?

Summers would of course encompass the annual peak in vegetation growth and cover, increasing from late spring onwards, with implications for hominin and other terrestrial animals' mobility, in terms of both ease and speed, lines of sight and the visibility of resources. Data on summer temperatures, like winter (Chap. 3), is generated by the mutual climate range method and similar approaches. The maximum summer temperatures associated with European Lower Palaeolithic hominin sites are typically between the high teens and low 20s °C (Table 5.1). Variations are again both chronological and spatial. Chronological variations appear to mainly reflect sub-stage variations: the relatively low summer temperatures at a number of British MIS 13 sites (*e.g.* High Lodge and Happisburgh I) reflect their position late in the warm stage, while MIS 11 summer estimates from Poland vary markedly across the full extent of the stage (Table 5.2). Comparison of the Early and Middle Pleistocene sites perhaps also suggests an increasing seasonality through time, as measured through summer/winter temperature contrasts (Table 2.3), although the site-specific ranges are large.

The other key trends are geographical: in general site summer temperatures associated with hominin occupations were highest in the Mediterranean (*e.g.* Gran Dolina and Aridos) and the continental interior (*e.g.* Bilzingsleben). Palynological and isotope studies from five lakes in eastern Poland (the Ortel Królewski, Hrud, Ossówka, Roskosz and Szymanowo) also provide an interesting continental contrast to the Atlantic-type conditions of Britain, with slightly warmer summer conditions in central Europe (Table 5.3). Mean July temperatures increased from 13°C to 16°C across the early stages of the MIS 11 warm stage, peaking at 20–22°C in the optimum phase (the present-day value is 18.3°C), with an accompanying shift from boreal to mixed forest and back to boreal conditions (Szymanek *et al.* 2016).

In light of these summer conditions, the potential range of benefits of controlled fires would be reduced (see also Chap. 3), with less need for artificial light to extend the day (Table 1.1) and less need for artificial heat to raise core temperatures. Adopting a cost/benefit analysis approach (Henry *et al.* 2018) Lower Palaeolithic lives during this season may therefore have been predominantly fire-free and focused on a raw food diet. Summer/winter contrasts might also apply to clothing: Chapter 3 emphasised the

Table 5.1: Selected summer temperature estimates (T_{max}) for European Lower Palaeolithic sites

Site	T_{max} (°C)
Early Pleistocene	
Barranco León D	+26.2
Fuente Nueva-3	+24.3
Sima del Elefante (Level TE9c)	+20.5
Happisburgh III (Bed E)	+16 – +18
Gran Dolina (TD-6.2)	+22.0
early Middle Pleistocene	
Pakefield (Bed Cii–Ciii)	+17 – +23
Cúllar Baza 1	+21 – +27
Boxgrove (Unit 4c & Freshwater Silt Bed ≈ Units 4b & 4c)	+15 – +20
Happisburgh I (Organic Mud)	+12 – +15
High Lodge (Bed C1)	+15 – +16
Waverley Wood (Channel 2, Organic Mud)	+10 – +15
Brooksby (Redland's Brooksby Channel)	+15 – +16
later Middle Pleistocene	
Barnham (Unit 5c; HoII)	+17 – +18
Hoxne (Stratum D; HoIIIa)	+15 – +19
Bilzingsleben II	+20 – +25
Aridos I	+20 – +28
Gran Dolina (TD-10 [sub-level T1])	+16 – +22
Schöningen 13 II-4	+16

See Table 2.3 for winter temperatures, temperature data sources, site ages and references

Table 5.2: Polish summer temperatures across the MIS 11 succession (Szymanek 2017)

MIS 11 Sub-stage	Vegetation zone	July temperatures (°C)
Pre-optimum	*Betula-Pinus*	+12 – +14
	Picea-Alnus	+16 – +19
	Taxus	+19 – +21
Climatic optimum	*Carpinus-Abies*	+20 – +22
Post-optimum	*Picea-Pinus-Pterocarya* &	+17 – +19
End of warm stage	*Pinus-Juniperus*	+15 – +17

See also Table 3.2

Table 5.3: Summer temperature estimates for selected non-hominin sites from central Europe

Site	Age (MIS)	Intra-stage phase	Proxies	July temperatures (°C)
Ferdynandów, Zdany & Łuków (Poland)	15	Not specified	Pollen, plant macrofossils	+18
	14	Not specified		+16
	13	Not specified		+19
Bilshausen (Germany)	13	Beginning of warm stage	Pollen	+15
		Interglacial maximum		> +17
		End of warm stage		+14
Dethlingen, Hetendorf & Munster-Breloh (Germany)	11	Beginning of warm stage	Pollen, diatoms	+16
		Interglacial maximum		+20/21
		End of warm stage		+17
Ossówka, Woskrzenice, Kaliłów & Wilczyn (Poland)	11	Early warm stage	Pollen, diatoms	+12 – +14
		Interglacial maximum		+20 – +22
		Late warm stage		+17 – +19
		End of warm stage		+15 – +17

data from Szymanek and Julien (2018); see also Table 3.3

potential need for it in light of winter cold. However the highest summer temperatures, of central and southern Europe in particular (*c.* 20–25°C; Table 5.1), fall only just below the lower critical temperature (LCT) estimates for *H. heidelbergensis* (adjusted for elevated BMR and enhanced muscularity; Table 3.6). Moreover, those estimates for Early and Middle Pleistocene environments are presented in the form of monthly averages – maximum temperatures may have been significantly higher. As a point of comparison, data from the weather station near Bilzingsleben has recorded an average monthly air temperature (at 2 m) of 17.5°C for July (1951–2017), but the average of absolute maximum July air temperatures over the same period is 30.8°C, with a highest individual value of 36.5°C. The summer might therefore have seen a marked reduction in the use of clothes, either in whole or in part, particularly during the daytime and on the warmest days. This would have obvious parallels with those many other animals which have summer and winter coats, and if hominins were reliant on body hair rather than clothing a similar response might have occurred (see also Chap. 3).

Water ... and wildfires?

These discussions of higher temperatures and possibly reduced fire and clothing needs might seem to imply that summers were a period of relatively reduced stress for hominins. But was that really the case? Although the discussions of climatic stress have so far been primarily concerned with coping with winter cold (Chap. 3), the impacts of summer heat and aridity may well have been a further significant challenge for hominins, particularly in the Mediterranean and the continental interior (Table 5.4). The key characteristic of Mediterranean vegetation is an ability to withstand long and drastic summer dryness (Moncel *et al.* 2018) and such conditions, and the need for water access, must also have impacted on hominins, particularly in their choice of local foraging landscapes and sleeping sites. The potential significance of this particular challenge is evident at Isernia la Pineta (MIS 15/14), where the pasture-rich, open woodland-steppe vegetation is associated with an essentially bi-seasonal climate: a long and arid 'dry' season and a short 'wet' season in which the annual rainfall is concentrated (Moncel *et al.* 2018). Blain *et al.* (2014) also identified a likely strong summer drought period at the site of Aridos 1 during the MIS 11 climatic optimum (probably MIS 11c), while a shift towards more arid interglacial conditions is evident after MIS 16 in the Tenaghi Philippon sequence in Greece (Tzedakis *et al.* 2006). 'Mediterranean' conditions (*i.e.* strongly seasonal precipitation with warm, dry summers) have also been suggested for the Pakefield 'rootlet' bed (Candy *et al.* 2006) in the UK. More generally, the problems of summer aridity in the Mediterranean and continental zones (probably in both warm and glacial stages in the former) would have been further enhanced by the increased seasonality of the Middle Pleistocene (Table 2.3). Summer occupations may therefore have sometimes been just as 'tethered' as those of the winter (Chap. 3), but with water (springs, streams, rivers, lakes), rather than food sources, being the most significant structuring factor, perhaps especially in the south and the east. The need for natural or artificial shelter in the heat of the day might also have been a factor, at least in more mosaic, open landscapes, and periods of peak activity might therefore have been concentrated into the early morning and late afternoon. Finally, extreme

Table 5.4: Summer precipitation data for selected Spanish Early and Middle Pleistocene sites (Blain et al. 2013; 2014; 2016)

Site	Month	Precipitation (mm)		
		Mean	SD	Range
	Early Pleistocene			
Fuente Nueva-3	June	22.0	15.0	10–60
	July	13.0	6.0	10–30
	August	14.0	9.0	10–40
Gran Dolina TD-6.2	June	67.5	4.6	-
	July	48.8	3.5	-
	August	60.0	0.0	-
	late Middle Pleistocene			
Aridos I	June	25.0	15.0	10–50
	July	10.6	2.0	10–20
	August	12.6	4.0	10–20

summer heat may have been as much of a threat to vulnerable individuals (*e.g.* the very young and very old) as extreme winter cold.

A further consequence of aridity would be its impact upon prey ranges: among semi-wild modern horses in Britain fresh water supply is one of four key factors impacting on range sizes (alongside grazing, shelter and 'shade' availability; Corbet and Harris 1991). Increasingly large prey ranges, with implications for search times, would presumably impact significantly on hominin dietary and mobility strategies and might even be a constraint on hominin survival in relatively dry summer landscapes (the possiblity of larger hominin ranges during arid periods has also been suggested in the broader context of initial dispersals and colonisations beyond Africa; Dennell 2003).

However, once again there is local variability amongst these regional trends. The precipitation pattern at Gran Dolina (Blain *et al.* 2012) was both wetter and more homogeneous than the present day, with increased summer rainfall and less winter precipitation indicated by the Pleistocene data. This might have favoured year-around, if not necessarily continuous, hominin occupations, and been a key factor in their sustained presence at Atapuerca during much of the late Early and Middle Pleistocene (although the general trend was towards drier conditions after the Mid-Brunhes Event [*c.* 450 kya]).

A further summer aspect that might have been of particular significance to hominins is wildfires. These are characteristic of the hot, dry regimes associated with continental climates, and would have been an obvious threat to hominin (and most other) life. However, the aftermath of wildfires would also have been a significant resource: a source of fire, access to charred foods (*e.g.* carcasses of large animals, but also other items such as suffocated burrowing animals; Clark and Harris 1985, fig 11), and a mechanism for habitat regeneration (*e.g.* Niklasson *et al.* 2010), with significant and varying impacts upon resource availability (Nelson *et al.* 2008).

Summer foods

Assessing wild plant food availability in the Early and Middle Pleistocene is generally difficult, for a number of reasons already discussed (Chap. 4). A further complication is that the availability of wild greens has bloomed since the spread of agriculture in Europe in the mid-Holocene, due to the relationships between wild weedy greens and agricultural crops (Leonti *et al.* 2006). This in turn has shifted the wild food balance from tree-based starchy resources (*i.e.* nuts) towards wild greens.

It is therefore likely, at least during the peak interglacials, that within dense woody ecosystems, starchy tree resources such as acorns and nuts (Chap. 6), and perhaps also mushrooms, were a more significant wild plant food resource than is evident in more recent times. While mushrooms are not a high-energy plant food, due to their low lipid and dry matter content, they are a good source of protein, carbohydrate and selected vitamins and minerals (*e.g.* Kalač 2009; Beluhan and Ranogajec 2011; Heleno *et al.* 2015). Moreover, these characteristics are consistent across a wide variety of modern species, from Iberia to Finland and Croatia to Poland, relevant both to the geography of Lower Palaeolithic occupations and to habitat change across warm stage intervals (*e.g.* from deciduous to coniferous woods). Their medicinal benefits (Ferreira *et al.* 2009; Wani *et al.* 2010) are similarly diverse: a study of selected species of wild mushrooms from north-east Portugal, including those associated with both oak (*e.g.* Golden coral) and pine habitats (*e.g.* Common funnel cap), highlighted their significant potential as a medicinal food, with their benefits including anti-oxidant, anti-inflammatory and anti-viral properties (Pereira *et al.* 2012). However, many other plants also have antiseptic or painkilling effects when parts of them are eaten or chewed (Bigga *et al.* 2015, table 1): *e.g.* alder bark (astringent); common bearberry berries (astringent, antiseptic); birch bark (astringent, antiseptic); hornwort leaves (stings); Scot's pine inner bark (antiseptic). As Bigga *et al.* observed, hominins were likely to face wounds, gastrointestinal disease and different parasites in their daily lives, and would certainly have had a need for natural medicines (Spikins *et al.* 2019). Whether they were aware of them is more difficult to say, but certainly they were present in the environment.

It is likely that, if consumed, mushrooms and other fungi were collected and eaten rapidly, given their short shelf-life due to high water content (Kalač 2009), and they may therefore have been eaten 'on the go'. Although many mushroom species are available between late summer and winter, others appear in spring and still others can survive though winters (Table 5.5), making them a potentially year-round, if

Table 5.5: Seasonality in the availability of modern British mushrooms (Mabey 2012)

Season	Species (incl. common names)
Winter (December–February)	C. cibarius (Chanterelle); F. velutipes (Velvet shank); L. saeva (Field Blewit); L. nuda (Wood Blewit); M. oreades (Fairy-ring champignon); P. ostreatus (Oyster mushroom)
Spring (March–May)	M. esculenta (Morel); F. velutipes (Velvet shank); M. oreades (Fairy-ring champignon); T. gambosum (St. George's mushroom)
Summer (June–August)	A. arvensis (Horse mushroom); A. campestris (Field mushroom); B. edulis (Cep); C. cibarius (Chanterelle); C. comatus (Shaggy cap); H. repandum (Hedgehog fungus); L. gigantea (Giant puffball); L. procera (Parasol mushroom); M. oreades (Fairy-ring champignon); S. crispa (Cauliflower fungus); T. gambosum (St. George's mushroom)
Autumn (September–November)	A. arvensis (Horse mushroom); A. campestris (Field mushroom); A. auricula-judae (Jew's Ear); B. edulis (Cep); C. cibarius (Chanterelle); C. comatus (Shaggy cap); F. hepatica (Beefsteak fungus); F. velutipes (Velvet shank); H. repandum (Hedgehog fungus); L. gigantea (Giant puffball); L. procera (Parasol mushroom); L. saeva (Field Blewit); L. nuda (Wood Blewit); M. oreades (Fairy-ring champignon); P. ostreatus (Oyster mushroom); S. crispa (Cauliflower fungus); T. gambosum (St. George's mushroom)

Species noted as not to be eaten raw by Mabey have been excluded. None of the above information should be used as a 'safety guide' with regards to the picking and consumption of wild fungi. Any readers wishing to do so are strongly recommended to consult an appropriate, dedicated guidebook, such as Mabey (2012) varying, component of a hominin diet rather than a purely summer and autumn food (Beluhan and Ranogajec 2011). Mabey (2012) emphasised that younger specimens are more palatable, and cooking typically improves taste, texture and nutritional benefits (Manzi *et al.* 2001). The latter point raises question marks over their dietary significance, in light of the mixed evidence for fire and cooking. Mabey also noted that some species (*e.g.* Wood blewit) can be indigestible when eaten raw, although it is obviously difficult to know if Lower Palaeolithic digestive systems may have been more amenable to them.

Discussion of mushrooms and other fungi inevitably raises the concern of poisonous species. While the proportion of the latter category is small (Mabey [2012] noted that only 20-odd of the 3000+ species of British large-bodied fungi are 'seriously poisonous'), it is true that the identification of particular species can be challenging, because of mushrooms' relatively few differentiating characteristics and intra-species variability.[1] This connects to the issue of how botanical knowledge was acquired and passed on, and whether language, physical corrections (*e.g.* preventing a child from picking a poisonous species, or taking an already-picked specimen from them and discarding it) or simply behavioural conservatism (*i.e.* only picking what you see others picking) was required.

A potential summer food that is sometimes overlooked in discussions of hunter-gatherer foraging is honey. Crittenden (2011) has highlighted its value as an energy-dense food, with 80–95% sugar in liquid honey, and additional sources of protein, fat, essential minerals and B-vitamins in bee larvae. Crittenden's review of ethnographic and primate exploitation of honey highlighted tools and technologies well within the compass of Lower Palaeolithic hominins: chimpanzees and orang-utans utilise stick tools and vegetation probes respectively, while the Worora, Wunambal and Ngarinjin tribes of western Australia utilise digging sticks, hammerstones and stone hatchets to access bee hives – only the last item would be exceptional. What is unclear is how risk, in terms of bee stings, could be minimised, although exposure of the vulnerable, especially the young, could be minimised by transporting honey, and the associated larvae, in large eggshells (Crittenden 2011). It would be a highly seasonal European food, with shorter production seasons than in the tropics, but nonetheless a potentially valuable addition to the dietary range during the summer and early autumn months (although its contribution to modern hunter-gatherers' energy intakes has also been argued to be modest; Eaton *et al.* 1997). If used it could have been an easily digestible and high energy food source for weaned young. Such foods, and other 'soft' dietary items such as semi-digested chyme from guts (Buck *et al.* 2016), might also be a valuable food source for older individuals (but *cf.* Bermúdez de Castro *et al.* 2003a).

There were also many other types of plant foods available in the summer months. The exceptional palaeoenvironmental evidence from Schöningen highlights a wide range of potential foods, including berries, seeds, stems and leaves and roots (Table 5.6), although the latter might best have been utilised from late

Table 5.6: Dietary potential, and seasonality data, for selected plant species available for summer collection and recorded at Schöningen

Botanical name	Common name	Edible parts	Nutritional & other features	Seasonal availability	Ecology
A. tripolium	Sea aster	Leaves, stems	–	April–October	Moist soils and saltmarshes (tolerates saline environments)
A. uva-ursi	Common bearberry	Berries	Vitamins A & C; carbohydrates	August–Winter (fruit)	Open pine woods & heaths
A. prostrata	Spear-leaved orache	Leaves, seeds	–	July–October (flowers)	Well-drained, dry or moist soils
C. demersum	Hornwort	Leaves	–	Young plant growth begins in spring; blooms in summer	Eutrophic, warm lakes
C. album	Fat hen	Flowers, leaves, seeds	Vitamins A & C; contains oxalate	June–October (flowers)	Pioneer plant; incl. lakeshores
C. avellana	Hazel	Nuts	Oil, protein, carbohydrates & linoleic acid	August–November (nuts)	Woods, esp. on hill slopes; prefers moist soils
M. spicatum	Water milfoil	Roots	–	July–September (flowers)	Eutrophic lakes (plant is submerged)
M. verticillatum	Whorled water milfoil	Leaves	–	Summer	Eutrophic lakes (plant is submerged)
N. lutea	Yellow water lily	Leaves, roots, seeds	Fresh roots are toxic	Fruit & seeds follow flowering (June–September)	Shallow water & wetland (aquatic plant)
P. australis	Common reed	Leaves, roots, seeds, shoots, stems	Starch & sugar (roots)	–	Moist soils, bogs or fen wood (max. water depth: 1m)
P. aviculare	Knotweed	Leaves, seeds	Rich in zinc	May–June (young leaves)	Pioneer plant
P. pectinatus	Fennel pondweed	Leaves, stems, roots	Bark must be removed from roots	–	Shallow standing or slowly flowing alkaline water (plant is submerged)
R. cf. caesius	Dewberry	Fruits, leaves	–	Late spring–early summer (fruits)	Fen woods and shores

(Continued)

Table 5.6: (Continued)

Botanical name	Common name	Edible parts	Nutritional & other features	Seasonal availability	Ecology
R. fruticosus	Blackberry	Fruits, leaves, roots	Vitamin C (esp. fruits)	August–October (fruits)	Forest pioneer; not extremely dry or flooded sites
R. idaeus	Raspberry	Fruit, leaves, roots	Vitamin C, tannins	July–September (fruit)	Forest pioneer & low herbaceous vegetation
R. maritimus	Golden dock	Leaves, seeds	Contains oxalic acid (reduced by cooking)	Spring–summer (leaves)	Mud-weed habitats along rivers and lake shores
S. sagittifolia	Arrowhead	Leaves, roots	Starch-rich tubers	June–August (flowers)	Still or slow flowing water and river banks
S. nigra	European elder	Flowers, fruits	Multiple vitamins, tannins	August–October (fruit); Late June onwards (flowers)	Moist woods & glades
U. dioica	Nettle	Leaves, stems, shoots, roots	Vitamins A & C, iron & protein	February–June (young leaves)	Woods and river valleys; widespread

Sources: Lippert and Podlech (2001); Mabey (2012); Bigga (2014); Bigga *et al.* (2015). Where a species is only edible when cooked (*e.g. R. aquatilis* [Common water crowfoot] or *R. repens* [Creeping buttercup]) they are not listed above. None of the above information should be used as a 'safety guide' with regards to the picking and consumption of wild plant foods. Any readers wishing to do so are strongly recommended to consult an appropriate, dedicated guidebook, such as Mabey (2012)

Table 5.7: Demonstrated and probable plant food consumption on European Lower Palaeolithic sites

Plant (species/taxa, where known)	Sites	Source
Celtis (seeds)	Arago cave; Gran Dolina; Le Vallonnet; Terra Amata	Allué et al. (2015); Hardy (2018)
Starchy plants (non-specific)	Sima de Elefante	Hardy et al. (2017)
Wild cherry (fruit)	Bilzingsleben	Mania and Mania (2003)

See also Chapter 6

autumn to early spring when other foods were in short supply (Hardy 2010; Bigga et al. 2015). Berry-yielding plants include raspberry, blackberry, dewberry, common bearberry, and elder, and would be a key source of vitamin C during summer and early autumn, probably supplanting the reliance earlier in the year on pine and bitch bark (Sandgathe and Hayden 2003), although wild fruits and berries are typically less sugar-rich than their modern domestic equivalents and may have had to be eaten in large quantities (Hardy et al. 2015). Rotted meat is a further potential source of vitamin C (Speth 2017). While direct evidence for fruit consumption is limited, the stones of sweet cherry recovered at Bilzingsleben offer a tantalising glimpse (Mania and Mania 2003; Table 5.7). In present-day Europe the fruit of sweet cherry (P. avium) matures in mid-summer, and is edible, although it is described as varying between sweet to astringent when eaten fresh. The fruits are consumed today by a wide range of birds and mammals, suggesting that there might have been competition for this resource, and it would presumably have only ever been a small dietary component (modern wild fruits are 1–2 cm in diameter), although one that hominins of nearly all ages could potentially have accessed. While some rodent and bird species crack open the stones to eat the kernels, imitating such behaviour would have been problematic for hominins, as all parts of the plant (except the ripe fruit) are slightly toxic.

Various other potential wild foods are highlighted by Mabey (2012; Table 5.8), although without direct evidence in a number of cases (e.g. wild strawberry), and the high vitamin value of the fruits of the Rosaceae genus/family have also been highlighted in Mediterranean landscapes (Tardío et al. 2005), along with the traditional sucking or chewing of selected flower species by children to access the nectar.

The widespread summer availability of fruit, fungi and other greens, and the modest technological (if not cognitive) requirements of their collection is likely to have made these foods an attractive resource for children. The long summer days would have enabled extensive foraging trips, as observed amongst some modern hunter-gatherers (e.g. Hadza girls and boys from age 6 upwards engage in up to 10 hours of berry collecting; Hawkes et al. 1995). It is obviously uncertain as to whether such foraging trips would have been in the company of adults, as in the case of the Hadza berry trips, or unaccompanied, as with modern Martu and Meriam children (Bird and

Table 5.8: Pleistocene records of potential summer plant foods (Mabey 2012), after Godwin (1975)

Species	Common name	Pleistocene?[1]	Earliest record[2]
A. ursinum	Ramsons	ND	ND
A. uva-ursi	Common bearberry	Yes	Ipswichian
B. vulgaris	Sea beet	Yes	Ipswichian
C. avellana	Hazel	Yes	Pre-Cromerian
F. vesca	Wild strawberry	No	Holocene
P. avium	Wild cherry	Yes	Cromerian
R. caesius	Dewberry	Yes	Ipswichian
R. canina	Rosehip[3]	Yes	Hoxnian
R. fruticosus	Blackberry	Yes (aggregate)	Cromerian
R. idaeus	Raspberry	Yes	Cromerian
R. nigrum	Blackcurrant	ND	ND
R. rubrum	Redcurrant	ND	ND
R. uva-crispa	Gooseberry	ND	ND
S. nigra	European elder (berries & flowers)	Yes	Hoxnian
U. dioica	Stinging nettle	Yes	Cromerian

[1]ND: No data available in Godwin (1975). [2]In light of Godwin's (1975, table 1) climate stage model (including the following sequence: Beestonian > Cromerian > Anglian > Hoxnian > Wolstonian > Ipswichian > Weichselian), 'Hoxnian' is cautiously interpreted as MIS 11 or MIS 9, and 'Cromerian' as spanning the early Middle Pleistocene. [3]Godwin (1975) lists this species as Dog rose. None of the above information should be used as a 'safety guide' with regards to the picking and consumption of wild plant foods. Any readers wishing to do so are strongly recommended to consult an appropriate, dedicated guidebook, such as Mabey (2012)

Bliege Bird 2000; 2005). Degrees of threat, *e.g.* from carnivores, might have influenced the composition of these foraging groups, as might the distances to be covered to acquire foods, and the cost/benefit balance if children did/did not accompany adults on foraging trips (Jones *et al.* 1994).

The seasonal changes of summer would impact on animal resources too, with ungulates in the pre-breeding period between birthing and the rut. Data from the BPF highlights the potential impacts on the availability of ungulate resources, in this case with reference to bison. From April–October bison bulls are mostly found in small groups of 1–3 individuals, before they join mixed groups for the August–October rutting season. If males were distributed along these lines in the Pleistocene they might be potentially difficult to locate in dense warm stage summer forests, although there would no doubt still be a range of visual, audible, and olfactory cues. By contrast BPF cows live in larger mixed groups (up to 39 individuals in the Polish portion of the forest), which would also include young after the late spring/ early summer calving (Daleszczyk *et al.* 2007). However, behaviours are clearly population-specific to some extent: based on living American and European bison Brugal (1999) suggested that late spring and early summer was a period of seasonal movement and aggregation for bovid herds, although without specifying the composition of the aggregations. It is noteworthy that larger mixed groups such as those in the BPF, with suckling calves and therefore high energy requirements for the cows, are more dependent upon abundant forage – comparable habits in their Pleistocene equivalents might have made their locations more predictable over the summer months. The possible tendency of cow's core areas to be smaller than those of the bulls (core area data varied in the BPF study depending on the methodology used, but was estimated as 16.1±8.7 km^2 when based on 50% relocation data), and to surround rivers, might also favour the locating of animals within their home ranges (Daleszczyk *et al.* 2007).

Modern studies highlight not just seasonal variations in range and group sizes, but also other changes in the social organisation and behaviour of prey species. In the case of red deer populations in the central Sierra Morena (Spain; Carranza and de Reyna 1987), female group organisation changed significantly across the year, reflecting broader seasonal changes (see also Chap. 1; Fig. 1.8). While females were found in larger harems during the rut (September–October), otherwise mean group sizes were 2.38 (pre-breeding; June–September) and 4.88 (post-breeding; December–May). These small groups consisted of the adult female, the calf (defined here as animals in the first year of life), and the previous year's young. A similar pattern is seen amongst modern British populations of red deer, where males and females tend to segregate for the majority of the year, divided into hind/calf groups and less stable stag groups (Corbet and Harris 1991). What should be apparent is that through much of the year (including summer), red deer females may only have been found in small numbers, and were perhaps more difficult to locate, although their physical summer condition was relatively healthy. Moreover the 'line order' of moving animals observed by Carranza and de Reyna (1987) highlights the potential importance attached to protecting the most vulnerable within the group – in the Sierra Morena red deer the youngest animals are found in the middle. In a Pleistocene context such behaviour might have encouraged hominins to target females and older young first, although the presence of calves is also associated with higher levels of hind alertness and aggression, with associated hiding or fleeing responses when disturbed (Clutton-Brock *et al.* 1982).

The challenges of locating animals in closed and mosaic woodland landscapes, particularly during periods of segregation and abundant summer vegetation, would presumably place cognitive demands on hominin awareness of, and alertness to, animal cues. While specific parallels should not be over-extended, tree 'thrashing' by modern red deer stags using their antlers (during August when cleaning velvet, and also in the rut) breaks side branches, *c.* 60–120 cm above ground (Corbet and Harris 1991). Similarly, modern male roe deer fray or rub saplings with their antlers, especially in summer (these traces are usually *c.* 30–70 cm from the ground). While their most well-known aural cue is rut roaring, red deer have many other calls, including a low 'moowing' sound made by hinds when they are trying to locate young calves. Similar Pleistocene behaviours might have been key visual and aural indicators for hominins of a local or recent animal presence, with the potential to distinguish between species.

Given the seasonal disaggregation of larger mammals and the widespread availability of other foods, it is uncertain how significant animal foods were for the hominin diet during the summer months. However, while late year breeding events (*e.g.* horse harems and the red deer rut) and/or migrations may have been the most predictable annual concentrations of ungulate prey, other types of aggregations may also have occurred at different times of the year. One such summer example amongst modern semi-wild horses is 'shading', during which horses stand together, largely inactive, and which may be an adaptation to minimise biting insect attacks (the behaviour is particularly common in mid- and late-summer; Corbet and Harris 1991). Interestingly horse 'shading' locations are maintained from year to year: if such behaviours had a Pleistocene equivalent then the locations might have been a reliable and repeated focus for summer game monitoring and hunting.

Long days, not lazy days ...

The longer days, warmer conditions, relative to the local winters, and relatively abundant food supplies raises the question of whether the summer months were a key opportunity for hominins to undertake other, more time-demanding tasks. In the archaeologically-detectable category such tasks might have involved raw material acquisition and tool-making, both lithic and organic. More speculatively, this might also have been the time of the year when hominins had the greatest opportunity to practice, and perhaps even be explicitly instructed in, the key skills of hominin life and survival.

Finding raw materials

Understanding of stone acquisition for tool-making in the Lower Palaeolithic period is inevitably linked to lithic transfer distance data, regional patterns and occasional insights at primary context sites. The former evidence typically suggests short distance transfers of less than 5 km (*e.g.* Féblot-Augustins 1999; Table 5.9), although occasional transfers of 20–30 km or further are known, as at Arago Cave (Wilson 1988). At Boxgrove the sources of raw material are extremely close, with flint nodules collected and tested at the base of the chalk cliff, and then moved to nearby activity locales such as the waterhole or the horse butchery site, where further knapping took place (Austin *et al.* 1999). At various sites in fluvial settings there is comparable evidence for the use of local sources, in the form of river gravels (*e.g.* at Pakefield, Swanscombe [Barnfield Pit] and Clacton; Ashton and McNabb 1996; Parfitt *et al.* 2005; McNabb 2007; Parfitt 2008). Immediately available material sources, in the form of outcropping flints, have also been reconstructed for sites in the Chalk upland landscapes of southeast Britain, such as Round Green (White 1998). This notion that Lower Palaeolithic hominins typically used the nearest available source, has also been inherent in, and supported by, key

Table 5.9: Lithic transfer data from the European Lower Palaeolithic

Site	Raw material sources (km)[1]			
	0–3/ 'Local'[2]	5–12	15–20	25–80+
Arago	> 80%	✓	✓	✓
Aridos I	100%			
Barnham[2]	✓			
Beeches Pit[2]	✓			
Bilzingsleben	96.4%	3.6%		
Boxgrove[3]	✓			
Brandon Fields[2]	✓			
Caddington[3]	✓			
Cagny La-Garenne I & II (Ca, Cxb, Cxv, I3, I4, J, Lg & Lj assemblages)[2]	✓			
Clacton	Majority			Very little
Elveden[2]	✓			
Ferme de l'Epinette (MS assemblage)[2]	✓			
Gaddesden Row[3]	✓			
Gran Dolina (TD-10)[4]	✓			
Happisburgh 1[2]	✓			
High Lodge[2]	✓			
Hoxne (Upper Industry & Lower Industry)[2]	✓			
La Celle[2]	✓			
La Noira (Lower level)[2]	✓			
La Noira (Upper level)[5]	✓			✓
Maidscross Hill[2]	✓			
Orgnac	95%	5%		
Round Green[3]	✓			
Rue de Cagny (Series 3)[2]	✓			
St-Pierre-les-Elbeuf[2]	✓			
Swanscombe (Lower Middle Gravels & Upper Middle Gravels)[2]	✓			
Vértesszölös	✓	✓		
Warren Hill[2]	✓			
Waverley Wood[2]	✓			✓

Sources: White 1998; Féblot-Augustins 1999; Fluck and McNabb 2007; Rosell *et al.* 2011; Moncel *et al.* 2015; Rodriguez-Hidalgo *et al.* 2017. [1]Distance categories as used by Féblot-Augustins (1999); [2]'Local' as used by Moncel *et al.* (2015); [3]Raw material source distances derived from White (1998); [4]Raw material source distances derived from Rodriguez-Hidalgo *et al.* (2017) & Rosell *et al.* (2011); [5]Iovita *et al.* (2017).

technological studies (*e.g.* White 1998), and can perhaps be characterised as a 'find it when we need it'-type approach. Material from such local sources could no doubt be acquired throughout the year, even on the shortest (and coldest) of winter days.

This impression of predominantly local sources can also be supported at a regional scale: in the case of Brittany for example, the Lower Palaeolithic record is character-ised by scarce handaxes and by a lack of evidence for the introduction of sufficiently large blocks of flint raw material, or finished handaxes, from the adjacent flint-rich landscapes of north-central and northeastern France (although handaxes and cleavers are present, for example at Menez-Dregan on the south Armorican shoreline, they are produced in local materials such as quartz and sandstone; Ravon *et al.* 2016). In southern Britain the Thames record is characterised by a relative rarity of flint handaxes in its upper reaches, despite their abundance in the adjacent Chalk land-scapes of the Middle and Lower Thames (Wymer 1968; 1999; Hardaker and MacRae 2000; Weston 2008). Further to the north there is a dominance of quartzite artefacts in the Trent Valley, despite their relatively low visibility compared to flint artefacts in a fluvial context (White *et al.* 2014). All of these patterns can be explained by a reliance on local materials, even when local materials are not ideal (*e.g.* the use of elongated burrow flint at Cuxton; Shaw and White 2003).

Yet the raw material insights from site and regional studies such as these may lead into an interpretive cul-de-sac that sees Lower Palaeolithic life purely as a here and now experience – lived by hominins who were perfectly effective tool-makers and foragers, but who had a limited cognitive ability to plan ahead and anticipate future needs. As we have already seen elsewhere, such a portrait does not quite seem to mesh with the challenges of a seasonal Europe. Might there be more to lithic pro-visioning in this period?

Recent studies from central Spain have offered a valuable new insight. Acheulean quarry sites, Charco Hondo I and II, have been excavated in the Madrid Basin, on the interfluve platform between the Manzanares and Jarama rivers (del Cueto *et al.* 2016). These sites exploit the platform's geology, with flint nodules up to 1–3 m in maximum dimension occurring in layers interbedded with clays. These 'quarry' sites are nothing like the open-cast quarries of the European Neolithic, such as those exploiting the volcanic tuff sources on the Langdales in the English Lake District (Edmonds 2012). It is an exploitation of surface resources – stream erosion exposed the flint layers and the nodules at Charco Hondo, enabling hominin access to them (Fig. 5.1). However, it is still clear that hominins were visiting these streams with very particular strat-egies and goals in mind. At Charco Hondo I (Ch-I) production was focused on biface shaping (*façonnage*), with a clear preference for tabular nodules, 3–10 cm thick. This was followed by the removal of the finished handaxes to other locations. At Charco Hondo II (Ch-II) a particular sequence of tasks was evident: (i) the initial splitting of large blocks, some up to 1 m in maximum dimension and more than 50 cm thick, into more manageable fragments; (ii) the knapping of flakes to use as 'blanks' for large tool production; (iii) the testing of the flake blanks and perhaps some initial

Figure 5.1: *Flint nodules and knapping debris at Charco Hondo II, level G1 (del Cueto* et al. *2016, fig 11c).*

bifacial working of them; and (iv) finally the discarding of inadequate blanks. What is also noteworthy is the general absence at Ch-II of the later stages of bifacial tool making: this is a 'quarry', or alternatively (but rather less evocatively) a raw material acquisition and primary or initial stages of working site. But most excitingly is the evidence at Charco Hondo for the hominins' ability to plan and prepare. This is expressed in the hammerstones found at the sites. They are almost the only foreign raw materials at the sites, and they must have been introduced from elsewhere, supporting the impression of deliberate visits to the 'quarries' and the anticipation of technological needs. The quartz and quartzite hammerstones were probably brought from the riverbeds or gravel bars of the Manzanares and Jarama rivers, *c.* 4–8 km (straight-line distances) from the Charco Hondo sites, although it is uncertain if they were picked up specifically en route, or had been collected at some point previously. The former is perhaps most likely however as the raw materials of the hammerstones vary across the two sites (there are only the denser and harder quartzite hammers at Ch-II, while there are 6 quartz hammers at site Ch-I), as do the properties of the flint (the blanks at Ch-II are larger, denser and harder). Finally, it is also noteworthy that the Manzanares gravels lack quartzite, while those of the Jarama are dominated by quartzite, further suggesting deliberately planned provisioning.

The Charco Hondo sites are not unique. Although not exclusively quarry sites there are other workshop sites which again highlight the exploitation of surface resources, such as the flint sources from the Chalk slope that meets the Somme floodplain at Cagny-la-Garenne (Moncel *et al.* 2015). At La Noira in the Middle Cher river valley, a tributary of the Loire, 'millstone' slabs (a silicified limestone) were gathered from slope deposits during the older occupation (Moncel *et al.* 2016). Double patina on some of the bifacial cores suggests repeated occupations at this site, although the full range of technological sequences which are present, combined with suggestions of domestic activities (on the basis of use-wear; see also B. Hardy *et al.* 2018), make it difficult to ascertain the relative importance of different factors in drawing hominins back to the site – how significant was the lithic raw material source? A related example of anticipation of future needs is evident at Gran Dolina TD-10.2: the lack of handaxe and cleaver manufacturing evidence in the assemblages suggests that these tools were imported (and curated) from elsewhere, although the time-lag between their production and use at Gran Dolina is uncertain (Rodriguez-Hidalgo *et al.* 2017). The later, MIS 13, occupations at La Noira also reveal complex provisioning behaviour, with diverse raw materials sourced both locally (millstone) and from areas 30–100 km away from the site (flint; Iovita *et al.* 2017).

Although a rare and invaluable insight for the European record, this type of planned lithic exploitation behaviour is not unique for the Lower Palaeolithic on a global scale. Exploitation of surface deposits has been demonstrated at various sites, including Mt Pua, Israel (where limestone blocks were smashed to enable flint extraction and where recurrent visits in both the Lower and Middle Palaeolithic resulted in hundreds of debris heaps or 'tailings'; Barkai *et al.* 2002; 2006), at Isampur Quarry in India (where limestone slabs were levered up from the bedrock after percussive flaking at natural joint surfaces; Petraglia *et al.* 2005; Petraglia 2006) and in the Upper Karoo region, South Africa (Sampson 2006). What is also notable in a European context is the chronology of such behaviour. The Charco Hondo sites are dated to the later Middle Pleistocene and this tallies with the rich series of Acheulean sites in central Iberia, associated particularly with the Tajo, Duero, Manzanares and Jarama rivers and thought to date mostly to MIS 11 (*c.* 400 kya) onwards (Santonja and Villa 2006). This is well within the expanded, or 'second', phase of the Lower Palaeolithic settlement in Europe, and perhaps also within the gradual 'Neanderthalisation' of Europe (Meyer *et al.* 2016).

But from a seasonal perspective, why might this sort of lithic provisioning have occurred preferentially in the summer months? Given the time demands of travelling to sources and extracting/initially working the materials it seems likely that it would be associated with periods when other demands on time were relatively low. This is unlikely to be spring or autumn, when high levels of food foraging are predicted to have occurred and when days were shorter and light levels lower. In the spring (Chap. 4) foraging of renewed resources would have been critical to the rebuilding, after winter shortages, of individual health, with a particular requirement potentially being to meet the needs of pregnant females entering the 3rd trimester. During autumn (Chap. 6) concentrated foraging would have sought to exploit animals in prime condition and the social aggregations associated with the rut and/or migrations and build up individual hominins' body fat reserves or external stores prior to winter. Winter itself would also seem unlikely, because of the shortage of daylight hours, weather conditions (knapping is both a relatively stationery activity and is dependent on finger agility and sensitivity, not numbness, as anyone who has knapped between late autumn and early spring will testify!) and the need to acquire whatever foods were still available. However, the summer would be characterised by lusher, more extensive vegetation, and thus identifying raw material outcrops and sources at a distance, and accessing nodules close at hand, would not have been straightforward (see also Wenban-Smith 1998). Nonetheless, exposures of nodules in fluvial channels, as at Charco Hondo, or in association with steep, partially vegetated slopes such as the Boxgrove cliff face or eroded upland gullies, would still be visible. Thus, long summer days might well have been the most appropriate times for dedicated lithic provisioning – and perhaps offered opportunities for adolescents (and children?) to gain/practice key skills and knowledge: distinguishing a quartz and quartzite hammerstone, or testing the quality of a flake blank or core.

How frequent might such dedicated resourcing trips have been? The majority of lithic transfer data and on-site evidence, while only relating to one category of resources, is not suggestive of resourcing from non-local distances (defined here as more than 3 km; Table 5.9). It thus seems likely that most provisioning of lithic raw material, at least, was embedded within local foraging activity, and/or that most lithic provisioning occurred on a similarly local scale, and thus presumably throughout much or all of the year. However, it is important not to under-estimate practical limitations: the quantity, and potentially awkwardness, of transported food resources (see also Chap. 4), whether plant or animal, would restrict opportunities to undertake other tasks and/or transport other substantial, bulky resources such as provisioned stone or, perhaps, firewood at the same time. The long daylight hours of summer might therefore have been significant in maximising the available time for foraging for a diverse range of resources.

Organic materials

Lower Palaeolithic tool-kits extended beyond lithics, with bone, wood and antler artefacts all known. As noted above (Chap. 4) shed antler may have been processed during its key collection period of late winter–early spring but for other organic materials summer may have been a critical time for their collection and processing. The range of wooden artefacts from this period is, unsurprisingly, very limited. Nonetheless examples do exist, in the form of spears and other artefacts at Schöningen (Thieme 1997; Schoch *et al.* 2015), the Clacton 'spear' (Warren 1911; Oakley *et al.* 1977; Allington-Jones 2015), and a range of wooden artefacts at Bilzingsleben (Mania and Mania 2003). As with antler artefacts, the shaping of wood is time and attention-consuming and would again presumably require a relatively safe space. Haidle (2009) reports times of 4.5–5.5 hours for the entire process, from tree-cutting to tip-shaping, associated with a reproduction of the Lehringen spear, while McNabb's (1989) spear point-shaping experiments (on yew staves 23–46 cm long) ranged from 30 to 265 minutes. In recent experiments at the University of Reading (Helme 2017) 20–80 minutes were taken to debark and shape *c.* 1.4–1.6 m branches of birch, pine and yew into spears (the shaping involved the removal of knots and tip-finishing).[2]

The Reading experiments highlighted the benefits of working with fresh or green wood, as debarking could be achieved by hand, effectively a peeling off process. In contrast, debarking 6 month-seasoned wood was a much longer process, requiring a stone tool edge, although point shaping on this material was easier. This might favour the acquisition and at least the initial processing of wood resources at broadly the same time, although the use of the tools could obviously be delayed. The most likely felling season for the spruce trunks used to make the Schöningen spears was summer (based on tree ring characteristics), with the timing varying between the individual artefacts from early to late summer (Schoch *et al.* 2015). The evidence from Schöningen and other sites also highlights the hominins' sophisticated awareness of wood's properties and their preferences when producing spears: the use of coniferous tree species (*e.g.* yew at Clacton, spruce and pine at Schöningen; Oakley *et al.* 1977; Schoch *et al.* 2015) exploited their durability and tendency not to split when worked. At Schöningen the selection of older, slow-growing spruce trees (50–60 years old), ensured both relatively thin trunks but also hard wood. The tips were produced in the hard, dense wood at the base of the tree, avoiding the softer central pith, as opposed to the thinner tips. Detailed technological studies of the spears have revealed a clear sequence of actions: felling; stripping of bark; removal of underlying fibre (inner bark); removal of knots (indicated by tool-marks); and finally, the polishing of the surface. The largest diameter is always found towards the front of the shaft, with implications for the centre of gravity and the mode of use: Schoch *et al.* (2015) drew parallels with modern javelins, as Thieme (2005) did previously, and suggested that the hard wood tips would prevent breakage when thrown. The associated palaeoenvironmental evidence also suggests planning and curation, with the scattered presence of *Picea* pollen in the spear horizon supporting the introduction of the spears or their raw materials from some distance (Bigga *et al.* 2015). Combined with the tree ring-indications of summer felling this might reflect a combination of planning for the late summer/early autumn migrations (although the Schoningen horse kills now appear to represent a series of multi-seasonal events; Julien *et al.* 2015), or at least for a trip to the lake edge and the possibility of a hunting opportunity, and the scheduling of a time-demanding and repetitive task during a period of warmer weather, longer days, and reduced pressure on the availability of food resources.

The multi-staged nature of such behaviours has been linked to various aspects of enhanced working memory in *H. erectus s. lato*, including interference control, inhibition, sequential memory, temporal-order memory and integration of memory across space and time (Nowell 2010). These are all argued to be necessary to sequence activities in their correct order to meet a goal (Coolidge and Wynn 2005; Nowell 2010) – a description which would certainly apply to spear production (Haidle 2009) but also to stone and bone tool making (and using), clothing or shelter production, fire-starting (and maintenance), and food foraging and processing.

While there is often a tendency for discussions of wooden artefacts in the Palaeolithic to focus on spears and other projectiles, the importance of digging sticks and other wooden artefacts should not be overlooked. The pointed 'throwing stick' from Schöningen has also been suggested as a digging stick or a bark peeler (a means of accessing carbohydrates and, in the case of birch and pine, vitamin C; Sandgathe and Hayden 2003; Bigga *et al.* 2015), although no conclusive interpretation of that particular artefact has yet been reached (Schoch *et al.* 2015). The Bilzingsleben wooden artefacts are more difficult to interpret due to the quality of preservation but their dimensions might suggest usage as digging tools (a number of the remains and imprints are in the size range 41–52 cm long, reported for ethnographic examples by Sandgathe and Hayden 2003; Mania and Mania 2003, fig. 16). Such a use might also be indirectly supported by the potential dietary importance of underground storage organs (Hardy 2010).

A third major source of organic material for technological items is bone and at sites including Schöningen, Gran Dolina (TD-10) and Boxgrove there is clear evidence for the manufacture and use of a diverse range of bone tools: retouchers, scrapers, anvils and hammers (Rosell *et al.* 2011; Smith 2013; van Kolfschoten *et al.* 2015b; Rodriguez-Hidalgo *et al.* 2017), while bone bifaces are also known from a variety of other sites (Zutovski and Barkai 2016). There is less evidence of a specific seasonal bias in terms of the material acquisition for bone tools, indeed at Schöningen there is use of both fresh bone from butchered carcasses (the majority), and dry, residual, bone. However, what is again evident is the careful selection, preparation and use of these tools. Moreover, the use of this material does not appear to be forced by lithic shortages (*e.g.* at TD-10.1; Rosell *et al.* 2011). Marks on the Schöningen knapping tools indicate the hominins' detailed appreciation of the bones' morphology, with the knapping marks concentrated in featureless areas of thick cortical bone – enabling a 'smooth' knapping strike supported by a suitable thickness of bone. Scrape marks on these tools also suggest that the bones were prepared for use, with periosteum removed from fresh bone, and sediments from dry bone, to produce a harder and more efficient surface for knapping (the absence of scrape marks on those that were only broken open suggests that these marks are not related to marrow extraction, unlike in some of the cases previously described by Binford 1981). The tools also reveal their careful usage, with evidence for changes in the location of the grip, the forces applied, and other handling characteristics (van Kolfschoten *et al.* 2015b). The Schöningen tools on fresh bone used both butchered and un-butchered specimens, suggesting that 'background scatter' bone was perceived as an everyday resource, just as stone, wood and other items would be.

While the knapping tools were mainly produced on long, heavy limb bones, the unusual marks on a horse innominate (hip bone) at Schöningen suggest that it was used as an anvil or support for bifacial knapping – a technique which may have been well suited to the working of small flakes. The latter is likely to have been a common technological scenario given the relative paucity of lithic raw materials at the site. Of particular interest is the lack of butchery marks on the innominate bone, which along with its general condition suggest that it was deliberately selected for use as an anvil from the scatter of naturally defleshed ('dry') bones on or near the butchery site (van Kolfschoten *et al.* 2015b).

The final category of bone tools are the hammers, with their distinctive flaking and percussion damage at the distal ends of the bone. The lack of diagnostic knapping damage and the absence of microscopic flint chips in the battered surfaces, suggests that these may have been used as hammers for cracking open bones to access marrow. This is supported by ethnographic observations (Binford 1978, 153–155), the lack of hammerstone-type marks on the marrow-extracted bones' impact notches, and the lack of stone cobbles suitable for use as hammers in the fine-grained sediments of the lake shore (van Kolfschoten *et al.* 2015b). This last point, in particular, highlights the hominins' ability to adapt their behaviours to local circumstances. If these hammers were used for marrow extraction, it is probably also noteworthy that they were made on horse and bovid metapodials – exploiting the dense properties of the distal epiphyses on these bones, especially in the case of horses (van Kolfschoten *et al.* 2015b). The Schöningen bone tools also highlight a flexible approach to tool function, with some of the metapodial hammers combining evidence for knapping (on the mid-shaft) with hammering (on the distal epiphyses).

The bone bifaces from sites in Italy, in particular, are argued to have similar design concepts and flaking techniques as their lithic equivalents, including the production of large bone flakes. Zutovski and Barkai (2016) have suggested that the bone was knapped when fresh and more elastic, but since fresh bone would have been potentially available throughout the year it is difficult to infer a seasonal bias to production. The bone artefacts are comparable in size to stone bifaces at the same sites, challenging the notion that bone biface making was due to small-sized or absent lithic sources and, instead, suggesting a deliberate choice. Zutovski and Barkai (2016) have suggested that bone bifaces reflect deeper connections between hominins and elephants: this is intriguing, and may be reinforced by the apparent limitation of bone biface making to elephant remains. However, the bones of other taxa are also flaked (but not into bifaces), and thus the elephant bone bias might simply reflect raw material size issues. Instead I find the most striking aspect to be the technological flexibility that is again evident, and while evidence for other uses of organic materials is scant, the wooden, bone and antler artefacts which have been recovered suggest the strong possibility of other uses of organic items (*e.g.* of scapulae as digging tools for accessing roots?). The use of such materials as expedient tools on temporary foraging sites (*e.g.* wetlands) would also mitigate against their appearance in the archaeological record.

Overall, while knowledge of the use of organic materials in the period remains highly partial, there are sufficient insights from bone and wood working to emphasise the likelihood of a strongly non-lithic life, as is known amongst both chimpanzees and extant hunter-gatherers (Sillitoe and Hardy 2003; McGrew 2010), and at least some seasonal aspects to the acquisition of materials and the production of artefacts. A complementary question to the suggestion of a summer focus for time-demanding artefact production is whether such artefacts were more intensively curated over the late autumn–early spring period?

Information gathering

The longer days and milder conditions of summer may also have facilitated resource monitoring and information collection across a wider range of landscape settings, potentially embedded within logistical trips (Kelly 1983). These expanded settings might well have included European uplands, which could provide opportunities to observe game movements and other resources (Kolen *et al.* 1999), although this would likely be complicated by the lusher summer vegetation during warm stage occupations, and perhaps also access springs. In contrast to spring (Chap. 4), the lower water levels, and perhaps slower speeds, of summer rivers, might also have facilitated, and perhaps even encouraged, wider-ranging summer movements. This might be especially evident in the Mediterranean south, where the evidence from a number of key sites and regions suggests very low levels of summer precipitation (Table 5.4).

Environmental knowledge of plants and animals is critical in seasonal landscapes, where food availability is seasonal or otherwise time-delimited. Although animal monitoring may appear to be a more challenging task, they can be monitored from distance (at least for larger prey) or remotely via tracks, whereas plants can only be observed through close inspection. A further difference concerns the 'breadth and depth' of knowledge: as the proportion of hunting, and therefore the size of foraging areas, is increased in a low effective temperature environment, the coverage of that larger area becomes less comprehensive and it is 'known' in less detail, because of the greater reliance on logistical over residential mobility (Kelly 1983; MacDonald 2007). Levels of knowledge might also vary seasonally, with fluctuations in food density and distribution impacting on the size of foraging territories and/or mobility strategies in, for example, winter and summer. Expanded environmental knowledge might also link to the general encephalisation trend that is evident in *Homo*: Kaplan *et al.* (2007) have noted that greater resource patchiness is associated with larger home ranges and that these place greater demands on spatial memory. The proportions of directly observed knowledge will also vary significantly between individuals (*e.g.* between hunters who undertake long logistical forays and those who mainly forage in the core residential area), highlighting the potential need for some form of spoken language. More generally, Kelly (1983) stresses that such information can be gathered for both immediate and future use, which also has implications for the scope and use of language, especially since other behaviours (*e.g.* sourcing of wood for the Schöningen spears) appear to be suggestive of an ability to anticipate future needs. Foraging trips may also have combined the meeting of dietary needs with other types of resourcing, such as fuel (adding a further complication to the cost/benefit approaches of Henry *et al.* 2018) and lithics, although much of the known lithic transfer data (Table 5.9) appears more local than the logistical mobility data (distances and days per trip) presented by Kelly (1983, table 1, 9 & fig. 6).

Learning in a Lower Palaeolithic childhood

What else may Lower Palaeolithic hominins have been doing during the longer days of the year? One key aspect may have been the opportunity for children and adolescents to learn the knowledge required for survival. This is strongly implied by the likely life history models for Lower Palaeolithic hominins (Chap. 4: Box M), which involve significant investment in delayed maturation and slower growth, while observations of human children today, and of infant primates too, would suggest that learning, of techniques and knowledge, is a central focus of this stage of life. Hublin *et al.* (2015) have emphasised human socio-cultural evolution through the concept of niche construction, and in particular the roles of protracted growth and delayed maturation/reproductive life in the development, maintenance and modification of those niches. From a European Lower Palaeolithic perspective, it seems likely that memory development, inhibition and attention, acquired through learning activities and experiences such as provisioning, cooperation and play, were the focus of those extended growth and development periods (Nowell 2010).

This immediately raises two questions: first, what do we mean by learning in a Lower Palaeolithic context? Secondly, what knowledge would be required? The question of learning mechanisms is difficult. Chimpanzee studies support the notion of social learning through simulation, facilitation and active teaching (Boesch 1991): might a similar model of social learning be applicable amongst Lower Palaeolithic hominins? The apparent paucity of change in that most iconic of Lower Palaeolithic artefacts, the handaxe, has recently been used by Corbey *et al.* (2016) to instead propose a significant degree of 'hard-wired' genetic control over artefact production (Box N). While objections to this view have been raised primarily on the basis of the artefact record (Hosfield *et al.* 2018; Wynn and Gowlett 2018), the sheer diversity of knowledge required by a Lower Palaeolithic life (Shea 2006; see also below) is another strong argument in favour of active social learning, as is the extended childhood implied by the available dental evidence (Chaps 2 & 4). But would this learning have involved spoken language? Predictions based around the social brain hypothesis would seem to favour some form of spoken language, at least for the latter half of the European Lower Palaeolithic (Dunbar 2003; Chap. 2), as might technology-based models (*e.g.* Stout *et al.* 2008; Arbib 2012; Stout and Chaminade 2012; Uomini and Meyer 2013; Morgan *et al.* 2015), and perhaps also the anatomical evidence for the expansion of Broca's Area in *H. erectus* (Zollikofer and De León 2013). Spoken language would also have been a valuable addition to a fission–fusion lifestyle (Grove *et al.* 2012; Grove and Dunbar 2015; Chap. 4: Box K), given that it facilitates the discussion of, and learning from, knowledge and situations which not everyone has experienced at first hand. At the same time, some experimental evidence for the role of language in Lower Palaeolithic-type knapping skills is ambiguous (*e.g.* Putt *et al.* 2014), and highlights the range of non-verbal mechanisms which can also be used to learn stone knapping: reverse engineering, imitation/emulation, basic teaching, and gestural teaching.

Box N: Handaxe making: not learned but inherited?

Corbey *et al.* (2016) have recently raised the possibility that the handaxe, that icon of the Lower Palaeolithic (at least from our perspective), was fundamentally under genetic control, rather than being a cultural object. This is an intriguing hypothesis, which seeks to address the longstanding interpretive challenge of handaxes: the combination of the apparent technical skill involved in their making, which might imply intention and design, and over 1.5 million years of apparent stability in form, which might suggest the handaxe as a prosthetic extension of the hominin body, akin to birds' nests and beavers' dams (Ingold's 'double bind'; 2013, 37). I certainly do not dispute that at least some of the patterning in handaxes reflects genetic inheritance – for the simple reason that individuals' hand-eye coordination, for example, would have a genetic basis, although it could and probably would be improved over an individual's life through practice and experience. However, I suggest that there are two problems with an over-arching genetic perspective (see also Hosfield *et al.* 2018). One is a fundamental aspect of the record which I think that Corbey *et al.* (2016) have ignored. In short, handaxes are not homogeneous. They can certainly look so when perceived through the long lens of 1.5 million years or more: there are few new artefact forms, with one or two possible exceptions such as ficrons, and there is relatively little evidence for clear inter-regional differences (*e.g.* Marshall *et al.* 2002). This global persistence reflects, I suspect, their inherent usefulness as a large butchery knife.

Yet at the level of individual assemblages, within regions, and even between regions, there is evidence for the diversity of handaxes and other large cutting tools (LCTs), and of their status as items of learned material culture. Variability occurs at the continental (*e.g.* the contrast in cleaver abundance between Africa and Europe), the regional (*e.g.* Acheulean/non-Acheulean trends across Europe from west to east) and the local or assemblage level (*e.g.* variations in modal tendencies; Gamble and Marshall 2001; Gowlett 2005; Moncel *et al.* 2015). Using the well-understood British record White (2015, fig. 9.4) has demonstrated differences in the dominant handaxe forms across MIS 13 through 9: given the wider palaeoenvironmental context of cold (uninhabitable)/warm cycles and relatively short periods of access across the English Channel and/or the southern North Sea, the most parsimonious interpretation of the pattern is of local traditions of handaxe making, being over-turned at each successive warm stage by the influx of new hominin populations. Similarly, shorter-term variations in material culture on MIS 11 sites have been seen as indicating brief incursions into Britain by different groups (Davis and Ashton 2019). This can also be seen at the site level, such as at Boxgrove where handaxe shape does not appear to be clearly related to either nodule/blank morphology or reduction intensity, but is rather the product of flexible knapping strategies and hominin choices (García-Medrano *et al.* 2019).

The second reason is reflected in, and responds to, Ingold's (2013, 44) statement: 'But unlike modern experimental archaeologists, who deliberately set out to produce exemplary replicas of the [handaxe] type, the original handaxe makers knew nothing of this taxonomy, and were not guided by it'. While I am happy to agree that young *H. heidelbergensis* were spared lessons in formal taxonomy I find this very difficult to agree with in the broader sense, since the hominin infants of the Lower Palaeolithic would have been exposed to the sights, sounds and touch of handaxes throughout much of their young lives. They would have been repeatedly exposed to the ways of doing, and the material products, of their elders and peers (see also Gosden 2005). Surely as they began to produce handaxes for themselves, they would have been guided by those experiences and by their own notions of what a handaxe was – the 'mental construct' of Ashton and McNabb (1994). Handaxes are obviously not identical, due to the vagaries of skill, raw material, resharpening and perhaps even the amount of knapping time that was available, and this is reflected in the 'noise' at the level of individual assemblages (Fig. N.1). But their production within, and reflecting, a particular social setting and a particular way of doing would seem to be the best explanation for intra-assemblage similarities such as at Boxgrove (Fig. N.2), and the inter-assemblage differences detected in studies such as White (2015).

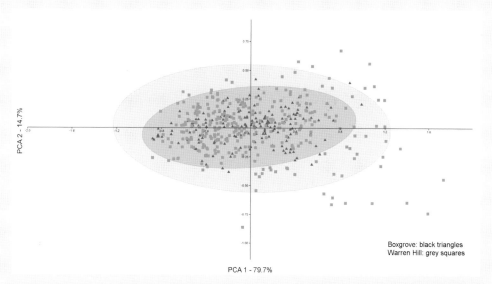

Boxgrove: black triangles
Warren Hill: grey squares

PCA 2 - 14.7%

PCA 1 - 79.7%

Figure N.1: Intra- and inter-assemblage handaxe shape variability in the UK assemblages of Boxgrove and Warren Hill. The greater degree of variability at Warren Hill probably reflects the palimpsest nature of the assemblage, in contrast to the time-constrained nature of the Boxgrove landsurface (probably less than 100 years; Hosfield et al. 2018, fig. 6).

Figure N.2: 50 silhouetted Boxgrove handaxes (selected by random number generator). O: handaxe with cutting edge all around the circumference, or nearly so; T: tranchet flaked (handaxes not to scale; Hosfield et al. *2018, fig. 1).*

In short, handaxes, and by extension other material items such as retouched flakes, were produced, used, and learned about, in a social context: this is material culture (see also the arguments in Hosfield *et al.* 2018).

Whatever the specifics of Early and Middle Pleistocene language, it seems likely that task learning occurred in a social context, through a blend of observation, imitation, trial and error, and teaching and, as children aged, through doing by participation: perhaps progressing through fuel collection, plant food foraging, animal observation, tool-making, fire-lighting and scavenging/hunting, among many other things. Acquiring language skills, whatever the exact form of that language (*e.g.* gestural and verbal; see also Chap. 2), would also be a key aspect of childhood learning and critical to wider socialisation.

What would a Lower Palaeolithic infant need to learn? This can be discussed both from the material archaeological record and on the basis of invisible, but likely, behaviours. The material record encompasses the making and using of stone (*e.g.* Stout *et al.* 2014), bone (*e.g.* Zutovski and Barkai 2016) and wooden tools (*e.g.* Schoch *et al.*

2015), and butchery techniques (*e.g.* skinning, disarticulation, defleshing, evisceration and marrow extraction; Parfitt and Roberts 1999; Voormolen 2008; Saladié *et al.* 2011; Huguet *et al.* 2013; Rodriguez-Hidalgo *et al.* 2015; 2017; van Kolfschoten *et al.* 2015a). These tasks all include the ability to recognise the properties of materials, either by their visual appearance or through physically testing them (*e.g.* tested nodules at Boxgrove; Austin *et al.* 1999), and it is noteworthy that the use of specific lithic materials for particular tools is evident from the earliest European sites. At the late Early Pleistocene site of Barranc de la Boella for example hammerstones are selected from the hardest and best quality varieties of quartzite, schist and quartz, LCTs are made on schist, and chert was predominantly selected for flake production (Mosquera *et al.* 2016): it seems likely that these preferences were both learned and passed on, rather than repeatedly re-learned. While the evidence is much rarer for organic materials, the Schöningen spears highlight a sophisticated appreciation of wood characteristics (Schoch *et al.* 2015). As I write this I am watching my daughter's swimming class, and the instructors are repeatedly demonstrating the arm and body motions for various strokes. It is surely not a great leap to envisage Pleistocene skills being learned in a similar manner: through observations of others' hand grips, cutting motions and knapping gestures, if not necessarily involving active demonstrations by adults or perhaps older children, followed by kinaesthetic learning.

But knowledge in the archaeologically invisible (or rarely visible) sphere extends much further. The following list is certainly not conclusive, but knowledge categories might include: animal behaviours (*e.g.* fight or flight responses, aggression indicators and predator avoidance strategies, reproduction cycles, preferred habitats, dietary strategies [including animal–animal and animal–plant relationships], denning habits, timings of migrations: Laughlin 1968; MacDonald 2007), animal cues, both direct (*e.g.* footprints and other marks, calls, scents, faeces) and indirect (*e.g.* the responses of other animals to predators), animal carcass properties (*e.g.* the nutritional value of specific components; the characteristics of safe vs. unsafe food; Speth 2017), plant properties (*e.g.* toxic/non-toxic, edible portions, medicinal properties, seasons of availability; Hardy *et al.* 2013), landscape and habitat characteristics (*e.g.* vegetation, topography, water sources, natural shelter, regular carcass accumulation points, weather cues and seasonality [including indicators of seasonal change]), social 'rules' (*e.g.* sexual politics, infant care, hierarchies of food access, recognising the moods and emotions of individuals), and, at least in the later Lower Palaeolithic, fire-making and maintenance skills (*e.g.* tinder and fuel selection, methods and materials for igniting and/or maintaining fires and for moving embers; Sorensen *et al.* 2014; Henry *et al.* 2018). More speculatively, knowledge might also include techniques for processing animal hides and constructing simple shelters. These categories mirror Mithen's (1996) domain-specific model of the hominin mind and its emphasis on social, natural history, and technical intelligence. Would any of the above be instinctive or hard-wired rather than socially learned knowledge? Possibly in some cases (*e.g.* a flight response to an aggressive predator; Adolphs 2013), but it is worth reflecting that many responses often assumed to be universal are in fact locally learned and culturally specific (as demonstrated for the disgust response to rotten meat; Speth 2017). Moreover, there is increasing evidence that our psychological capacities for cultural learning evolved as an adaptation to spatio-temporally variable environments (Richerson *et al.* 2010), a description that is certainly applicable to Early and Middle Pleistocene Europe, although it is unclear exactly when this particular mode of gene-culture co-evolution started to become important.

It is likely that children had a significant period of acquiring foraging skills and knowledge and this learning may well have begun during the middle childhood phase which has been argued to be a part of Early and Middle Pleistocene hominin life history (Nowell and White 2010; Thompson and Nelson 2011). Middle childhood, starting between roughly 6 and 9 years old in modern children, is specifically associated with a suite of cognitive developments (*e.g.* abstract reasoning, independent task organisation, acquiring and using new knowledge; Collins *et al.* 2002), near-completion of brain growth and eruption of the first permanent molar (Bogin and Smith 1996), and thus seems a likely point at which children would become actively involved in foraging. This early initiation of learning might have been especially important given the evidence for relatively short life expectancies (Table 4.6), although the fossil record may be missing examples of older individuals (see also Chap. 4). In contrast to Neanderthals (Thompson and Nelson 2011), Europe's Lower Palaeolithic fossil record is currently insufficient to demonstrate shared trauma and, by extension, shared dangerous activities between adults and non-adults. Nonetheless, the evolutionary costs of non-participation (burdening all food provision for weaned non-adults onto adults) make a strong argument for child participation in foraging, although the evidence from modern hunter-gatherers is mixed (Jones *et al.* 1994; Hawkes *et al.* 1995). Moreover, foraging groups would have provided intense and valuable learning environments for children, during a critical phase of brain plasticity. As to which foraging activities, and by extension which skills and knowledge, were the focus of pre-adult learning, it is likely the nature and content of learning varied with age – as both the mind and the body developed (MacDonald 2007).[3] However, while an extended childhood enhances learning opportunities, it also has implications for the ages at which non-adults can actively contribute to particular foraging activities, by extending the duration of their mental and physical development.

I suggest an emphasis by Lower Palaeolithic children on plant foraging and collection of static animal resources (*e.g.* birds' eggs) rather than hunting mobile prey in the pre-adolescence stage. This is for three reasons (see also Chaps 4 & 6). First, while hunting weapons and lesions are scarce, available evidence from both the Lower and Middle Palaeolithic is currently suggestive of short-range hunting, possibly with an ambush-style component (Schoch *et al.* 2015; Gaudzinski-Windheuser *et al.* 2018), although debates remain as to the effective range and use-mode of the known spears (Churchill 1993; Milks 2018a; Milks *et al.* 2019). Whether thrown or thrusted, strength and skill would be required for the effective use of spears and the meeting of these requirements might be more likely to start developing significantly in adolescents. Close-range hunting also exposes the participants to significant physical risk. Yet, as noted elsewhere, hunting episodes would have been about more than just spear use. Could younger children have contributed to other aspects of the hunt, such as monitoring or tracking the prey? A key consideration is that whether effective weapon ranges were a few tens of metres or less, discipline (*i.e.* stealth and silence) and efficient mobility would presumably be a critical attribute for the successful hunter prior to the attack. Modern studies of the Martu suggest that their children can be silent while men are tracking and pursuing animals but it is notable that Martu men never took children on foot hunts (prior to the introduction of vehicles), while Martu women would remark that children were too slow to keep up with them while they were searching and tracking (Bird and Bliege Bird 2005). While European warm stage habitats were very different to those of the Martu in Australia, collectively the above points seem to make a case against the involvement of sub-adolescents in adult hunting (and perhaps in other types of adult foraging as well). Differences in mobility are also relevant to the second point: namely that pursuit of an injured animal (if it was not killed in the initial strike) would require speed and endurance, to ensure that the prey was accessed before other carnivores could reach it (and efficient mobility would also be critical to successfully access and scavenge other predators' kills). Thirdly, notions of risk and value might also extend to children, given the dental evidence for an extended childhood and thus a clear evolutionary investment in slow growth and development and a relatively long productive life. An early exposure to risky activities therefore seems unlikely, and hunting risks could include exposure to other scavengers, not just the prey itself. Overall, it therefore seems likely that pre-adolescents were not habitually part of hunting parties.

However, these younger children could still make significant dietary contributions, both for themselves and potentially also the wider group. Their own feeding could occur both during (*e.g.* 'snacking' on leaves, flowers, berries and other foods; Hawkes *et al.* 1995) and after their foraging trips. Amongst the Martu for example children hunt lizards and small birds, pick fruit, collect birds' eggs and dig-up plants and immediately consume much of their foraging (Bird and Bliege Bird 2005). While the specific foods would obviously vary, comparable foods would certainly have been available in the warm stage habitats of Lower Palaeolithic Europe. Children may have foraged in the company of those not involved in the hunting of mobile prey (*e.g.* adult females in the latter stages of pregnancy and/or older individuals?) but may have also foraged unaccompanied (Crittenden *et al.* 2013). Such child/adult foraging groups may have been highly productive: amongst the Martu, older women (35+) and children up to 15 tend to be the most active foragers, with inter-generational foraging partnerships quite common (Bliege Bird and Bird 2008). Alongside food returns children would also be learning and practising critical skills and knowledge and elements of this learning would probably be relevant to both the tasks at hand and to the adult-style foraging of later life. The learning of visual, auditory and olfactory signs of different animals might also start at this younger age, as might participation in small-game hunting.

Furthermore, in some modern hunter-gatherer groups it is apparent that children undertake independent foraging, with their skills and decisions more strongly influenced by their peers than by adults (Bird and Bliege Bird 2005). This challenges traditional notions of 'growing up' (*i.e.* that childhood is solely about gaining the skills required for successful adulthood). While it is difficult to know the extent to which this applied in Lower Palaeolithic communities, the seasonal, patchy and diverse nature of the available food resources, combined with significant provisioning needs, might suggest that at least some sub-adult foraging behaviour was child-specific rather than proto-adult in character. Possible support can be found at the Sima de los Huesos, where the consumption of specific, less abrasive, foods by older children has been suggested based on differences in dental striations, although the degree of overlap between the sub-adult and adult samples means that this cannot be established confidently (Pérez-Pérez *et al.* 1999). However modern hunter-gatherer perspectives, including amongst societies where child foraging is significant, suggest that such 'work' might be motivated as much by play and enjoyment as by food-getting (Gray 2009).

By contrast, involvement in sustained tracking and ambushing/stalking activities of larger animals would likely await late childhood/adolescence and the requisite physical endurance and perhaps also sustained task focus and concentration: MacDonald (2007) highlighted that boys tend to join hunts as adolescents, while amongst the Martu both girls and boys aged 13–16 begin to adopt adult hunting strategies and hunt with women and men respectively (Bird and Bliege Bird 2005). The tasks of hunting to be learned would be many, covering locating, tracking, disadvantaging/catching, injuring/killing, butchery and potentially carcass transportation, while Milks *et al.* (2019) have emphasised the years of practice required to develop expertise in hand-held spear throwing. Overall it seems probable that a relatively gradual shift from playful to real participation in hunting and other adult tasks (Gray 2009) would have characterised an Early/Middle Pleistocene childhood. The likely communal and cooperative nature of hunting (Chap. 6) would further influence skills learning, with the presence of older hunters probably resulting in a highly conservative 'many-to-one' mode of transmission (Hewlett and Cavalli-Sforza 1986).

Might the transmission of different types of knowledge have differentially involved adult males and females depending on the nature of the associated task? This is evident amongst chimpanzees (*e.g.* Boesch and Boesch 1984; Boesch 1991) and would seem likely, if not inevitable, in a Lower Palaeolithic context *if* degrees of male/female bias existed in terms of habitual tasks (see also Chap. 4). Such bias is evident in chimpanzees with respect to frequency of hunting activities (Gilby *et al.* 2017) and is well known amongst modern hunter-gatherers (*e.g.* Bliege Bird *et al.* 2009). The archaeological record for the Lower Palaeolithic period is frustratingly limited on such issues but Gilby *et al.* (2017) noted that low female chimpanzee hunting rates and a female focus on low-cost prey (*e.g.* terrestrial/sedentary) may stem from risk-averse foraging strategies (*i.e.* reducing the energetic demands and the levels, and consequences, of failure) and the potential for losing carcasses to males, rather than any constraints of maternal care (*i.e.* clinging offspring). Potential loss of carcasses to males might be minimised by female/male pair bonding and associated changes in sharing and provisioning behaviour amongst Lower Palaeolithic *Homo* (although the evidence for pair bonding is ambiguous; Chap. 4), and maternal care practices and constraints are clearly not directly comparable between chimpanzees and Lower Palaeolithic hominins. However, a female emphasis on risk-averse hunting/foraging may have deep evolutionary origins that pre-date offspring provisioning (Gilby *et al.* 2017) and may well be relevant to both foraging behaviours and knowledge transmission mechanisms in small Lower Palaeolithic communities where the inherent value, and nutritional demands, of female individuals and their offspring may have been further accentuating factors.

Another key experience during older childhood, and particularly adolescence, might have been observation and participation in infant-care practices, alongside other social and sexual behaviours. Kennedy (2003) has argued that adolescence was an especially key period, when youngsters who were close to maturity could observe and participate in the social, sexual and infant-care practices that are critical to success in adulthood, while Bogin and Smith (1996) have also emphasised adolescence as a period when parenting skills could be practised, perhaps resulting in the greater survival of the practitioner's own offspring later in life. These participatory contributions from adolescents might also be critical to the success of the alloparenting model discussed in Chapter 4, although adolescence may have been relatively short in Early and Middle Pleistocene *Homo* in comparison to modern humans (Thompson and Nelson 2011; Schwartz 2012). Thus alloparenting and other contributions may have begun during middle childhood, at least in the very earliest Europeans. However post-1 mya the available dental evidence is suggestive of more prolonged maturation and clear stages of both childhood and adolescence (Chap. 4: Box M; Bermúdez de Castro *et al.* 2003b), and this might have delayed participation in alloparenting (although it could also have extended the period over which children could provide alloparenting). The potential role of older infants in childcare also has interesting parallels with another social carnivore: amongst the wolves of the BPF Jędrzejewski *et al.* (2001) recorded that the daily movement distances of sub-adult females remained depressed into July, suggesting that they acted as carers while the adult females, after birthing in May–June, were hunting away from the dens and their pups.

The impact of a potentially shorter childhood has also been considered by Hopkinson *et al.* (2013), with reference to the apparent conundrum of the lack of long-term cumulative or directional change in handaxes and other LCTs. In short, how can this rather 'un-modern' technological characteristic be married to the broadly modern life-history of Acheulean hominins? They noted that if childhood and adolescence was relatively short (Chap. 4: Box M), then opportunities to experiment or innovate in technical skills might have been relatively limited, not least because of the other pressures on time (*e.g.* learning the sexual and social politics of Lower Palaeolithic life). A further cause may well have been the social and demographic context of the technology, and in particular how changes came to be, and how they came to be shared more widely (or not). The arena for innovation might have been limited by relatively small group sizes, both because of the relatively limited sources of 'inspiration' but also because small intimate and effective social networks (*c.* 5 and 20 individuals respectively; Gamble 1998b) would be likely to favour the learning and refining of existing skills rather than innovation, given these small networks' emphasis on stability, confidence and trust. Hopkinson *et al.* (2013) argued that small social groups are generally hostile to novelty and that most novel behaviours were probably quickly suppressed in the early Palaeolithic (although if they *were* adopted then that self-same small group conservatism would probably have helped to maintain it). In short, the focus was on reproducing the existing society as opposed to teenage rebellion in the Early and Middle Pleistocene.

More broadly, Hopkinson *et al.* (2013) emphasised the relationship between the local population and the meta-population (the population of populations) and three key factors: first, that the transmission of innovations between local populations will increase the longer each local population survives, with smaller groups more vulnerable to age structure and sex ratio variations and therefore to extinction; secondly, that the risk of extinction for local populations is lessened by larger foraging territories, as these will enable the group to benefit from the asynchronous histories of mosaic landscapes (*i.e.* that the different landscape components will respond differently to the same climate changes); and thirdly, that large local populations encourage immigration, whereas small populations do not. This led Hopkinson *et al.* (2013) to relate the Acheulean characteristics (local, short-lived 'novelties' and global 'stasis') with a familiar life history, albeit with a reduced-duration childhood, small, dispersed local populations, limited foraging ranges, and low migration. This socio-demographic perspective is to some extent supported by the handaxe record (Box N), which at certain sites shows evidence of a modal form – the material expression of a shared and persistent 'way of doing' (*e.g.* Boxgrove; Roberts and Parfitt 1999). Alongside this it seems likely that young hominins would also learn a suite of 'base-line' or universal knapping skills, which could underpin all Lower Palaeolithic core and flake working (White 2000). Assessing and dating the appearance of extended childhood and adolescence, and modelling demography, is thus critical (although difficult; Box O), and might perhaps be a key factor in the shifting intensity of European occupation over the course of the Lower Palaeolithic.

Socialisation in Lower Palaeolithic Europe might be further complicated both by the relatively large communities (although seemingly not the day-to-day groups), as predicted by neocortex size and the social brain hypothesis (*e.g.* Dunbar 1998; 2007; Gamble *et al.* 2014), and perhaps also by periods of individuals' absences. As noted elsewhere short fission–fusion events (*i.e.* dF–F; Chap. 4: Box K) would probably be a regular occurrence in the seasonal mid-latitudes throughout the year (Grove *et al.* 2012), arising from dispersed resource distributions, dietary demands, and lower

Box O: Estimating populations: a guessing game?

Group sizes, whether at the level of the day-to-day band or the regional community, would be controlled by the need to maintain ecological viability, *i.e.* sustaining healthy reproduction and not imperiling the group. Yet specific population estimates are notoriously difficult to calculate in archaeology, due to factors such as excavation and survey bias, temporal (and spatial) palimpsests and the uncertain relationships between material remains (artefacts, buildings, food residues) and group numbers. These problems are certainly applicable to the Lower Palaeolithic and are perhaps exacerbated by the relative shortage of occupation sites, the large error-margins on absolute dating techniques, possible differences in hominin life history and the limited availability of genetic data-sets.

Nonetheless various attempts have been made (see also French 2016). Dennell *et al.* (2011) suggested European population sizes of 3000–5000 (warm stages) and 1500–2500 (cold stages), with effective breeding populations of 1200–2000 (warm stages) and 600–1000 (cold stages), based on estimates of 40% of the population as reproductive (also suggested by other estimates of age profiles in bands of 25; Kelly 1995). However, these figures were partially based on previous estimates for the Upper Palaeolithic (Bocquet-Appel *et al.* 2005), which were calculated using modern hunter-gatherer densities and numbers of sites. Such hunter-gatherer demographic data should be used cautiously, given the significant differences between contemporary global contexts and the Pleistocene 'world of hunter-gatherers' (French 2016). Is it possible to suggest figures on the basis of Lower Palaeolithic data?

As a starting point, 'Europe' provides a potential maximum occupation area of *c.* 2,165,000 km² (rounded up to the nearest thousand). This includes all countries west of a 'line' linking Turkey–the western edge of the Black Sea–Moldova–Ukraine–Belarus–Lithuania and south of Denmark and excludes mountainous territory,[1] although it is acknowledged that relatively high altitude Lower Palaeolithic occupations are known, *e.g.* at Atapuerca, and that relative altitudes, and land extents, would have varied in response to sea-level fluctuations. Linking this with hunter-gatherer population density estimates for *c.* 40–50° latitude (Grove *et al.* 2012, fig. 1; 0.1–0.2/km²) generates overall warm stage population estimates of 216,500–433,000. For cold stages a maximum occupied core territory of *c.* 779,000 km² is suggested here (including the Iberian and Appenine peninsulas and the Balkans; Dennell *et al.* 2011), and this produces population estimates of 77,900–155,800. These figures are over an order of magnitude larger than published figures for European Upper and later Middle Palaeolithic populations (*e.g.* 4400–5900 inhabitants from the Aurignacian to the LGM; Bocquet-Appel *et al.* 2005; Bocquet-Appel and Degioanni 2013), and clearly highlight the problems of applying modern hunter-gatherer densities to Pleistocene populations.

What other options are available? If foraging band sizes are assumed to be 43 (derived from estimates in Gamble *et al.* 2014), and a territory of 3318 km² (diameter: 65 km) is suggested (utilising the maximum lithic transfer distances from Arago Cave; Wilson 1988), population densities of 0.01/km² are generated. While the Arago lithic transfer data may well be a palimpsest of mobility behaviours, a territory of 3318 km² falls within the estimates of *c.* 1400–5400 km² [*c.* 40–80 km diameter] suggested for Neanderthals using a 'wolf model' (Churchill *et al.* 2016), and could be walked across in 2–3 days. Given the area estimates of habitable Europe presented above and a population density of 0.01/km², the overall hominin population numbers are now, respectively, 21,650 (warm stage) and 7790 (cold stage). These are still roughly 4–7 and 3–5 times larger than the estimates of Dennell *et al.* However, these numbers assume that hominins occupied *all* the available space *all* the time, with core band areas effectively forming a series of tessellated tiles. This seems unlikely for two reasons: first, the level of competition from other predators, especially in the Early Pleistocene (*e.g.* Rodríguez *et al.* 2012; Rodriguez-Gomez *et al.* 2017); and secondly, because inherent in hunter-gatherer mobility models is the notion of regular residential moves, in order to stay ahead of the problem of resource exhaustion within the local or core foraging area (Chap. 3: Box I).

The frequency of core area moves (and of returns to a previous location) is difficult to estimate, but a potentially interesting temperate forest perspective is offered by the BPF wolves. Their core areas and home ranges were significantly different in size: 11–23 and 173–294 km² respectively (Okarma *et al.* 1998). Thus, the wolves' core area was just 5–13% of their overall home range, a ratio of roughly 1:10. Such a ratio might be less applicable to hominins however, given the carnivorous nature of wolves and the proposed omnivorous diet of hominins. If, following Cordain *et al.* (2000), a *c.* 35:65% balance of plant:animal foods is suggested (with the latter a mix of lean meat and fats to avoid exceeding the protein ceiling), then a ratio of 1:6 might be more appropriate. If, as seems likely for the Lower Palaeolithic, lithic transfer data reflects the hominins' core foraging area, then an overall home range could be *c.* 20,000 km² (*c.* 160 km diameter; using the Arago data, adjusted using the suggested 1:6 multiplier). This larger area might be the landscape through which the hominins moved residentially across the year, while the majority of foraging, for plants and animals, occurred at any one time within the core foraging area, although longer hunting trips (*e.g.* associated with mobile or low-density prey) might occur across the entire home range.

This generates a revised population density of *c.* 0.002/km² (and significant 'hominin-empty' landscapes at various points in the year), and by extension population numbers for habitable areas of Europe of *c.* 4330 (warm stages) and 1558 (cold stages). Effective breeding population estimates, using the suggestion that *c.* 40% of each group were reproductive, are therefore *c.* 1732 (warm stages) and 623 (cold stages). Interestingly these figures are broadly comparable with those of Dennell *et al.* (2011).

Using Birdsell's 'magic number' of 500, these data suggest the presence of roughly nine and three regional populations in the warm and cold stages respectively (if using the mega-band [n=384] of Gamble *et al.* 2014 the numbers change to roughly 11 and 4). However contact and genetic exchange between many of these populations would be unlikely at any particular moment, with only home range 'boundary' zones being active areas of exchange. This is based on the notion that ecological sustainability (*i.e.* the dangers of resource exhaustion) would effectively prevent, by behavioural choice or extirpation, significant encroachment into temporarily 'empty' areas of another groups home range: limited home range overlapping is also evident amongst the wolf packs of the BPF (Okarma *et al.* 1998, fig. 1). However increased genetic exchange *would* periodically occur as fragmented cold stage populations re-expanded from southern core areas, leading to recombinations (Fig. O.1) and avoiding long-term population isolation (Dennell *et al.* 2011).

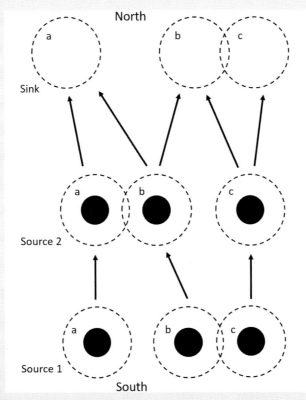

Figure O.1: Population expansion and demic inter-mixing (Dennell et al. 2011, fig. 6).

One last, chronological, caveat. The question of prey and predator density adds a further layer of complication. While modelled prey biomass (kg/km²) in the BPF is broadly comparable to that at Amalda V, Atapuerca-Galería IIa/IIb and Venta Micena, the number of predators in the BPF is approximately half that at Venta Micena (although the BPF is broadly comparable to the estimates for Amalda V and Atapuerca-Galería IIa/IIb; Rodríguez *et al.* 2012). Home ranges (and perhaps also core areas) might therefore be even larger in the Early Pleistocene – this would be in-keeping with the high level of carnivore competition inferred by various studies (*e.g.* Rodríguez-Gómez *et al.* 2016; 2017), and might be equally applicable irrespective of the balance of hunting and scavenging within the hominin strategy. Purely as an illustrative exercise, this might change the core area and home range to 6600 km² (diameter: *c.* 90 km) and *c.* 40,000 km² (diameter: *c.* 225 km), with likely implications for the frequency and scale of group (*i.e.* residential) mobility, and perhaps also for long-term hominin sustainability in the Early Pleistocene.

All the above is intended primarily as a 'number thinking' exercise and the suggested numbers should certainly be treated accordingly. They do, however, highlight the potential of combining lithic, carnivore and modelling data as one possible means of approaching the vexed question of Palaeolithic demography.

---

[1] This occupation area estimate is derived from data in the European Commission's *Mountain Areas in Europe - Final Report* (table 3.2; https://ec.europa.eu/regional_policy/sources/docgener/studies/pdf/montagne/mount4.pdf; European Commission 2004), where mountainous areas were defined by a combination of altitude and local relief. Therefore, this altitudinal 'cut-off' does not ignore the likely importance of lower elevation 'uplands' (*e.g.* the Chalk landscapes of northwest Europe; Blundell 2020), which could have offered a range of resources and opportunities to hominins, including raw material sources, spring-lines, and vistas for observing game movements and carcass resources (*e.g.* Wymer 1999; Kolen *et al.* 1999).

population densities. From a seasonal perspective, longer temporary group fragmentations (over multiple days?) might be differentially associated with the milder and longer days of late spring and summer.

Finally, the nature of all this learning would vary across the year, reflecting nothing more than annual ecological rhythms – the timings of the deer rut and the seasonal fruiting of trees for example (Chap. 1). But would the hominin year simply 'dance to the rhythms of the Pleistocene' (Gamble 1999, 125)? This is difficult to address with certainty, but it seems likely that the frequency of any social learning, from direct instruction to simply observing what others do (and trial and error learning), would vary with the seasons and the ecological stresses faced by hominins – short winter days (see Table 1.1), an urgent need for food, and the potentially fatal costs of hunting failure might well be a less conducive learning environment than long summer hours of sporadic game observation. Thus, the acquisition and practising of key skills in lower-stress environments may have been an important aspect of Lower Palaeolithic summer life, although such occasions would likely be defined by place as well as time. For example, in highlighting the impact of predation threats on carcass transport decisions and thus on the duration of processing time at the kill/butchery site, Saladié *et al.* (2011) may be indirectly demonstrating the value of butchery skills being learned and practiced by older children at home bases or during child-only foraging trips, prior to employing them in the potentially time-limited scenario of a fresh kill. Can we see evidence of such learning directly? This is again difficult, but high rates of knapping errors (*e.g.* step and hinge fractures, hammer marks, core rejuvenation strategies; Finlay 2015) may be indicative of this aspect of Lower Palaeolithic life. However, to date there have been very few attempts to look for such material in the early Palaeolithic or even explicitly consider what it might look like (Shea 2006), although Stapert (2007) has proposed a number of potential examples of children's artefacts, both learning pieces and possible 'toys', in the European Middle Palaeolithic.

Childs' play ...?

While focusing on Neanderthals, Spikins *et al.* (2014) have emphasised a secure childhood and internal bonds, based on factors including small, isolated groups, the importance of attachment security in 'normal' cognitive/emotional development (Fig. 5.2), and the strong visibility of children in burial evidence. While the latter is absent in the Lower Palaeolithic, the former seems likely based on ecological and lithic transfer data (Table 5.9). Spikins *et al.* also stress the likelihood of mother/carer–infant and infant–infant play (it occurs widely in humans, primates and social carnivores; *e.g.* Bekoff 1974), with its benefits including exploration, safety and advanced group cohesion. They suggest the possible importance of small artefacts as a means of both learning techniques and also becoming familiar with the various uses of such artefacts, and include selective examples. Security and social cohesion might also have been built and reinforced through reliable food sharing, while repetitive and familiar artefact forms (*e.g.* handaxes; Box N; Figs N.1 & N.2) might have contributed to a wider sense of conformity and belonging (see also Hopkinson *et al.* 2013). Pleistocene play, alongside socialisation, may also have played a significant role in developing attention, memory and inhibition, as part of an enhanced working memory capacity (Nowell 2010).

Gray (2009) links play and its key components (*e.g.* self-chosen and self-directed; intrinsically motivated; structured by mental rules; imaginative; and produced in an active, alert but non-stressed frame of mind) to various, oft restated, qualities of many, but not all, hunter-gatherer societies: autonomy, equality, sharing and consensus. While specific parallels are difficult to make, two arguments in particular may well have applied to Early and Middle Pleistocene children in Europe: age-mixed play (likely to be the case in small bands, with a probable inter-birth interval of 2–3 years)

Figure 5.2: H. heidelbergensis *child and adult female (© Mark Gridley).*

tends to be a less competitive and more nurturing environment; and the presence of such mixed-age groups would also support a scenario, common amongst modern hunter-gatherers, in which much childhood learning is through self-directed exploration and play. A diverse set of adult behaviours could thus be mimicked through their incorporation into play (*e.g.* caring for infants and other social interactions, imitating animals, tool use, stalking and tracking).

It is also very likely that Lower Palaeolithic children (and adults!) laughed as well. Gray (2009) emphasised the widespread presence of humour (to 'prick' egos and ensure humility) and play (as a lifelong mechanism for coping with, not escaping from, the dangers and difficulties of life) amongst modern hunter-gatherers: in short, play is not just for children, although it is more common. The role of laughter from early *Homo* onwards has also been considered by Gamble *et al.* (2014) in the context of the social brain hypothesis, as a possible means of promoting endorphin release (and thus building/maintaining relationships) amongst larger groups than can be reached by one-to-one grooming alone (see also Chap. 2).

From a material perspective it is obviously difficult to identify children's artefacts in the Palaeolithic, not least because it is unclear what we are seeking. There are certainly examples of small artefacts in the archaeological record (Fig. 5.3) but their size may reflect many other factors, such as raw material size, resharpening intensity or specific functional needs, rather than being a simple proxy for smaller children's hands. Perhaps unsurprisingly, a comparison of selected quantitative 'refinement' attributes (degree of symmetry, cortex percentage, percentage of circumference worked and the thickness/breadth [cross-sectional thinning] index) for 'small' (length < 80 mm) British handaxes against the remainder of the sampled assemblages (Table 5.10) suggests no significant differences between the 'small' and 'larger' handaxes, indeed if anything the smaller handaxes were 'better' made. This doesn't mean that children's artefacts are absent in the Lower Palaeolithic but it is likely that detailed technological (Stapert 2007), rather than quantitative, studies will be needed to detect them.

Figure 5.3: Selected 'small' (length [L] <80mm) chert and flint handaxes from Broom, UK (images © Jennifer Chambers). Left: chert handaxe (L: 67.4 mm; artefact no. 49); Middle: chert handaxe (L: 73.8 mm; artefact no. 57); Right: flint handaxe (L: 76.4 mm; artefact no. 96). Scales: 10 mm intervals.

Table 5.10: Comparison of 'small' and 'large' LCTs from Bowman's Lodge, Broom and Warren Hill (Marshall et al. 2002)

Site	Size[1]	n	Th/B[2]	% circum. worked[3]	% cortex	Symmetry[4]	
						Planform	Profile
Bowman's Lodge	L	16	0.466	86.5	4.4	0.030	0.035
	S	13	0.411	94.3	0.8	0.014	0.030
Broom	L	239	0.442	84.1	2.6	0.019	0.030
	S	14	0.412	91.9	1.5	0.014	0.029
Warren Hill	L	250	0.439	93.1	2.4	0.019	0.026
	S	91	0.389	96.8	1.3	0.009	0.017

[1]L: 'large' LCTs (length > 80 mm); S: 'small' LCTs (length < 80 mm); [2]Th/B: thickness/breadth; measure of cross-sectional refinement, where lower values = thinner cross-section; [3]Percentage of the LCT's edge which has been deliberately shaped through secondary retouch; [4]Recorded using the Continuous Symmetry Measure (CSM); values of 0.000 = 'perfect' symmetry.

A world beyond the horizon?

While the possible concentrating of distant foraging and/or raw material collecting in the longer days of summer may seem logical, it raises a key question: what were the scales of Lower Palaeolithic hominins' landscapes of habit or local [hominid] networks – the stages for the habitual, day-to-day routines of hominin life (Gamble 1996a; 1996b; 1999)? The great challenge in answering this question is that methods are heavily reliant on lithic transfer data, which clearly has the potential to underestimate the size of habitual landscapes (if lithic material, due to its weight, was transported as little as possible) but also to potentially overestimate them (if the material found at a single site represents, through yearly or seasonal time-averaging, the intersection of multiple landscapes associated with different groups or with the same group but at different times). The former issue may explain one of the curiosities of the Lower Palaeolithic record: the apparent disconnect between the strong ecological arguments for relatively large territories or home ranges, in line with those seen for other carnivores, and the limited archaeological evidence for comparably large habitual landscapes. This is especially true in mid-latitude regions such as Europe, where the lower effective temperature, greater seasonality, greater reliance on animal foods and more clumped and dispersed resources all appear to favour a mobile strategy (*e.g.* Kelly 1995, table 4-1 and figs 4-7 & 4-8). Such strategies are evident amongst other social carnivores such as wolves, including at an intra-continental scale: their home ranges increase from *c.* 80–240 km^2 in southern and central Europe to *c.* 415–500 km^2 in northern Scandinavia (Okarma *et al.* 1998). Yet much of the lithic transfer distance data from the European Lower Palaeolithic is very small in scale (White 1998; Féblot-Augustins 1999; Table 5.9), particularly when comparing wolf pack sizes (typically a dozen or fewer) with suggested hominin band sizes (n=43, after Gamble *et al.* 2014, table 2.1).

In some specific areas (*e.g.* southern Britain and northern France) the disconnect may also reflect the analytical difficulties of sourcing a widely used but difficult to distinguish raw material: Cretaceous flint (although Pettitt *et al.* 2015 have achieved recent success in tracking Late Upper Palaeolithic lithic transfer and mobility patterns). More broadly however the pattern may simply reflect the physical demands of carrying raw materials, and perhaps even finished tools, around the landscape: in other words, while hominins may have been highly mobile for other foraging purposes, their tools were made, used and discarded at a local scale. Thus, perhaps stone was often casually acquired from the immediate locality as and when required – and just as rapidly returned to the ground surface after usage. In that regard the recent teaching-inspired observations of Shea (2016), that a skilled user can deploy a stone tool to cut almost anything and that any associations between form and function are likely to be highly transitory, are notable and might help to explain brief artefact lifespans and short lithic transfer distances. Such transfer patterns might also fit with a scenario in which the principal tools carried were organic (a digging stick or spear perhaps[4]), with many stone tools made, used and discarded expediently as needs dictated (*e.g.* for rejuvenating a wooden tool). A long spear or digging stick might well have been a preferable weapon over a stone tool if unexpectedly faced with an aggressive predator in the woods ...

However, there are occasional exceptions to this pattern of very local material sources, most notably La Caune de l'Arago in southern France. Here, while the majority of the raw materials in the 10,000+ artefact assemblage was sourced from the Verdouble river below the cave, there is also evidence for more distant sources, located 7–35 km from the site (Wilson 1988). These non-local sources vary in quality and were described by Wilson as 'poor' to 'excellent', suggesting that their transportation around the landscape was structured by more than a simple decision for the best knapping material – although the fact that the exotic materials are more common amongst the flake tool component of the Argo assemblage suggests that knapping quality was at least one factor in their acquisition and use. These exotic materials were both brought back to Arago and knapped and introduced as finished tools. This gives a clear sense of Arago as part of a wider hominin landscape across which hominins sometimes acquired and transported raw materials to be knapped at Arago and, on other occasions made and maybe also used tools at other locations and then transported those tools to Arago. Wilson (1988) has suggested, on the basis of the range of exotic lithic raw materials in each occupation level, that the minimum exploitation range was 65 × 30 km. If calculated as a hypothetical circular territory, encompassing both Roquefort-des-Corbières (30 km to the northeast of Arago) and Vinça (35 km to the southwest), these transfers suggest a range of 3318 km². In contrast to estimates generated from other sites' lithic transfer data this is an order of magnitude larger than European wolf territories (Okarma *et al.* 1998), and is comparable in scale to the average residential move distances for a number of the hunter-gatherer groups reviewed by Binford (2001, table 8.04; selected data in Table 5.11) and Kelly (1995, table 4-1; selected data in Table 5.11). It is also broadly comparable with Steele's

Table 5.11: Selected territorial area and mobility data for hunter/gatherer groups in (1) subarctic and continental mid-latitude forests of North America and Asia (n=37; data from Binford 2001, table 8.04) and (2) temperate and boreal forests (n=24; data from Kelly 1995, tables 4.1 & 4.3)

Sources	1: Binford (2001)			2: Kelly (1995)[1]		
n	37			24		
	Mean	*Min.*	*Max.*	*Mean*	*Min.*	*Max.*
Residential moves/year	11.2	0.10	20.0	17.6	1.0	60.0
Average distance/move (km)	33.8	1.60	55.4	19.2	4.3	64.0
Total distance (km)	425.3	0.16	793.6	160.1	8.6	510.0
Total area (km²)	ND			829.5	32.0	3385.0
Logistical mobility (days)	ND			34.0	27.0	48.0

[1]Data calculated without Crow values, due to extreme size of total area (61,880 km²). Data calculations excluded cases where value ranges were quoted.

(1996) predictions of home ranges, using body mass and group size, which suggested range diameters from *c.* 30–90 km for Neanderthals and *H. erectus* (with community sizes varying from 25 to 300).

So is Arago the exception or the rule when it comes to the scale of habitual landscapes? One key point is that it is not unique as even longer transfer distances are suggested by the flint artefacts at Waverley Wood, in the UK (the nearest sources are *c.* 100 km to the east; Moncel *et al.* 2015) and the upper level at La Noira (Iovita *et al.* 2017), but Arago's overall atypical-ness probably is evidence that lithic material was generally sourced from as nearby as possible. These much shorter distances are also evident at both open air and other cave sites (*e.g.* Gran Dolina TD-6.2). Nor should this be surprising, given the weight of lithic nodules and cores, the varying impacts of terrain on ease of journey (Wilson 2007), but also the potential risks involved in long Pleistocene journeys for a relatively slow biped without sharp claws and teeth. These generally short distances also offer little support for the idea that lithic provisioning was regularly embedded into longer foraging trips, or that material was transported during residential moves. To some extent this is also evident at Arago, where there is no clear evidence that the proportions of exotic materials varied in response to shorter (seasonal) and longer-term occupations (Table 5.12).

If the Arago-based estimates are broadly representative of hominin home range sizes then one potential cause of the contrast with the European wolf data is likely to be differences in group size: Okarma *et al.* (1998) suggested a mean pack size of 4–5 individuals. Hominin group estimates are fraught with difficulty (Chap. 2), but the social brain hypothesis (Dunbar 1998; 2007; Gamble *et al.* 2014) and ethnographic studies of local group sizes in nomadic hunter-gatherers (Kelly 1995, table 6-2) might suggest bands of 25–50 individuals. The inevitable increase in animal food demands, which would be significant at European latitudes, despite the likely mixed diet of hominins, could explain the order of magnitude difference between the wolf data and the Arago estimate, as might possible differences between the predatory efficiencies of Lower Palaeolithic hominins and wolves.

Table 5.12: Comparison of occupation and exotic raw material data for Arago Cave (Wilson 1988; McNabb 2007, table 8.2)

| Arago unit | Length & timing of occupation | Exotic[1] raw materials | | | | | |
| | | Flakes | | Small tools | | Choppers | |
		%	n	%	n	%	n
D	Long-term	0.0	2	100.0	2	40.0	5
E	Long-term	35.2	329	58.3	24	13.0	23
F	Seasonal, spring to summer	35.3	1203	53.1	567	8.9	124
G	1 year or more	27.0	2799	47.2	411	2.5	445
J	Seasonal, autumn	40.5	37	13.3	15	100.0	1

[1]Exotic raw materials include sources from a minimum of 9–35 km from the cave (Wilson 1988)

Summer conception: hunting with benefits?

Alongside foraging and provisioning there may have been one final complication to mobility in the summer. Conception has been suggested to be concentrated in late summer on the basis of annual cycles in food availability and individual condition (Mussi 2007). But if day-to-day group sizes were at the band scale (*i.e.* perhaps *c.* 40 individuals), or smaller, how was genetically-viable reproduction maintained? A fission–fusion model that periodically brought together the community (*c.* 130 individuals) or the 'mega-band' (*c.* 400 individuals) appears central to this (Grove *et al.* 2012), but what stimulated such aggregations? Seasonal instinct? Tradition? Perhaps bands encountered one another on the fringes of the late summer/autumn animal aggregations that were a focus of hunting activity (Chap. 6)? The tendency of modern deer and horse populations to maintain traditional rut territories and 'stand' locations might make such hominin encounters more likely (if potentially unintended).

Another possibility is that site occupations might have been shorter, and residential moves longer, during the late spring, summer and early autumn periods, as hunting and/or scavenging debris would be likely to attract a variety of insect life (Yellen 1977, 67) and carnivores (Grove 2009). Larger-scale residential moves might enhance the possibility of encounters with other bands at the margins of a territory, especially if residential moves were structured by discrete resources (*e.g.* watercourses, food patches, fixed shelters and/or stone outcrops).

Considering the nature of such hominin band encounters is inevitably speculative but insights from other social carnivores (*e.g.* wolves) suggest that they would likely be a mixture of inter-group aggression (*e.g.* Cassidy *et al.* 2015) and perhaps also the occasional dispersal of individuals between bands (*e.g.* Lehman *et al.* 1992). Moreover, it is likely that such encounters might have been unpredictable, noisy affairs, perhaps with a parallel to wolf howling during aggressive encounters (Harrington 1987). Aggressive encounters with 'other' hominins (and territory defence) are suggested by the exo-cannibalism at Gran Dolina TD-6.2 (Carbonell *et al.* 2010; Saladié *et al.* 2012). While the demographic profile of those cannibalized remains suggests a focus on infants and juveniles (*i.e.* low-risk targets), with implications for the reproductive abilities of that other group, this particular approach to inter-group relations cannot have been a sustainable strategy for genetically viable populations unless reproductive females or males were also taken as part of such encounters, whether willingly or not.

Males and/or females may also have moved voluntarily between groups during these encounters: *i.e.* an exchange of individuals as well as genes. However while this might have been motivated by individual rivalries or tensions within specific groups, I suspect it may have been the exception rather than the rule. This is suggested by the evidence for locally-focused, intimate lives (*e.g.* artefact conservatism, small-scale lithic transfer distances, small site 'footprints'; see also Hopkinson *et al.* 2013), in contrast to the 'social alliance' archaeology of the Upper Palaeolithic (Gamble 1982). This might well have been a social environment in which 'others' or 'outsiders' were generally excluded and in which attempts to relocate to another group were rare and unwelcome.

Much of the above is inevitably speculative, but the issues are important, since the small-scale site evidence (and ecological implications of permanent large communities) present a clear contradiction to the question of population viability and the widely cited, albeit problematic, numbers for breeding populations of 500 (Birdsell 1953) and 175–475 (Wobst 1974; see Kelly 1995 for a summary of the various caveats). What does seem likely is that any such aggregations broke up before moving on, since long moves for large groups are costly, especially when there are many young (Grove 2009), and presumably longer moves would be necessary for individual bands to relocate to their own, non-overlapping, foraging areas.

Long days ... and dry days

Summers were therefore long days, rich in plant foods. But they were not necessarily always 'easier' days, with a probability of dispersed prey and, especially in the

Figure 5.4: A summer strategy.

Mediterranean and central Europe, potential problems of heat, aridity and water availability. Nonetheless those longer days may have been a favourable time of year for time-demanding tasks such as lithic resourcing and organic tool-making: possible indicators of an ability to anticipate, and plan for, future needs (Fig. 5.4).

Long summer days, with an associated relative abundance of resources, may also have been a significant time for children to participate in a wide, probably age-dependent, range of tasks: these were opportunities to acquire knowledge, gain experiences and practice skills. Late summer may also have been the start of a socially dynamic, perhaps fraught, time of year: this was probably structured by traditional animal aggregations and hunting opportunities, but perhaps was also the focus for inter-group encounters, which seemingly ranged from breeding, to conflict and cannibalism, and perhaps some cooperation.

Notes

1. None of the above discussion should be used as a 'safety guide' with regards to the picking and consumption of wild mushrooms or other fungi. Any readers wishing to do so are strongly recommended to consult an appropriate, dedicated guidebook, such as Mabey (2012).
2. During her experiments at Reading Helme (2017) also favoured the use of an unmodified fresh flake as the ideal tool for tip-shaping.
3. Alongside ethological knowledge Laughlin (1968) emphasised the 'body training' exercises undergone by Aleut children in preparation for kayak hunting, although within a strongly pro-hunting framework and a dismissive attitude to plant foraging ('the amount of information which must be exchanged between plant eaters is small compared with that needed in group-hunting of large animals'; *ibid.*, 318).
4. The extensive embedding of microscopic flint fragments in the heavily worn deer antler soft hammers from Boxgrove is suggestive of *their* curation, and probable transport (Pitts and Roberts 1997, fig. 54; Stout *et al.* 2014).

Chapter 6

Autumn – rich in food and colour

Wild harvests … and shorter days

Autumn in mid-latitude Pleistocene Europe would, in many ways, have been the opposite of spring: shortening days, declining temperatures, increasing precipitation, and the gradual 'opening up' of the landscape. But autumn would also have provided an abundance of food resources: animal breeding aggregations and, potentially, migrations, and a range of plant foods: nuts, late-year fruits and fungi. From a hominin perspective this season of change and colour would have been both dynamic and bountiful and perhaps also critical to survival through the upcoming winter. While climatic conditions again varied over time, latitude and local topography, the general trends from September to November of decreasing temperature and increasing precipitation (Table 6.1 & Fig. 6.1; with the partial exception of Gran Dolina TD-6.2) are familiar, with the marked differences between September and November being especially notable.

Ruts and nuts

The importance of dietary fats, as well as carbohydrates, was highlighted in Chapter 2. In that context, the seasonal fluctuations in mid-latitude foods from summer/autumn to winter (*e.g.* animals lose their body-fat reserves, plant foods become less available) are highly significant. Speth (1991a) suggested a number of strategies to address these seasonal variations in resources, including a selective approach to the animals hunted at particular times of year (*e.g.* avoiding ungulate females when they are pregnant/nursing, but targeting them in the summer and autumn; targeting males in the late winter/spring). It is difficult to test this model against the Lower Palaeolithic record due to the paucity of seasonality data, but it can perhaps be considered in the context of ungulate behavioural cycles, and the concentration of breeding in later summer and autumn (see Table 4.5 & Fig. 1.8).

Table 6.1: Autumn temperature and precipitation data for selected Spanish Early and Middle Pleistocene sites (Blain et al. 2013; 2014; 2016)

Site	Month	Temperature (°C)			Precipitation (mm)		
		Mean	*SD*	*Range*	*Mean*	*SD*	*Range*
		Early Pleistocene					
	September	21.9	2.8	16.0–24.0	35.0	12.0	30–80
Fuente Nueva-3	October	16.3	2.5	10.0–18.0	78.0	21.0	60–130
	November	12.6	2.8	6.0–15.0	110.0	28.0	70–150
	September	18.3	1.7	–	70.0	0.0	–
Gran Dolina TD-6.2	October	12.5	2.0	–	92.5	12.8	–
	November	8.3	1.7	–	70.0	9.3	–
		Middle Pleistocene					
	September	21.8	2.6	16.0–24.0	30.6	12.0	20–60
Aridos 1	October	15.4	2.8	10.0–18.0	71.5	18.0	50–110
	November	11.8	2.8	6.0–14.0	96.8	22.0	60–140

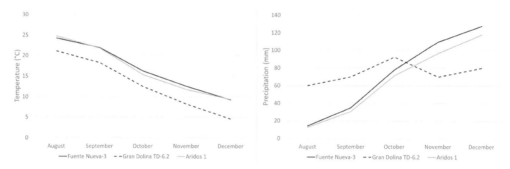

Figure 6.1: Autumn temperature (left) and precipitation (right) trends for selected Spanish Early and Middle Pleistocene sites (Blain et al. 2013; 2014; 2016).

The nature of rutting behaviour appears to be variable: in the case of modern red deer, examples include harem defence (*i.e.* males joining and following a group of females, restricting their movements and defending the harem against other males) and territory defence (*i.e.* settling in an area, defending it against neighbours and not courting females outside the territory; Carranza *et al.* 1990). Different rutting behaviours would presumably impact on the mobility requirements of hominin foraging, *e.g.* following a harem or targeting a local territory. The scale of resource associated with deer ruts would also likely vary: in the case of modern British red deer, harem sizes alter according to habitat (Corbet and Harris 1991). However, as deaths are not uncommon during red deer ruts, this might be an additional potential source of carcasses and a further cause for hominins to 'shadow' a rut. Locating the resource may have been aided by the tendency of modern red and roe deer rutting areas to follow traditional locations (although this may in part reflect the characteristics and management of modern landscapes and habitats) and these locations could well have structured residential moves by hominins (and range shifting by other predators) in late summer/early autumn. Residential moves of the entire band rather than logistical forays would seem more likely, given both the length of modern red deer ruts (the oestrus cycle of red deer is *c.* 18 days; Corbet and Harris 1991) and the number of potential hunting opportunities. If different Pleistocene deer species' rut periods were staggered, as is the case for modern British red (September–November) and roe deer (mid-July–end of August), then there may even have been opportunities for groups to move from one resource to the next during late summer and autumn.

Tracking of red deer during rutting would be further aided by the roaring of the stags. In modern populations this is loud and repeated, although at a variable rate, during the 4–5 week reproduction period and is associated with the herding of the hinds (Reby *et al.* 2001). The question of whether rutting deer would be more vulnerable to hominin hunting is difficult to resolve. At one level, the group size during the rut could offer security through numbers – simply a case of more eyes, ears and noses. However, Bartoš *et al.* (2007) emphasised the various tactics used by the Březka deer park's fallow deer to minimise fighting at the rut – *e.g.* groaning and parallel walk displays. The key question is whether this concern with social politics might make the deer, or at least individuals, more vulnerable (or, as a counterpoint, more aggressive)?

The modern behaviours of various species highlight the potential risks and dangers associated with hunting during animal mating seasons. Amongst semi-wild modern horses stallions are especially concerned with the maintenance of their harems and territories during the mating season (Corbet and Harris 1991) and such heightened awareness and behaviour (*e.g.* circling the mares and rounding up stragglers) might extend to aggressive defence against potential predators. While bison aggression in BPF populations is generally low, males are more aggressive during the rut (Haidt *et al.* 2018). It therefore seems likely that hominins were attuned to the threat displays and aggression indicators of a range of species (*e.g.* the ear and head-threats of horses and various pre-attack warnings of bison, including head swinging, hoofing the ground and tail swinging; Corbet and Harris 1991; Haidt *et al.* 2018).

While breeding aggregations such as the deer rut would have been key events in the annual cycle, various other behaviours in the latter part of the year would also impact on resource availability. The study of non-migratory red deer in the BPF (see also Chaps 2 & 3) highlighted significant seasonal differences in mobility patterns: home ranges were largest in the autumn for males (23.0 ± 3.6 km^2), although female ranges were largest in winter (Kamler *et al.* 2008). While home range sizes can be influenced by a number of factors, including population density and food resources, a further factor is the presence of large carnivores. From a Palaeolithic perspective these range size data highlight the potential degree of mobility required to monitor and scavenge or hunt red deer, particularly towards the end of the year, in light of the diverse range of Early and Middle Pleistocene carnivores.

The question of mobility also raises the possibility of animal migrations and their interception by hominin hunters. However, it is unclear to what extent key ungulates in warm stage woodlands and forests were migratory. Modern European red deer populations appear to be both sedentary and migratory and the impacts of modern land-use and habitats on this behaviour are complex (Szemethy *et al.* 1999; Kamler *et al.* 2008). Isotope studies have also challenged migratory models for some species (Julien *et al.* 2012), while confirming them in others (Britton *et al.* 2011). It is thus difficult to assess the potential opportunities for intercepting ungulate migrants (even before considering the cognitive and behavioural demands of such strategies upon hominins).

Alongside animal foods, European forests would have offered a significant autumn bounty in the form of nuts. While direct evidence is limited, there are examples. On the margins of Europe, at the Qesem Cave site in Israel (*c.* 300–400 kya), the essential fatty acids (linoleic and alpha-linoleic acids) which are present on dental calculus samples must have originated from dietary sources (Hardy *et al.* 2016). These fatty acids are abundant in pistachio and linseed but the seeds/nuts from *P. halepensis* (the Aleppo pine) are suggested to fit especially well with the Qesem evidence. Oak and hazel nuts have high carbohydrate and fat contents (Divišová and Šída 2015) and, in the case of Pyrenean mixed forest habitats today, oak and beech nuts enable wild boar to store fat for the critical end of winter period (Herrero *et al.* 2005).[1] It is certainly possible that they served a similar purpose for Lower Palaeolithic hominins (there is evidence for acorn cracking with pitted stones at Gesher Benot Ya'aqov on the margins of Europe; Goren-Inbar *et al.* 2002), although raw acorn kernels are strikingly bitter due to their concentration of tannic acid, and can be toxic if eaten in large quantities (Šálková *et al.* 2011). Nut roasting, as in the Mesolithic (Mithen *et al.* 2001; Divišová and Šída 2015), is a possibility, but while fire traces have been reported at Gesher (Goren-Inbar *et al.* 2004; Alperson-Afil 2008) the general rarity of European 'fire sites' suggests that such behaviour may have been atypical (see also Chap. 3). Nonetheless, if nut collection was widespread, this is an activity that children, including young children, could usefully have partaken in (see also Chap. 5). Moreover, the timing of nuts' availability probably varied by species, enabling a changing, time-staggered focus on different resources: while modern beech nuts ripen in early autumn, hazelnuts tend to drop at the end of the season.

Autumn migrations and winter in the sun?

As noted in Chapter 3, available climatic reconstructions for southern European sites make a strong case for the south of the Alps and the Pyrenees as preferred winter landscapes: January temperatures from Spanish Middle (and Early) Pleistocene sites contrast markedly with those from northwestern and north-central Europe (Table 3.1). Over the longer-term the role of the Iberian and Apennine peninsulas and the Balkans as cold stage core population areas is widely accepted (Dennell *et al.* 2011). But on the short-term, annual scale, was winter residence in the landscapes of Burgos or Madrid part of a much broader, highly mobile, strategy of landscape exploitation, or just one particular season for a 'permanently' Mediterranean dwelling hominin population? In short, did Lower Palaeolithic survival involve significant annual migrations (see also Chap. 4)?

As Kelly (1995) observed, not only do hunter-gatherers move around a lot, but they also move in many different ways (Chap. 3: Box I). Might high levels of mobility have been a Lower Palaeolithic response to the turning of the year and the onset of a European winter (as numerous cohorts of students have suggested to me)? It is important to be careful here with terminology. Migration is defined as both the movement of people or animals from one region, place or country to another, but also as a journey between different areas at specific times of year. It is clearly the latter definition that applies here and is referred to below as an *annual migration*. It can be broadly linked to the hunter-gatherer concepts of the seasonal round and residential mobility (*e.g.* Jochim 1981, 148–155; Kelly 1995), but are the scales of mobility comparable (annual migrations invariably conjure up images of caribou moving hundreds of kilometres between winter feeding grounds and summer calving habitats)? Although the average residential moves of hunter-gatherers are in the order of a few tens of kilometres at most (see Table 5.11), total distances moved over a year can total hundreds of kilometres (Kelly 1995, table 4-1 & fig. 4-7). While these distances are typically not linear (as shown by the ratios between total distance and total area covered in Table 5.11), they do highlight the human potential for large-scale movements, albeit split-up into several individual journeys.

Evaluating the feasibility of an annual autumn migration as a (pre-) winter survival strategy in a European Lower Palaeolithic context requires consideration of the scale of movement necessary for the strategy to be effective (see also Ashton and Lewis 2012; Ashton 2015). Attention must also be given to the practical factors which would have influenced and impacted upon migratory behaviour: for example, resource knowledge, group composition and assessment of risk (Kelly 1995, 144–148). These issues are divided below into four categories: the gradients of climate change and habitat change in the Lower Palaeolithic landscapes of Europe; rates of hominin movement; resource knowledge; and the motivations for long distance mobility (in short, what encourages migrations to occur?).

If they occurred annual migrations would surely have been driven by the need for habitats with one, or more likely both, of the following characteristics: milder climatic conditions (*e.g.* warmer, with reduced snow cover and/or precipitation) and enhanced availability of winter foods (animal and/or plant). However, these criteria are not always mutually overlapping: for example, modern and comparative data from the Stage 3 Project (Van Andel and Davies 2003) suggests that the Atlantic West tends to be milder with reduced snowfall, but is also characterised by higher levels of rainfall. It is not assumed that hominin migrations would have pre-set 'destinations' in mind, either specific landscape locations in the style of modern hunter-gatherers (*e.g.* Binford 1980), or the winter/summer feeding grounds and spring/early summer calving grounds that structure the migrations of herd animals such as reindeer (*e.g.* Burch 1972, 345). However regular 'destinations' may nonetheless have been a possibility if yearly animal migrations were followed by hominins, although it has been argued that humans on foot have little or no ability to move 'with' long-distance migrants such as reindeer (Burch 1972). Either way, hominin movements on a day-to-day basis would be structured by landforms (*e.g.* surface drainage and relief), local conditions (*e.g.* thickness of woodland, ground surface conditions under-foot), local resource availability (*e.g.* plant and animal foods, fuel), predatory threats and environmental cues of changing conditions (*e.g.* vegetation characteristics and animal behaviours).

As introduced in Chapter 2, one of the major challenges to understanding survival strategies in the Lower Palaeolithic concerns the nature of the available palaeoenvironmental and palaeoclimatic data-sets. While individual sites provide rich records of fauna and flora (*e.g.* Coope 2006; Messager *et al.* 2011; Rodríguez *et al.* 2011; Ashton and Lewis 2012), it is frequently very difficult to identify contemporary records across sites, especially on a Europe-wide scale. Consequently, there is currently little specific data on spatial patterns and gradients of change in climate and habitats in the Lower Palaeolithic (but *cf.* Candy and Alonso-Garcia 2018). Latitudinal and longitudinal gradients of climate and habitat change are therefore explored here with reference to the Stage 3 Project data for MIS 3 (*c.* 60–24 kya) 'warm' intervals (Van Andel and Davies 2003). Those project data are not used as exact estimates for Early and Middle Pleistocene conditions. They are used instead to provide insights into likely degrees of latitudinal and longitudinal change in Europe, and into the differences and contrasts between multiple palaeoclimatic measures.

Annual migrations in Europe during the Early and Middle Pleistocene would also face significant potential barriers in the form of major landscape features. This is evident for example in northwest Europe, where global sea-level changes and regional isostatic processes resulted in the fluctuating status of Britain as both an island and a peninsula of Europe (see also Ashton and Lewis 2002; White and Schreve 2000; White 2015; Fig. 6.2). This chapter follows the general recent consensus, namely that Britain had a permanent peninsula status prior to MIS 12, but that post-MIS 12 hominin movements between Britain and the Continent were reliant upon relatively narrow time 'windows' which combined habitable conditions with relatively low sea-level stands and/or relatively elevated sea-bed heights (*e.g.* during the MIS 11 sub-stages in the southern North Sea Basin; Ashton and Lewis 2002; Ashton *et al.* 2011; White 2015; Fig. 6.3).

Many other potential barriers would also have impacted on hominin movements. Most obvious are the high uplands of Europe. At a continental scale the Pyrenees, Alps and Carpathians essentially divide the southern peninsulas from the northern European Plain while, at a regional scale, uplands such as the Apennines, Massif Central and Dinaric Alps would also have structured hominin movements (Fig. 6.4). The impacts of such landscapes are obvious, but no less important for that, with higher altitudes impacting on air temperatures, outgoing night-time thermal radiation

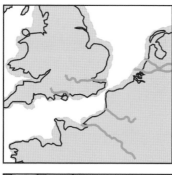

Post-MIS 12 high sea-level
stand (i.e. warm stage)

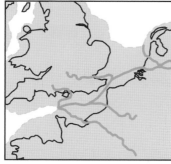

Post-MIS 12 low sea-level
stand (i.e. cold stage)

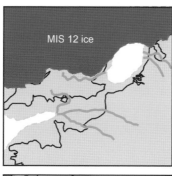

MIS 12: note ice sheet(s),
and pro-glacial lake in the
southern North Sea

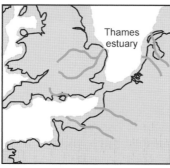

Pre-MIS 12: note permanent
land mass between southern
North Sea and English Channel

Figure 6.2: Britain's changing palaeogeographical status, prior to and post an MIS 12 breaching (redrawn after Ashton 2017, figs 36, 80, 85 & 108). Note the scale of the low sea-level stand Channel River in the post-breach period.

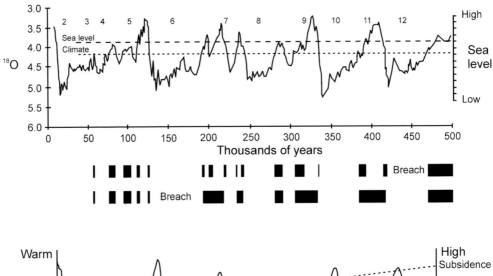

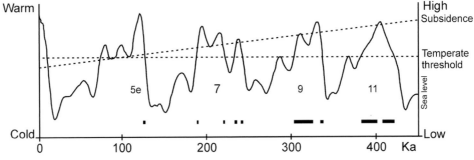

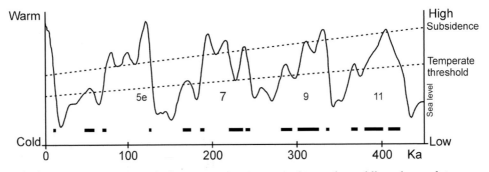

Figure 6.3: Colonisation 'windows' for hominin access to Britain during the Middle and Late Pleistocene (Ashton and Lewis 2002, fig. 3; Ashton et al. 2011, fig. 4.2). Top: Note the contrasts in 'window' duration before and after an MIS 12 breaching of the Weald–Artois anticline; middle: progressive subsidence of the North Sea basin further narrows the 'windows' over the course of the later Middle and Late Pleistocene; bottom: a progressively reduced hominin temperature threshold (e.g. through the introduction of complex clothing) partially counteracts the impacts of progressive subsidence of the southern North Sea basin.

Figure 6.4: Key European uplands (© Google Earth 2019).

(in cloudless conditions) and temperature driven-seasonality and productivity, although other factors, *e.g.* precipitation and wind speed, are not generally altitude-specific (*e.g.* Baldwin and Smithson 1979; Körner 2007). But even at low altitudes, significant potential barriers exist in the form of Europe's major river systems (Fig. 6.5; see also Chap. 4). Specific dimensions for Pleistocene rivers can be difficult to estimate (but see also Chap. 4). However useful indicators can still be derived from the width and depth of today's rivers, despite their recent histories of management and modification. The UK's Thames for example averages 305 m wide and 9 m deep (Ackroyd 2008), although there are course substantial variations from source to mouth, while the Rhine varies between *c.* 150 m and 450 m wide in its *Oberrhein, Mittelrhein* and *Niederrhein* sections (Frings *et al.* 2019). While these modern parallels are only illustrative, it seems unlikely that the Early and Middle Pleistocene interglacial/ warm stage equivalents of these rivers would have been substantially different, given the probability of meandering, relatively deep, single channel river types (Chap. 4; Fig. 4.3). Depending on where they were encountered such water bodies would have presented significant, if not practically impassable, obstacles. This is not simply because they could not be waded in parts due to the depth of water, but also because of the vulnerability of hominins both during and after crossings, respectively from fauna and currents and as a result of reduced body temperature, particularly given late year water temperatures. Rivers may therefore have significantly structured hominin migrations, either delaying crossings until shallow/narrow sections were

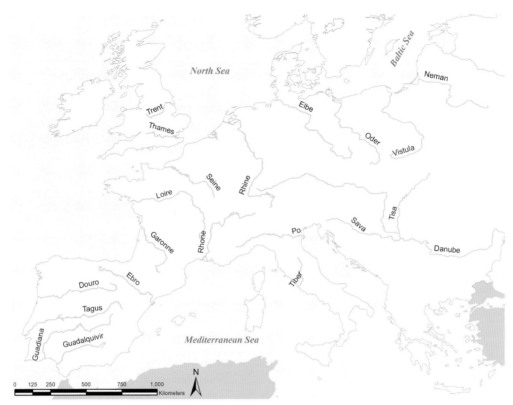

Figure 6.5: Selected modern European rivers (created in ArcGIS 10.5.1, ©1999–2017 Esri Inc.; river data from European Environment Agency: https://www.eea.europa.eu/).

reached or perhaps even preventing them entirely, *if* river crossings were conscious choices rather than the instinctive behaviours of, for instance, migrating reindeer. The paucity of Lower Palaeolithic occupation to the east of the Rhine, at least until the later Middle Pleistocene, is noteworthy in this context. Might it in part reflect the significant difficulties presented by the river to hominins approaching from a more populated Atlantic west?

A further river consideration relates to the potential significance of coasts and estuaries in hominin mobility (Cohen *et al.* 2012). These are considered here with reference to annual migrations but are also relevant to smaller scale logistical and residential movements (see also Chap. 3), and perhaps also to larger-scale dispersals. Cohen *et al.* (2012) highlighted the potential richness of coastal resources (*e.g.* sea-weed, shellfish, beached marine mammals), and the possibility that northern species dispersals from glacial refugia occurred via coastal zones/near-coastal river reaches,[2] as this would require minimal re-adaptations during dispersals. However, the major 'Atlantic' rivers of western Europe, such as the Garonne, Loire and Seine all have substantial deltas and would not be easily passable if they were encountered during a 'linear' migration.

Cold stage rivers are likely to have been very different, probably shallow and braided in type (Fig. 4.3), although the likely water temperatures would have presented a significant crossing challenge, even in southern Europe. Moreover, the major rivers of the glacial stages' exposed coastal plains would have been even more substantial. Thus the low sea-level stand landscapes between southern Britain and the continent would still not have been easily navigable, even before considering the likely low temperatures, with a major river system dominating the 'English Channel' region that was fed, after the breach of the Dover–Calais landmass, by the Thames, Rhine–Meuse, Somme, Seine and Solent (Gibbard 1988, fig. 5; while differently configured, the pre-breach riverine landscapes would also have been significant in scale: Fig. 6.2).

How far do we have to go …?

What is evident from the predictions and mapping of the Stage 3 Project (Van Andel and Davies 2003) is that only relatively limited environmental benefits would be gained from both north–south and east–west migrations of hundreds of kilometres. Mean winter surface temperature data (Barron *et al.* 2003, fig. 5.7; Fig. 6.6) indicate only small differences along a north–south transect: from −4–0°C (at 52°N, 0°E ≈ north London) to 0–+4°C (at 45°N, 0°E ≈ Bordeaux) over a distance of *c.* 780 km 'as the crow flies'. Those latter temperatures still fall within the range, albeit the upper end, evident at

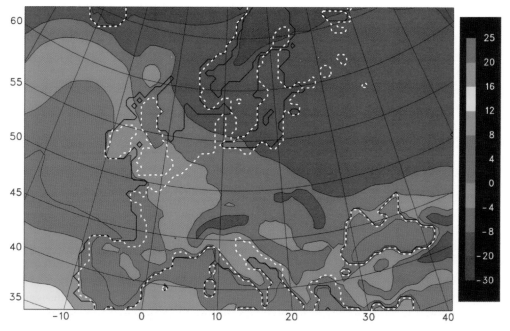

Figure 6.6: Mean winter surface temperature data (°C) from the Stage 3 Project's MIS-3 'warm' simulation (Barron et al. 2003, fig. 5.7 [Stage 3 Warm Phase DJF]). Dashed white line: Modern European coastline.

northern European Early and Middle Pleistocene sites in Germany and Britain (see Fig. 3.1 & Table 3.1). Indeed such 'soft' gradients, suggested for a south–north transect in western Europe in MIS 13 during the Middle Pleistocene, may instead be a factor in range expansion (Candy *et al.* 2015; see also Chap. 2).

Interestingly, following a 'coastal' route southwards from 52°N, 0°E offers improvements in air temperature over shorter distances. This reflects the topography of western Europe and the ameliorating effects of maritime climates, a factor which may also explain the estuarine locations of some of the earliest sites in northwestern Europe (Parfitt *et al.* 2005; Parfitt *et al.* 2010; Cohen *et al.* 2012; Ashton 2015). This particular coastal zone trend might also limit the need to cross major river estuaries (see above), by providing tolerable coastal micro- and/or meso-climate habitats in closer proximity to the archaeological 'core' of northwest Europe during the warm stages of the later Middle Pleistocene (southeast Britain and northern France). By contrast, the distinct northeast–southwest trend for the main temperature gradient (Barron *et al.* 2003, 64) highlights the particular challenges of winter occupation in the northern European continental interior and may help to explain the particular patterning of the Lower Palaeolithic record east of the Rhine (see also Chap. 3) and the rich records of the Iberian Peninsula.

This coastal/continental interior contrast is also evident in two of the Stage 3 Project's other palaeoclimatic measures: summer and winter contrasts in mean surface temperatures (Barron *et al.* 2003, appendix 5.1; Fig. 6.7) and snow-cover characteristics (snow depth and the number of days with snow cover; Barron *et al.* 2003, fig. 5.9; Figures 6.8 & 6.9). The former data again highlights the ameliorating effects of coastal settings, with less marked summer/winter contrasts in the coastal west of the continent, although the day/night temperature contrasts for the winter months follow an essentially north–south trend. In the case of snow, the coastal zone offers both a shallower cover, and a reduced number of 'snow days'. This would be relevant to hominins in terms of their own movements, with regards both to the increased energetic costs of moving through snow and the potential exposure to frostbite and hypothermia, and with regards to the timings of any major moves such as seasonal migrations. However snow cover is also highly significant in terms of resource distribution (and acquisition), as snow cover is a key limiting factor for particular species (*e.g.* 50–70 cm+ for *C. elaphus* and 60 cm+ for *R. tarandus*; Gamble 1986, table 3.12).

However, this model of mild coasts and harsh interiors is by no means the full picture. While there is a distinct northeast–southwest trend for the winter temperature gradient and for wind chill (Barron *et al.* 2003, appendix 5.1; Fig. 6.10), precipitation shows a clear increase from the interior to the coast (Barron *et al.* 2003, appendix 5.1; Fig. 6.11). Thus, the western coasts, while relatively warm and with reduced wind chill and snow cover (Figs 6.6 & 6.8–6.10), might also present the challenges of regular winter rainfall, with implications for the availability of dry fuel. Moreover, the Stage 3 project models of net primary productivity (gC/m2/year) and annual growing days (above 0°C and 5°C) indicate higher values, and therefore longer and more productive summer growing seasons, for inland areas of western Europe (eastern

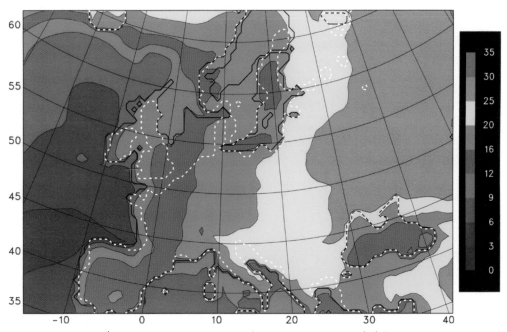

Figure 6.7: Summer/winter contrasts in mean surface temperature data (°C) from the Stage 3 Project's MIS-3 'warm' simulation (Barron et al. 2003, appendix 5.1). Dashed white line: Modern European coastline.

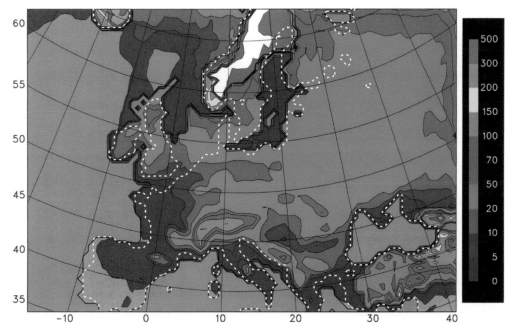

Figure 6.8: Snow depth (cm) data from the Stage 3 Project's MIS-3 'warm' simulation (Barron et al. 2003, fig. 5.9). Dashed white line: Modern European coastline.

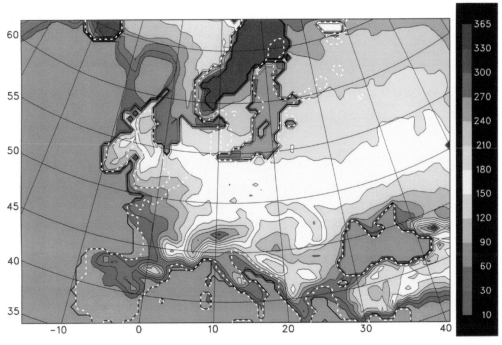

Figure 6.9: Number of days with snow cover data from the Stage 3 Project's MIS-3 'warm' simulation (Barron et al. 2003, fig. 5.9). Dashed white line: Modern European coastline.

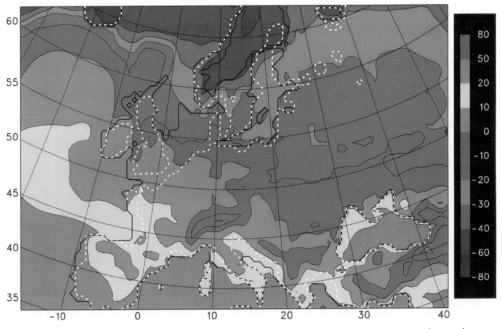

Figure 6.10: Winter wind chill (°F) data from the Stage 3 Project's MIS-3 'warm' simulation (Barron et al. 2003, appendix 5.1). Dashed white line: Modern European coastline.

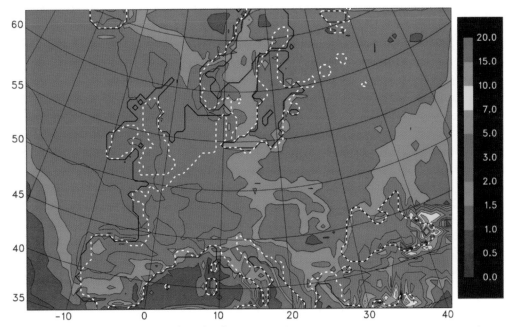

Figure 6.11: Winter precipitation (mm/day) data from the Stage 3 Project's MIS-3 'warm' simulation (Barron et al. *2003, appendix 5.1). Dashed white line: Modern European coastline.*

France and southern Germany; Barron *et al.* 2003, appendix 5.1). This would have clear implications for regional variations in plant (and therefore animal) food availability.

Are we nearly there yet?

The hypothetical distance of nearly 800 km outlined above highlights a further key aspect of any annual migration 'strategy': group composition. If migrations involved entire groups, then how did the very young move? Were they carried? Were they walking (where able to)? The former would generate significant energetic costs for the carriers (Wall-Scheffler *et al.* 2007). Either way, this would seem to impose obvious restrictions on distances covered, as would the relatively short daylight hours associated with late autumn–early winter and, assuming the migrations were bi-annual, early spring periods (see Table 1.1). Even at an ambitious average of 20 km/day (after Kelly 1995, 133; but adopting a lower estimate from Kelly's 20–30 km/day range to reflect group composition and short late autumn–winter–early spring day lengths), the 780 km (straight-line) distance from 52° to 45°N would still take 39 days. Moreover, this estimate assumes that the migration was effectively one long, continuous residential move and does not take into account local relief, vegetation cover and ground condition or larger obstacles such as major rivers. While the 'speed of the slowest member' problem can be reduced by simply abandoning the slow, whether sick, old or young (see also Pettitt 2000), this would seem to be an unlikely evolutionary solution if it habitually extended to the group's young. A further factor potentially impacting on movement rates is the presence of pregnant females. If Mussi's (2007, 170–173 & fig. 3) suggested peak in conception during late summer/early autumn is correct, then implications for autumn mobility, broadly equating to the first trimester, would be modest, but spring mobility would be much more demanding (Wall-Scheffler and Myers 2013). The numbers of any such groups are extremely difficult to estimate but a band seems much more likely than a community given the costs of food provision.

These problems of group composition can be significantly reduced if the migrating group consists only of adults. Examples of such groups could include seasonal hunting or foraging parties. Similar arguments have been made, albeit at a smaller scale, for Neanderthals, with reference to the British MIS 3 record and its distinctive signature of bout coupé handaxes (White 2006). Temporary task-specific foraging groups (*i.e.* fission–fusion) have also been emphasised as a key response to patterns in resource distribution, territory sizes and mobility in mid- and high latitudes (Kelly 1995; Roebroeks 2001; 2006; Grove *et al.* 2012). However, the Lower Palaeolithic archaeological record offers little clear evidence for task-specific sites (beyond occasional raw material provisioning sites; Chap. 5) and/or tool-kits, suggesting that fission–fusion probably operated over much shorter-term timescales (*e.g.* daily), and was embedded within intra-seasonal, and probably relatively local, mobility strategies (see also Chap. 3: Box I and Chap. 4: Box K). The typically local character of raw material sources in the European Lower Palaeolithic (Féblot-Augustins 1999; see Table 5.9) is also potentially an argument against habitual migration strategies and therefore very large annual ranges, although the relatively short life history of lithic artefacts is probably a further complicating factor here (*e.g.* Pope 2004; Wenban-Smith 2004; Hallos 2005).

Alongside the issue of individual and group mobility, days would presumably also be shortened by the need to acquire resources: fresh water, food, and possibly fuel. Intra- and inter-seasonal mobility has often been discussed, for various periods and different hunter/gatherer groups, in terms of animal migrations and resource interception (*e.g.* Spiess 1979, chap. 4; Loring 1997). But Gamble (1987, 87) has argued, with specific reference to the Lower and Middle Palaeolithic, that 'the mobility of prey far exceeded the capacity of the human predator to keep up'. This point has also been made more generally by Burch (1972; but *cf.* Spiess 1979, 137–139). Practical Lower Palaeolithic possibilities might therefore include the acquisition, both *ad hoc* and more targeted, of animals encountered over the course of the hominins' migration (see Spiess 1979 for various examples), or perhaps a 'lag' pursuit of migrating herds. Both strategies would require notable 'natural history' knowledge (Mithen 1996; MacDonald 2007): whether in terms of a knowledge base concerning the timing and locations of available animal (and plant) foods along the migration 'route' (and suitable hunting grounds), or in the form of 'reading' the tracks of a migrating herd (Burch 1972, 351–352; Haynes 2006). The 'lag' pursuit model would, in particular, also require the finding of alternative food sources en-route. While the frequencies of hunting encounters with animals are largely dependent upon the animals' density, harsh conditions would also have introduced other complicating factors:

> in the boreal forest, heavy snow, especially if it has crust, reduces moose mobility, making them easier to pursue, but making them harder to find since they move less and do not leave long trails. (Kelly 1995, 88)

Migrations would presumably also involve, on each separate occasion, habitats which although essentially familiar in terms of their flora and fauna and the presence of 'readable' cues (*e.g.* tracks, sounds, dung), would also contain fixed resource locations which would need to be found/learned (*e.g.* lithic outcrops, pools for ambush hunting). This application of locational and limitational knowledge and information from existing, familiar landscapes to new locations has been highlighted by Rockman (2003, 4 & 19). The transferability of knowledge may vary according to how fixed a resource is and to the scale of its distribution. The ranges of large animals on one hand and plants and lithic materials on the other are an obvious example, with the distributions of the latter respectively controlled, and subject to modification, by local variations in climate and topography and geological history. This would presumably add to the demands and costs of resource acquisition during periods of late autumn–winter–early spring shortages, as might the processing of those resources within an essentially highly mobile lifestyle. A further option would be to mimic many ungulates, and 'fatten-up' before the journey. The quantities of fauna on later Middle Pleistocene sites such as Schöningen and Gran Dolina TD-10.2 (which includes an autumn seasonality signal in the 'bison bone bed' assemblage) would permit such a strategy. The problem is that this extra weight must then be carried over significant distances, with resulting metabolic costs (*e.g.* Grabowski *et al.* 2005).

Any such movements would also not have occurred in isolation. Long-distance autumn migrations would only have been required within a wider context of twice-yearly movements (*e.g.* northwards in spring and southwards in autumn), and as Ashton and Lewis (2012, 60) have noted such migratory behaviour would appear to be an entirely new adaptive strategy for early hominins dispersing beyond 45°N, in contrast to the well- and long-established 'southern' solutions that hominins had developed in the Mediterranean during the Early Pleistocene. Comparable doubts have been expressed by Orain *et al.* (2013) over the feasibility of long-distance mobility between central/southern and northern Italy, due to the need for hominins to modify their behaviours fundamentally to adapt to the different environments. While the long-distance tracking of herbivore migrations is a possible answer (but see Burch 1972), the evidence for such large-scale migratory behaviour among the warm stage ungulate fauna of the Early and/or Middle Pleistocene is very limited.

Occasional southerly (or westerly) migrations might be undertaken by a previously residential 'northern' group, in response to markedly deteriorating climatic conditions and/or resource availability at the beginning of a stadial sub-stage or more short-lived cold event. There is increasing evidence of the latter (*e.g.* the YHO event in MIS 11; Chap. 2 & Box A), but such movements would be exceptional rather than annual occurrences. They are also migrations as 'permanent' relocations from one region to another (at least from the perspective of a hominin lifespan), rather than annual migrations. Moreover, any group migrating southwards would presumably face competition, and perhaps active resistance, from the existing occupants of those territories (Hublin and Roebroeks 2009): as Dennell *et al.* (2011) have argued, these southern 'refugia' were lifeboats for local groups, not arks for all.

Returning to the four factors outlined above, the challenges of group mobility and resource acquisition, combined with the relatively shallow gradients of climate and habitat change, would seem to argue strongly against the feasibility of annual migrations as an evolutionary survival strategy in Lower Palaeolithic Europe (but *cf.* MacDonald *et al.* 2012).

However there are other types of movements. Local residential movements to habitats with favourable micro- or meso-climates may have been a key strategy for surviving winters (see also Chap. 3). Such habitats would presumably be selected for small but critical advantages in climatic conditions (*e.g.* temperature, windchill) and/or resources (*e.g.* plant and animal food sources, shelter). Coastal environments might have met some of these requirements (Cohen *et al.* 2012). It seems likely that such local moves would have occurred in autumn, before the full cold and darkness of winter, but they may have been delayed by the need to exploit rich, albeit patchy, autumn resources first …

Preparing for winter?

Aggregations of ungulates, either for the rut or in migrations, would have presented opportunities for hominins to accumulate substantial quantities of animal foods prior to the relative shortages of a local winter, in terms of the number and/or condition of available animals (Chap. 3). Modern red deer for example are in peak condition in late September (extending to late November for yeld [calf-less] hinds), deteriorating to an annual nadir in late winter/early spring (Corbet and Harris 1991). Yet if such a 'stockpiling' strategy was followed, two important requirements must have been met: acquisition, presumably through killing, of multiple animals over a relatively short time interval; and storage of the resources, either internally or externally, for winter.

The first question, the nature of autumn hunting strategies, is difficult to answer, both due to the limited number of clear-cut hunting sites, and the paucity of seasonality data associated with those sites. While Thieme (2005) suggested autumn for the 'spear horizon' horse exploitation at Schöningen (13 II-4), this interpretation has not survived the recent re-analysis, which has interpreted the horse remains as the products of multiple hunting events, with possible inter-event differences in the seasons of death (Julien *et al.* 2015; Starkovich and Conard 2015).

However, a clear indication of autumn/early winter hunting can be found at Gran Dolina TD10.2 (see also Chap. 4), where the dental eruption, use-wear and microwear data indicates a significant seasonal peak late in the year (Rodríguez-Hidalgo *et al.*

2016). The hunting of both yearlings/young bison and prime age adults is indicated, respectively by dental eruption data (*e.g.* individuals dying around 5–6 months old) and microwear data (with high numbers of scratches argued to reflect an abrasive autumn/early winter diet). A deliberate, rather than ecologically-forced, hominin focus on bison at Gran Dolina is also suggested by the low-level presence of other prey, which has been modified by other carnivores, but not by hominins (Rodriguez-Hidalgo *et al.* 2017). It is possible that this autumnal hunting reflects the interception of a migratory herd, although Julien *et al.* (2012) have questioned whether Pleistocene European steppe bison (*Bison priscus*) were migratory.

Such large-scale hunting has clear implications for the numbers of individuals involved. Rodriguez-Hidalgo *et al.* (2017) suggested that up to ten people may have been needed to systematically butcher one bison: at the very least an entire band-sized community (*c.* 40–50 individuals?) would likely be involved in the butchery (if not also the procurement) of the carcasses. In light of the need to focus on complementary food resources across the year, this communal event would likely have involved many who were not frequent hunters and may well have included adolescents (see also Chaps 4 & 5; Fig. 6.12). Moreover, Gran Dolina (TD10.1 and TD-6.2), Boxgrove,

Figure 6.12: Communal butchery (© Chris Crump; http://www.chriscrumpartist.com/).

Schöningen and Soucy also enable questions about broader hunting strategies to at least be considered, if not convincingly answered.

Boxgrove

The inter-tidal mudflat habitat associated with the horse butchery site at Boxgrove (GTP-17) would offer little or no vegetation to facilitate an 'ambush from cover'-style approach (Roberts 1999a). 'Stalking in plain sight'-type approaches might therefore be necessary. Milks *et al.* (2019) have suggested a minimum range for wooden throwing spears of 15–20 m (although with evidence for reduced accuracy at the upper end of this range), offering an estimate for the distances to which hominins might have needed to close with their prey. The importance of not startling the prey can be seen in modern horse responses to threats: rapid flight, associated with pantherine predators and more open ground, and a standing defence, using foreleg strikes, associated with canid predators and habitats that are not conducive to rapid flight (Goodwin 2003). A rapid flight response would therefore seem likely in the Boxgrove landscape and stalking from up-wind directions would presumably have been a key component of avoiding or minimising such a response by potential horse prey. It also seems likely that any stalking was of horse herds or smaller family groups (where possible), to maximise opportunities for killing or wounding at least one individual.

Having closed within range, a killing or severely disabling throw would also be critical, since bipedal hominins would be limited in their ability to pursue fleeing prey, at least over short distances (Fig. 6.13). As a point of comparison, sprinting speeds of wolf, hyaena and modern humans are respectively 56–64 km/hour (Mech *et al.* 2015), 65 km/hour (Bro-Jørgensen 2013) and 37 km/hour (and this can be only be sustained by humans for less than 15 seconds, unlike other carnivores; Bramble and Lieberman 2004). By contrast, domestic horse can run at speeds of up to 70 km/hour (Garland 1983), although lifestyles and locomotion are likely to have varied slightly for all these species' Pleistocene equivalents[3] (*e.g.* van Asperen 2010). The availability of, and physical ability to use, throwing weapons would also reduce hominin vulnerability to kicking and biting from horses and other potential prey species.

It is possible that an already injured horse with limited or no mobility was targeted by the hominins at GTP-17 and potentially finished off with a close-range spear throw or thrust (although the original 'impact wound' interpretation has recently been critiqued; Smith 2013; Milks 2018a; 2018b, 183). This view has further implications however: was the earlier injury due to hominin activity, an attack by another carnivore, or some other event? If either of the latter two, then close monitoring of the local habitat (visually and aurally), to rapidly identify and then intercept vulnerable potential prey, is implied: the top of the *c.* 60 m high cliff might well have served as a significant observation 'point' (but only as long as the hominins could then rapidly access the mudflats). If an initial carnivore attack had occurred then an ability to drive them off and/or appropriate the prey is also implied, since carnivore modifications on the GTP-17 horse carcass are very limited and only ever overlay

Figure 6.13: A Boxgrove horse hunt (© Chris Crump; http://www.chriscrumpartist.com/).

the hominin traces (Smith 2013). This also implies control of the carcass once it had been acquired, as does the comprehensive butchery (skinning, disarticulation, meat removal and bone fracturing for marrow), which seemingly left little or nothing for other carnivores when the carcass was sporadically re-exposed by subsequent inter-tidal cycles (Smith 2013). This view is also supported by the evidence that carcass control was maintained while flint was acquired from the cliff for handaxe production around the carcass (carcass defence is a persistent element of classic Boxgrove visual reconstructions; Fig. 6.14). Unsurprisingly the spatial and technological patterns of artefact production suggest cooperative butchery, and perhaps also hunting, by at least six or seven hominins (Roberts 1999b, 373 & figs 279–280), while further refit-ting analysis has suggested the possibility that a larger number of individuals were present (Pope *et al.* in press).

While the palimpsest archaeology associated with the Boxgrove palaeosol (Unit 4c and the equivalent units at the waterhole) makes the specifics of individual hunt-ing (or scavenging) events more difficult to reconstruct, many of the key indicators

Figure 6.14: Rhinoceros hunting and carcass defence at Boxgrove (© John Sibbick).

are similar. Carnivore traces are minor in comparison to those of hominins, and the implied grassy plain with few trees and shrubs (Roberts 1999a, fig. 89) would again have limited the potential for 'ambush from cover'-type hunting. However occasional large trees are suggested, and these could potentially have facilitated short sprints that sought to span the distance from cover to spear-throwing range as quickly as possible, although the distances from any such tree patches to the waterhole are unclear.

Schöningen

The habitat at the 'spear horizon' (13 II-4) is also relatively open. Urban and Bigga (2015) describe a period of falling lake levels and a local fauna of aquatic and riparian species (*e.g.* pondweeds and sedges), set against a regional context of dry, steppic, open woodland. Various models have been proposed, and some rejected, for the accumulation of horse remains and spears. An early interpretation proposed a single hunting event, preserved *in situ* on a dry lake shore (Thieme 2005). However more recent analysis has concluded that there is no evidence of a dry surface, instead describing the context as a constantly submerged area near the edge of the palaeolake (Stahlschmidt *et al.* 2015a). A second model argued for several hunting events that exploited the soft, wet muds of the lake shore to disadvantage the horses (Voormolen 2008). However, the depositional environment required by this model is also not supported, with the find horizons underwater and not accessible for butchery (Stahlschmidt *et al.* 2015a). A third model argued for the concentrating of the materials through fluvial displacement, but while small-scale reworking is plausible there is no evidence for large-scale displacement. Hunting and butchery on the frozen lake surface has also been proposed but this is not supported by isotopic data suggesting that the horses died in multiple events at different times of the year. The final model argued for anthropogenic discard in the shallow waters near the lake edge, to reduce predator attraction, the massing of insects and/or to avoid scaring future prey (Stahlschmidt *et al.* 2015a). This last option looks particularly intriguing in light of Speth's rotten meat model (see also Chap. 2). The site is highly complex, and perhaps unsurprisingly:

> As none of the models explains every property of the archaeological assemblage, it seems that the archaeological assemblage at Schöningen 13 II-4 represents a palimpsest of different behaviours that occurred in a restricted, localized area at different points in time. (Stahlschmidt *et al.* 2015a, 89)

It nonetheless seems very likely that the hunting events occurred along the foreshore, again raising questions as to the particular strategies used.

Spears and other hunting strategies

Schoch *et al.* (2015) have argued that the Schöningen spears had sufficient accuracy when thrown to hit larger animals at ranges up to 35 m (although the size of such 'larger animals' is undefined), while Milks *et al.*'s (2019) experiments suggested that taller or standing prey may have been easier to hit than prey lower to the ground. The range suggested by Schoch *et al.* is a significant reduction on the effective range of up to 60 m suggested by Rieder (2000; cited in Thiem 2005), but would nonetheless enable 'remote' hunting of horse, deer and other prey, as would the effective ranges of 15–20 m suggested from experimental trials by Milks *et al.* (2019). This would potentially tally with the relatively open lakeside described by Urban and Bigga (2015), although there was probably more cover than on the Boxgrove mud-flats. However, penetration depths of 22.5–25.5 cm, argued to 'significantly' harm an animal, were based on a 5 m experimental range (Schoch *et al.* 2015), perhaps suggesting that close-range hunting might have been most appropriate with spears of this sort. This is in-keeping with ethnographic examples, where throwing ranges were also very short (7.8 m on average [n=14], reduced to 5.7 m if exceptionally long throwing distances from Tasmania are excluded; Churchill 1993).

Churchill (1993) highlighted the strong association between spear use and disadvantage hunting in ethnographic case studies, with a lesser link between spears and ambush hunting (Table 6.2). The data also suggested the domination of thrusting rather than throwing, in-keeping with the interpretation of the Clacton spear point by Oakley *et al.* (1977), although the two modes of use may not be mutually exclusive (Milks 2018a). If these patterns are applicable to Lower Palaeolithic hunting, then a key requirement is to find evidence for terrain or other features that would disadvantage the prey.

While the exact location and nature of the bison hunting associated with the assemblage at Gran Dolina TD-10.2 remains uncertain, it is possible that local landscape features, including narrow valleys and limestone escarpments, were exploited (Rodriguez-Hidalgo *et al.* 2017). These might well have provided opportunities to disadvantage the bison. However, this interpretation is complicated by the distances from these features to the cave (*c.* 150 m from the valley and *c.* 500 m from the escarpment). These distances appear highly significant given the size of the carcass (an adult steppe bison, *B. priscus*, can reach *c.* 900 kg). Given the evidence for primary access at the cave, is it perhaps not more likely that the killing grounds were immediately adjacent to the cave mouth?

The Boxgrove cliff line and the Schöningen lake edge would also provide 'disadvantaging' features as a result of their linear form, since they would limit any animal escapes to certain directions, although the specific trapping of animals in muddy lake shores has been critiqued (Stahlschmidt *et al.* 2015a). However the open mudflats at Boxgrove, and to some extent the Schöningen lakeshore, suggest very limited cover to support ambush hunting (although a much more wooded habit is suggested for the riverine Soucy localities; Lhomme 2007). Although animals drinking at the waterhole could perhaps be surprised by a slow, careful approach from a downwind direction, such approach hunting (Table 6.2) is rarely associated with spears amongst modern hunter-gatherers (Churchill 1993). For it to have been used at Boxgrove then the need for both throwing spears and an effective throwing physiology (argued to extend back to early *Homo*; Roach *et al.* 2013) would seem to be paramount.

Churchill's (1993) data also suggested a bias towards the hunting of large prey with spears (mean and median prey weight values of 504.1 and 202.5 kg are reported for the hand spear-disadvantage method), and these weights are in-keeping with some, although not all, of the Lower Palaeolithic faunal evidence. Moreover, Churchill argued

Table 6.2: Key hunting strategies (Churchill 1993, 16)

Hunting strategy	Definition
Ambush hunting	Hunters wait in hiding, whether behind man [*sic*]-made blinds or natural features, for animals to pass within effective range of their weapons.
Approach hunting	Includes stalking free-moving animals to within effective weapon range. The object of approach hunting is to avoid evoking the prey's flight response before the hunter is within effective weapon range.
Disadvantage hunting	Any technique that limits the escape of an animal or exploits an animal naturally disadvantaged to gain time or access so that a weapon can be employed.

that the disadvantage hunting technique works best with larger prey, as smaller prey can more easily escape traps and other obstacles. Finally, a thrusting spear can be repeatedly used to kill an animal that cannot escape (the durability of spears was suggested in experiments where yew spears were thrust or thrown, the latter from 6 m, into a lamb carcass target, totalling 40 hits each for each mode of use; Smith 2003).

A reliance on short-range disadvantage and, to a lesser extent, ambush hunting with spears would potentially expose individuals to kicks and/or contact with antler or horns (Lieberman *et al.* 2007), further highlighting the potential importance of social care (Spikins *et al.* 2019; Chap. 4). Moreover, the above options ignore persistence hunting as a possible strategy. This has been widely discussed and debated for low latitude hominins and environments (*e.g.* the very open and hot environments of the southern Kalahari in the present; Liebenberg 2008), yet persistence hunting would seem very unlikely in the temperate woodlands of much of interglacial/warm stage Europe. This is in light of both their relatively closed and mosaic nature, and perhaps also their relatively low temperatures (in persistence hunting animal hyperthermia is encouraged by keeping it above its trot–gallop transition, as most mammals cannot pant when galloping; Lieberman *et al.* 2007). The strategy might however have been more feasible in the cold stage open steppe habitats of the Mediterranean.

A further complication concerns the relative speeds and endurance of hominins and their prey. Bramble and Lieberman (2004) observed that higher speeds of human endurance running (2.3–6.5 m/s [albeit the upper values reflect elite athletes]) overlap with the average daytime galloping speed of horses (5.8 m/s), highlighting the potential for persistence hunting. Hominins could also minimise the length of the pursuit, and thus boost their effective speed, through their knowledge of animal behaviour, enabling them to run 'as the crow flies' wherever possible (Laughlin 1968). However, horses' top speeds (*e.g.* 8.9 m/s for elite race horses), which would presumably be a part of their flight behaviour when faced by predatory hominins, can be sustained for 10–15 minutes and would likely enable the 'prey' to reach cover and/or evade the hunters (at speeds of 8.9 m/s a straight-line distance of over 5 km can be covered in 10 minutes). A further potential problem is that humans are not equally efficient at running at all speeds, and therefore may not have been able to exploit the gait optima of specific prey (Steudel-Numbers and Wall-Scheffle 2009). Steudel-Numbers and Wall also highlighted the extremely high energetic costs of persistence hunting by running, as opposed to walking – although Lieberman *et al.* (2007) have strongly criticised the feasibility of persistence hunting by walking (albeit in hot, dry conditions).

However, endurance running, which is suggested to have emerged in early *Homo*, might nonetheless have enabled hominins to reach carcasses before (or at least around the same time as) other predators, after responding to cues such as the visual presence of scavenging birds (Bramble and Lieberman 2004). This running might only have been undertaken by a small group, who then secured and defended the carcass until other hominins joined them.

In summary, a widespread reliance by European Lower Palaeolithic hominins on endurance running and persistence hunting seems unlikely, due to environmental conditions (see also Steudel-Numbers and Wall-Scheffler 2009). This is perhaps also indicated by the evolutionary trajectory in body shape of Middle and Late Pleistocene European hominins, which seems most suited to short range, encounter and ambush-style hunting, and thus power locomotion and sprinting (Stewart *et al.* 2019), although specific evidence regarding the running capacity of European Lower Palaeolithic hominins is currently limited. Nonetheless it seems possible that some pursuits would have occurred, perhaps in those instances where an initial spear strike was not fatal, or to reach newly available carcasses. Presumably such pursuits could have used a combination of walking and running, depending on the severity of the injury and the speed of the animal, or the presence/absence of competitors for the carcass.

Spears and carcass defence

While the Schöningen evidence strongly argues for the use of spears as hunting weapons, at least by the end of the Lower Palaeolithic, we should not ignore the spears' potential role in other areas of Lower Palaeolithic life. Alongside the extensive Early Pleistocene carnivore guild (Chap. 4), the identification of sabre-toothed cat remains (*H. latidens*) at Schöningen highlights the specific nature of some of the competition also faced by later Middle Pleistocene hominins: a withers height of 0.9–1.1 m, 1.5–2.0 m in total length, and up to 200 kg (Serangeli *et al.* 2015b). The specific presence of such large felids and/or canids at the site when or shortly after the hominin-killed carcasses were accumulating is evident in the various punctures, bites, digestion damage, and teeth drag marks, recorded on 15.7% (n=426) of the Spear Horizon South faunal assemblage (Starkovich and Conard 2015). Spears may have played a key role in keeping these and similar animals, *e.g.* bear and hyaena, at rather more than arm's length – the Schöningen spears range in length from 1.84 m to 2.53 m (Schoch *et al.* 2015). Similar competition is evident at other Middle Pleistocene kill/butchery sites, for example wolf, lion and spotted hyaena at Boxgrove (Parfitt 1999a) and wolf at Soucy (Lhomme 2007).

Competition at the kill is also likely to have impacted upon hominin social structures: in, for sake of argument, a foraging group of 7–8 individuals, how many would have been required to 'keep watch' while others engaged their attention in the demanding task of carcass butchery? What would have been the implications of the presumably extended time spent around the carcass? Could more vulnerable group members, such as young children, have been safely introduced to the carcass for the purposes of feeding (see also Chap. 4)? The kill might not intuitively seem a safe place, and yet zooarchaeological data from the Spear Horizon South at Schöningen does not argue in favour of the removal of parts of carcasses, and only limited evidence for the removal of large equid bones by carnivores (Starkovich and Conard 2015).

Starkovich and Conard (2015) have seen this as evidence of the hominins' ability to defend a kill from other carnivores (as also proposed at GTP-17), and suggested that there was no need for the hominins' to remove meat to a safer place (see also Serangeli *et al.* 2018). The evidence for horse skinning (Voormolen 2008; Van Kolfschoten *et al.* 2015a) is also suggestive of significant time being spent at the site, as is the fact that the green bone knapping tools (Chap. 5) were broken after use, to extract their marrow content (Van Kolfschoten *et al.* 2015b). However, it is worth considering whether possible meat removals would have to occur on-the-bone. Filleting of meat would reduce the transported weight, but it would also increase the awkwardness of the package. One possible solution would be the moving of filleted meat in a hide 'bag', drawn together by the hooves, or perhaps in a stomach (Buck *et al.* 2016, fig. 2). It is noticeable that foot elements are under-represented at both the main Spear Horizon site and the Spear Horizon South (Voormolen 2008; Starkovich and Conard 2015) and that could indicate complete removal of the skins (with the caudal vertebrae and phalanges left attached to the skinned hide).

The transport to another location of meat cut off the bone at the kill site is an interesting possibility (see also Chap. 4) – and there is certainly extensive evidence for filleting at both Schöningen and Boxgrove (*e.g.* Parfitt and Roberts 1999). But what of the bone marrow? Its value at the Schöningen Spear Horizon South and at the main 'spear site' is evident by the distribution of marrow processing across all classes of animals, with the exception of the small ungulates (Starkovich and Conard 2015; van Kolfschoten *et al.* 2015a). Yet the absence of evidence for the significant transportation of limb bones elsewhere would suggest that marrow was accessed, and presumably eaten, at the kill site. Was this particular treat restricted to the foraging party, or did other individuals join the kill?

Multiple events ... and managing the resource

One of the key debates surrounding the Schöningen horses concerns the length of time over which the assemblage accumulated. In contrast to the original suggestion of a single event, targeting a single herd (Thieme 2005), Voormolen (2008), Starkovich and Conard (2015), Julien *et al.* (2015) and Rivals *et al.* (2015) have all favoured multiple events (although with specific variations in the different taphonomic models), with horses killed at the same spot over many seasons and/or years. An important implication of this latter model concerns the nature of hominin memory – it could imply landscape and animal behaviour memory, and repeated visits, as does the likely timespan represented by the waterhole assemblage at Boxgrove (Roberts 1999a; Roberts and Pope 2009). Starkovich and Conard (2015) also discuss the kill events in the context of the hominins' predictive ability. However, I would be cautious on the latter point, since the archaeological record does not tell us about failed visits to the location. In light of dental seasonality indicators at Gran Dolina TD-10.2 (Rodríguez-Hidalgo *et al.* 2016) and the mixture of young (n=21) and prime age adults (n=36) Rodríguez-Hidalgo *et al.* (2017) similarly suggest multiple seasonal hunting events and occupations, and multi-animal predations (although not always, given the relatively small number of individuals overall). Those multiple predations at Gran Dolina may have targeted cow herds (which dominate bison social structures), although autumn rut herds are also possible.

The most significant impact of the time-depth of the zooarchaeological assemblages for the current discussion is that it reduces the quantity of animal food generated by each hunting event. However, even if each Schöningen event involved a horse family group (one stallion, 2–6 mares, and their foals), as suggested by Voormolen (2008), then the meat yield alone would be c. 400–1000 kg of meat. Applying the calorific values for horse in Cole (2017) generates calorie values (muscle) of 276,000–690,000 kcal (based on 1150 calories/kg [muscle], with muscle weight estimated to be 60% of total body weight [*i.e.* muscle weights of 240–600 kg for the horse family group suggested above]). Combined with a suggested calorific requirement for *H. heidelbergensis* of 3783 kcal/day (based on the mean of the 'Schöningen' values for males and females; see Table 2.11), the horses would sustain eight adults for 9–22 days (this ignores the consumption of other foods, both animal and plant, and infant needs). The question of how such quantities were managed therefore remains relevant. One possibility is that they were gorged upon, with hominins building up their body fat reserves for winter (*i.e.* internal storage). Chapter 4 explored the question of 'group to carcass' or 'carcass to group' strategies, partly with reference to the suggestion that different groups of individuals may well have pursued different foraging tasks. If autumnal gorging on migratory herds or rut harems, to build up internal fat reserves, was part of an annual strategy, then the scale of kills such as the Schöningen 'horse events' would likely make a 'carcass to group' approach unfeasible. This might imply both the involvement of all 'independents' (*i.e.* adults and older children) in the hunt itself, but also the nearby presence of dependents. Gorging might be especially likely if the communal hunt involved multiple groups, along the lines of a fission–fusion model (Grove *et al.* 2012), given the former's occurrence in various hunter-gatherer societies (Speth and Spielmann 1983). However, the mechanisms by which such social aggregations might have occurred remain unclear. The scale of a large 'bounty' might have reduced inter- (or intra-) group tensions, while low hunting yields would be likely to have heightened social tensions: Blumenschine (1991) has argued that small quantities of easily defended, energy-rich foods, such as bone marrow, might encourage intra-group competition rather than promoting cooperative sharing.

An alternative is that some of the meat was stored to allow it to rot, with various arguments in favour of rotten meat storage over smoking or drying (*e.g.* there is no evidence for fire at either Gran Dolina or Schöningen; see also Chap. 3). The potential importance of storage has long been recognised. Binford (*e.g.* 1980; 2001, fig. 8.04) argued that storage should increase in line with more marked seasonality, as it would enhance the potential for winter residency, and Speth's (2017) rotten meat model offers a practical and low-technology solution that is not dependent on habitual fire use. This seems an attractive option given uncertainties about hominins' ability to produce fire (Chap. 3). It is intriguing in light of the lake and waterhole settings of the Schöningen and Boxgrove kill-butchery sites, while raising further questions about the spatial relationships between such stores and residential sites. In-water storage methods similar to those in Fisher's pond experiments (as Speth explicitly suggested with reference to Schöningen) might potentially reduce the attraction of carnivores to the kill site (in contrast to air-drying methods), although it raises questions as to how the carcasses were secured. Perhaps the deposition of complete or near complete carcasses (*i.e.* utilising the natural weight of the animal) would be the most likely solution. Localised food stores could offset the need for long winter foraging trips, since amongst ethnographic groups Kelly (1983) has observed that logistical trip distances, already relatively high in the low effective temperature landscapes of Europe, would be further increased during winter by a reliance on live fauna (Chap. 3). Storage would also reduce the dangers of spending more residential time in one place (which otherwise increases the exhaustion of the local fauna).

Prey selection and complex cognition?

The seasonal, communal hunting at Gran Dolina TD-10.2 (the 'bison bone bed'), and perhaps also at Schöningen, clearly requires knowledge of environments, prey behaviours and annual cycles. In broader terms, a variety of cognitive and social attributes are implied, including anticipation, social integration and cohesion and articulate communications (Rodriguez-Hidalgo *et al.* 2017). Yet complex group hunting patterns are also evident in other social predators, such as wild dogs (*e.g.* Carbone *et al.* 1997) and wolves (*e.g.* Peters 1978). Were Lower Palaeolithic hominins really any different in terms of their foraging behaviours and if so how? Hunting-related aspects of hominin anticipation are likely to have included the locating of some autumn (and spring?) residential sites in order to monitor key locations, which is necessary when the exact timing of prey's arrival at an identified place in the landscape (*e.g.* a major river) is unknown (Kelly 1983). The counter strategy of logistical mobility (in essence, searching for the game), used if the location rather than the timing of animals is unknown, is a similar example of anticipation.[4] Yet both are perhaps no more than the hominin-equivalent of wolves travelling to regions where prey is likely to be found or to places where kills have been made in the past (Peters 1978). A key difference can possibly be found in the occasional hominin evidence for longer-term anticipation, planning to meet future needs and delayed returns, such as the summer sourcing of wood for spears in the landscapes around Schöningen (if not their production as well; Chap. 5). Yet perhaps the strongest indicator of a different type of predator is the occasional evidence for highly selective hominin hunting in the final stages of the European Lower Palaeolithic, most notably at Schöningen 13 II-4 (94% horse, n=782; based on all hominin modifications to fauna; van Kolfschoten 2015a, table 1) and Gran Dolina TD-10.2 (the 'bison bone bed'; 100%, n=1019; based on cut-marks; Rodriguez-Hidalgo *et al.* 2017). This is in contrast for example with the wolves of the BPF today, for whom red deer prey, while clearly preferred, made up 68.6% of wolf kills between 1985 and 1996 (Jędrzejewski *et al.* 2000).

However, such strong prey selectivity is extremely rare in the European Lower Palaeolithic record as a whole, with intra-site mixtures of butchered animal species characteristic of the majority of sites (*e.g.* Boxgrove, Soucy and Gran Dolina TD-6.2). It is also notable that, at a European scale, shifts in habitat, whether spatial or temporal, appear not to have significantly impacted on the range of foods selected by hominins. Comparisons of anthropogenically-modified fauna at Schöningen (*c.* 52°N), Boxgrove (*c.* 51°N), Soucy (*c.* 48°N) and Atapuerca (*c.* 42°N) suggest a predominance of medium and large ungulates at all four locations (*e.g.* horse at Schöningen, horse and red deer at Boxgrove, horse, red deer and bovids at Soucy, red deer at Gran Dolina TD-6.2 and bovids at TD-10.2; Parfitt and Roberts 1999; Lhomme 2007; Huguet *et al.* 2013; van Kolfschoten *et al.* 2015a; Rodriguez-Hidalgo *et al.* 2017), across both the Early and Middle Pleistocene, although this may partially reflect preservation bias. An interesting comparison can be drawn with modern lynx, whose dietary patterns across Europe reflect changing prey abundance in forests and woods along a north–south transect. Hare dominate their diet in the north, being replaced by ungulates from 52–54°N southwards, with tetraonidae birds (*e.g.* grouse) only becoming a significant component in boreal and montane forests (Jędrzejewski *et al.* 1993). All of the Lower Palaeolithic sites listed above fall within the modern lynx's 'ungulate range' – does the apparent northern limit to Lower Palaeolithic site distribution, which extends over the entire period, reflect in part relative prey abundance at different latitudes (as well as hominins' climatic tolerances)?

The apparent concentration on medium/large animals may also reflect an optimal payoff between food 'package sizes' and hunting risks (see also Gamble 1986), since hunting during these periods would have brought challenges as well as opportunities. The potential physical dangers presented by even larger prey may therefore explain why there is relatively little, and typically ambiguous, evidence for megafaunal hunting in the European Lower Palaeolithic. For example, at the Schöningen spear site there is no evidence that hominins generated the elephant and rhinoceros remains through hunting (van Kolfschoten *et al.* 2015a), and the relatively greater risks associated with those larger species may explain why horse was a strongly preferred prey. Interestingly, there is also no evidence of hominin exploitation of the elephant and rhinoceros carcasses either (van Kolfschoten *et al.* 2015a). This may be due to the potential for butchery of thick-hided animals not to leave traces on the bones (Frison 1989). Another possibility is that the lack of exploitation simply reflects timing, *i.e.* hominins not being present at the same time as the carcasses. However, the data on bison carcasses in the BPF suggest that even larger, megafauna-sized, carcasses might have been present on the lake-shore for several weeks. The respective manageability and defensibility of the horse and elephant/rhinoceros carcasses may instead have been a factor, as, more speculatively, might a genuine preference for horse meat.

Other sorts of stockpiling?

Alongside animal foods, the autumn landscape would still offer a range of plant foods. Mabey (2012) highlighted various nuts (*e.g.* hazelnuts, beech nuts) and fruits (*e.g.* elderberries, hawthorn berries, service tree fruits, raspberry and blackberry), but also other foods (*e.g.* dandelion roots), many of which are known from Pleistocene sites (Table 6.3). A range of potential autumnal plant foods are also documented at Schöningen (Bigga *et al.* 2015; Table 6.4). While direct evidence is again limited, the presence of hackberry (*Celtis*) remains has been documented at a small number of European Lower Palaeolithic sites (*e.g.* Arago Cave, Terra Amata and Kärlich; see Table 5.7), although the question of consumption has remained debated at those sites (Allué *et al.* 2015). However, the particular circumstances at Atapuerca TD-6.2 (*e.g.* the abundance of the seed remains and the associations with other hominin material)

Table 6.3: Pleistocene records of potential autumn plant foods (Mabey 2012), after Godwin (1975)

Species	Common name	Pleistocene?	Earliest record[1]	Comments
A. petiolata	Jack-by-the-Hedge	No	Roman	–
B. vulgaris	Sea beet	Yes	Ipswichian	–
C. avellana	Hazel	Yes	Pre-Cromerian	–
C. monogyna	Hawthorn	Yes	Hoxnian	Berries
C. sativa	Sweet chestnut	Yes?	MIS 7	Based on possible identification at Crayford
F. sylvatica	Beech	Yes	Cromerian	–
F. vesca	Wild strawberry	No	Holocene	–
J. communis	Juniper	Yes	Hoxnian	–
J. regia	Walnut	Yes?	Cromerian?	Possibly only introduced during Roman times (Godwin 1975, 248)
M. germanica	Medlar	No	Holocene	–
P. rhoeas	Common or Field Poppy	No	Bronze Age	–
P. spinosa	Sloe	Yes	Cromerian	–
R. canina	Rosehip[2]	Yes	Hoxnian	–
R. fruticosus	Blackberry	Yes	Cromerian	Identified to genus level
R. idaeus	Raspberry	Yes	Cromerian	–
S. aucuparia	Rowan	Yes	Weichselian	–
S. media	Chickweed	Yes	Cromerian	–
S. nigra	European elder	Yes	Hoxnian	Berries
S. torminalis	Service tree	Yes	Hoxnian	–
T. officinale	Dandelion	Yes	Hoxnian	Identified to *Taraxacum* genus
V. myrtillus	Bilberry	Yes	Weichselian	–

[1]In light of Godwin's (1975, table 1) climate stage model (including the following sequence: Beestonian > Cromerian > Anglian > Hoxnian > Wolstonian > Ipswichian > Weichselian), 'Hoxnian' is cautiously interpreted as MIS 11 or MIS 9, and 'Cromerian' as spanning the early Middle Pleistocene. [2]Godwin (1975) lists this species as Dog rose. None of the above information should be used as a 'safety guide' with regards to the picking and consumption of wild plant foods. Any readers wishing to do so are strongly recommended to consult an appropriate, dedicated guidebook, such as Mabey (2012)

Table 6.4: Dietary potential, and seasonality data, for selected plant species available for autumn collection and recorded at Schöningen

Botanical name	Common name	Edible parts	Nutritional & other features	Seasonal availability	Ecology
A. tripolium	Sea aster	Leaves, stems	–	July–September (flowers)	Moist soils & saltmarshes (tolerates saline environments)
A. uva-ursi	Common bearberry	Berries	Vitamins A & C; carbohydrates	August–Winter (fruit)	Open pine woods & heaths
A. prostrata	Spear-leaved orache	Leaves, seeds	–	July–October (flowers)	Well-drained, dry or moist soils
C. album	Fat hen	Flowers, leaves, seeds	Vitamins A & C; contains oxalate	June–October (flowers)	Pioneer plant; including lakeshores
C. avellana	Hazel	Nuts	Oil, protein, carbohydrates & linoleic acid	August–November (nuts)	Woods, especially on hill slopes; prefers moist soils
H. vulgaris	Common marestail	Leaves, shoots (incl. overwintered spring stems)	–	May–August (flowers); Autumn–Spring (ideal harvesting)	Aquatic conditions (shallow standing or slowly flowing water)
M. spicatum	Water milfoil	Roots	–	July–September (flowers)	Eutrophic lakes (plant is submerged)
N. lutea	Yellow water lily	Leaves, roots, seeds	Fresh roots are toxic	Fruit & seeds follow flowering (June–September)	Shallow water & wetland (aquatic plant)
P. australis	Common reed	Leaves, roots, seeds, shoots, stems	Starch & sugar (roots)	–	Moist soils, bogs or fen wood (max. water depth: 1 m)
R. fruticosus	Blackberry	Fruits, leaves, roots	Vitamin C (esp. fruits)	August–October (fruits)	Forest pioneer; not extremely dry or flooded sites
R. idaeus	Raspberry	Fruit, leaves, roots	Vitamin C, tannins	July–September (fruit)	Forest pioneer & low herbaceous vegetation
S. lacustris	Bulrush	Rhizomes, shoots, seeds	Starch & sugar-rich (fresh spring roots & autumn)	Spring & autumn	Slow flowing water, river banks & lakeshores
S. nigra	European elder	Flowers, fruits	Multiple vitamins, tannins	August–October (fruit) Late June onwards (flowers)	Moist woods & glades

Sources: Lippert and Podlech 2001; Mabey 2012; Bigga 2014; Bigga et al. 2015. Where a species is only edible when cooked (e.g. R. aquatilis [Common water crowfoot] or R. repens [Creeping buttercup]) they are not listed above. None of the above information should be used as a 'safety guide' with regards to the picking and consumption of wild plant foods. Any readers wishing to do so are strongly recommended to consult an appropriate, dedicated guidebook, such as Mabey (2012).

have been interpreted in favour of consumption of this fruit by *H. antecessor*. The species is most likely *C. australis* (Mediterranean hackberry), the fruit of which now ripens in October (Fern 1995–2010), and potentially stays on the trees into the winter and early spring. They consist of a thin sweet skin around a hard seed, and are high in oils, proteins, fibre and minerals. The seed is particularly rich in protein and fats, but is very hard – its regular exploitation may have been made easier by the use of percussion tools (although recent experimental work focusing on the artefacts from Barranco León has emphasised stone knapping and bone breakage; Titton *et al.* 2018). While the seeds could potentially have been consumed by birds and rodents, the lack of rodent tooth marks and the limited evidence for bird inhabitation of caves argues against this, while the absence of the seeds in hyena coprolites suggests that they were not unwittingly consumed by the carnivores (as part of the stomach portions of their herbivore prey).

While the nature of some plant foods, *e.g.* shelled nuts, might have offered the potential for external storage, this would have been complicated by their small 'package sizes', residential mobility (*e.g.* late autumn relocations to habitats with favourable local micro-climates), and therefore the likely need for containers of some sort. Immediate consumption, perhaps including gorging, therefore seems more probable, especially in the case of those plant foods that were not available throughout the autumn period.

An ability to plan? Preparing clothing ...

The availability of large numbers of animals, either rutting or migrating, combined with pelts in their best conditions/the growth of winter coats (Table 6.5) and relatively long early autumn days, would also offer a key opportunity for the preparation of clothing, prior to, but in anticipation of, the challenges of winter. As argued in Chapter 3, modelling of hominin physiology combined with palaeoenvironmental estimates suggests that clothing (*sensu lato*) may well have a key requirement for Lower Palaeolithic survival, but the practicals of clothing production may well have been better suited to autumn (or late summer).

In terms of the raw materials for clothing, Lower Palaeolithic sites provide evidence for a wide range of fauna with usable hides, such as various deer species, horse and *Bos*/Bison, alongside smaller species, including a number of carnivores (*e.g.* Parfitt 1998; Parfitt 1999a; Lhomme 2007; Huguet *et al.* 2013; van Kolfschoten *et al.* 2015a). The range of documented species (Tables 3.7–3.8 & 6.6) indicates that hominins could have accessed pelts with various different properties: thicker and warmer, such as fox, or more durable, for example otter. With reference to modern species, Hammel (1955, table 1) reported a total insulation value (clo) of 6.8 for *V. fulva* (Red fox), compared to 6.6 for *C. lupus* (wolf), 5.4 for untanned *R. arcticus* (Caribou), and 5.2 for *L. canadensis* (Canadian lynx).[5] While an absence of tailored clothing would partly reduce the benefits, the insulation gain of such pelts should be clearly apparent in light of the 1 clo advantages outlined in Table 3.6. Moreover, those pelts are often in prime condition during the autumn and winter months in the case of contemporary animal populations (Table 6.5), and there is little reason to imagine a different scenario in Pleistocene Europe. In the case of modern beavers for example, the dense fur (12,000–23,000 hairs/cm^3 of skin) is in prime condition in the latter part of winter, while the dead wood in beaver territories, both chips from gnawing and dry wood from standing dead trees, would be a further valuable resource for the starting and maintaining of fires (Coles 2006, 48 & 54; although not

Table 6.5: Hair density and other properties for selected mammal fauna

Species	Hair density (hairs/cm^2)	Further description[1]	Sources
B. bonasus (European bison)	641	–	(Sandel 2013)
C. lupus (wolf)	–	Underfur: fine, usually in tufts; Guard hair: c. 60–100 mm long; Winter fur is dense & fluffy	(Gronquist 2013; Mech 1974)
C. capreolus (roe deer)	–	Winter coat: 55 mm hairs; moult: mid-March–early June; summer coat: 35 mm hairs	(Corbet and Harris 1991)
C. elaphus (red deer)	–	Winter coat growth: September–December; 60 mm hairs with thick underwool (20–25 mm long); moult starts April–May; summer coat: 50 mm hairs with little or no underwool	(Corbet and Harris 1991)
D. dama (fallow deer)	–	Moult: May–June, winter coat growth: September–October	(Corbet and Harris 1991)
E. ferus (horse)	–	Winter coat growth: from late autumn; winter coat: thick insulating underfur grows into summer coat; winter coat shed April–May.	(Corbet and Harris 1991)
L. lutra (Eurasian otter)[2]	c. 70,000	–	(Kuhn et al. 2010)
M. erminea (short-tailed weasel or ermine)[2]	–	Underfur: short, even; Guard hair: slightly longer (c. 50–100 mm); Spring & Autumn moults (Spring moult delayed in low temperatures); Winter coat: denser	(Gronquist 2013; King 1983)
M. meles (badger)[2]	320	–	(Sandel 2013)
M. putorius (European polecat)[2]	6388	–	(Sandel 2013)

(Continued)

Table 6.5: (Continued)

Species	Hair density (hairs/cm²)	Further description[1]	Sources
O. moschatus (musk ox)	4480	Fine underwool (qiviut): mean diameters: 13–17.3 µm; Guard hairs up to 58 cm long, with mean diameters > 80 µm; Spring moulting (early April).	(Lent 1988; Sandel 2013)
U. arctos (brown bear)	–	Guard hairs: 100 mm long; underfur: *c.* 80 mm long; Prime pelt: Autumn	(Pasitschniak-Arts 1993)
V. vulpes (red fox)[2]	3780	Underfur: *c.* 40 mm long; Guard hair: *c.* 90 mm; Prime pelt (i.e. long, dense hairs): December; Summer fur is shorter & sparser	(Larivière and Pasitschniak-Arts 1996; Sandel 2013)

[1]Specific timings for autumn and spring moults will vary depending on local conditions and are included here as broad indicators. [2]The small body size of selected species (*e.g. M. erminea*; body length: 187–325 mm [♂] & 170–270 mm [♀]) would obviously limit uses of their fur, *e.g.* as hand or foot wraps for children.

necessarily in the case of the giant Pleistocene species *T. cuvieri*). The territorial habits of the modern equivalents of the selected species in Table 6.6, excluding *C. lupus*, also suggest that their furs may have been available without extensive, energetically-expensive searching, although their catching may have been rather more demanding. The good autumn condition of many species' hides would also have enabled them to be acquired prior to the winter foraging challenges of snowfall, low temperatures and shorter days (Chap. 3).

The particular value of mustelid (*e.g.* otter, badger, wolverine) and canid (*e.g.* wolf and fox) furs as cold-weather clothing has been recently emphasised with reference to the Middle and Upper Palaeolithic by Collard *et al.* (2016), on the basis of their hair properties (*e.g.* their mixture of long and short hairs which reduce air velocity and heat loss at the edges of clothing where skin is exposed). A number of such animals are well documented on Lower Palaeolithic sites, including cut-marked specimens of *V. praeglacialis* at Gran Dolina TD-6.2 (Huguet *et al.* 2013; Table 6.6). The fact that no individual mustelid and canid is large enough to provide a human body-sized pelt has further implications however, *if* their hides were used: either some form of rudimentary tailoring was used, there was a complementary emphasis on species with larger individual hides (*e.g.* cervids and bovids, or occasional large carnivores such as the lion, *P. leo fossilis*, skinned and butchered at Gran Dolina TD-10.1; Blasco *et al.* 2010), or there was an absence of all-over clothing. The skinning evidence at TD-10.2 is suggestive of the use of bison hides, although this does not immediately imply clothing (other possibilities include shelter and/or sleeping covers, and containers/carriers). The possible use of hand and foot-wraps, particularly in winter (Chap. 3), is also intriguing in light of the observation by Collard *et al.* (2016) that the effectiveness

Table 6.6: Potential non-ungulate clothing sources, with modern distribution data for comparison, documented on European Early and Middle Pleistocene sites

Species	Home range[4]	Density[4]	Mobility[4]	Site examples
M. martes (pine marten)	3–82 km²	1/0.8–10 km²	Solitary; not highly territorial; hunting trips up to 28 km	Swanscombe (LL)[7]
F. sylvestris (wild cat)	0.6–3.5 km²	1/0.7–10 km²	Sedentary; nomadic	Boxgrove[6] Swanscombe (LG)[7]
C. fiber (beaver)	500m–5.5 km (along river)	1.0–1.8/ km²	Family movements within territory	Arago[2] Bilzingsleben[5] Boxgrove[6] Hoxne (Beds C & E)[8] Soucy[3] Swanscombe (LL)[7]
C. lupus (wolf)	100–10,000 km² (food-dependent)	1/50–80 km²	Territorial (and correlating with prey migrations)	Bilzsingsleben[5] Swanscombe (LL/LG)[7]

Early Pleistocene species (by site) without clear modern equivalents

Canis mosbachensis/arnensis; Lynx cf. *issiodorensis; Mustela* cf. *palerminea/praenivalis; Panthera gombaszoegensis; Vulpes* cf. *alopecoides*	Sima del Elefante (TE9–TE14)[1]
Canis mosbachensis; Lynx sp.; *Panthera gombaszoegensis; Vulpes praeglacialis*	Gran Dolina (TDW4)[1]
Canis mosbachensis; Lynx sp.; *Vulpes praeglacialis*	Gran Dolina (TD-6.2)[1]

Other documented species include: *L. lutra, L. spelaeus, M. erminea, M. lutreola, M. putorius, P. leo,* and *V. vulpus.* [1]Huguet *et al.* (2013); [2]Lebreton *et al.* (2017); [3]Lhomme (2007); [4]Macdonald and Barrett (1993; modern European data – it is fully acknowledged that Early and Middle Pleistocene species' ecology would not have been identical to their modern equivalents: see also Chap. 3: Box H); [5]Mania and Mania (2005); [6]Parfitt (1999a); [7]Schreve (1996); [8]Stuart *et al.* (1993). Site units: Swanscombe (LL): Lower Loam; Swanscombe (LG): Lower Gravels; Mobility characteristics of a sample of potential hide-producing ungulates are presented in Table 3.7.

of clothing would impact on the feasibility of foraging tactics involving long periods of inactivity (*e.g.* ambush hunting or animal observations), as well as on the length of foraging time windows and on the latitudes and altitudes at which such foraging was possible.

What is critical is that at least some of this fauna has yielded clear evidence for skinning, although this may also have just occurred to enable access to the meat and the bones (Blasco *et al.* 2010). The Schöningen spear horizon's horse remains reveal extensive traces, with the almost complete absence of caudal (tail) vertebrae, and the under-representation of phalanges being strongly suggestive of skinning (Voormolen 2008; van Kolfschoten *et al.* 2015a). Voormolen (2008) suggested that both elements could be under-represented due to remaining attached to the horse skins after removal. The specific utilisation of the horse hides at Schöningen is uncertain, although use as a raw material for shelters and/or clothing is clearly possible. More speculatively, the phalanges could then serve as 'handles' to draw the hide together into a simple, if rather large, skin container, perhaps for the carrying (dragging?) of other butchery products away from the kill site? That the horse skinning evidence is not clearly repeated on the red deer and large bovids at that site, despite clear evidence for the dismemberment, filleting and defleshing of both these latter species, is potentially an argument against a 'carcass bag' interpretation, although there is no reason why the products of different animals could not be combined (*e.g.* deer flesh in a horse 'bag'). However, the overall pattern might suggest an exclusive use of horse hide for cultural insulation at Schöningen, an interesting observation given the widespread ethnographic use of bovid and deer hides (*e.g.* Gramly 1977; Creel 1991). Skinning of the latter animals is clearly documented elsewhere, at Gran Dolina TD-6.2 and TD-10 (Saladié *et al.* 2011; Rodriguez-Hidalgo *et al.* 2015; 2017). Perhaps of most significance is the more occasional evidence for animals whose hides have enhanced insulation properties. The warmth and waterproofing values of beaver fur are well known and Lebreton *et al.* (2017) documented butchery, albeit in low proportions compared to the other modified fauna, on beaver bones at Arago Cave. While a beaver carcass has other potential benefits to hominins, including the fat reserves in its winter tail (Aleksiuk 1970) and the sources of castoreum[6] in the castor sacs (Pincock 2005), the Arago butchery marks include suggested evidence for skinning. The butchery traces on the bear (*U. deningeri*) skull at Boxgrove have also been interpreted as skinning marks (Parfitt and Roberts 1999). Both bear and beaver remains are present at Bilzingsleben (Mania and Mania 2003) and cut-marks on bear and fox are also recorded at TD-6.2 (Saladié *et al.* 2011). If these animals were being targeted, in whole or in part, for their hides, then primary access through hunting is obviously implied, so that complete hides could be secured.

If hides and furs were targeted for clothing, then a seasonal focus to this activity is likely, given the evidence for significant seasonal changes in the insulation properties of fur amongst Arctic and North temperate zone mammals (Hart 1956) and the quality of hides. Summer parasites can also reduce the quality of hides, such as the warble-fly whose larvae burrow through reindeer skin (Lantis 1950). These seasonal variations are significant, in part because of the markedly differing life spans of treated and untreated hides which, in turn would impact on the frequency of clothing replacement. The limitations of untreated or improperly treated hides are that they will quickly become hard, dry and stiff (effectively rawhide) or, if wet, will be soft but will rapidly rot, probably within just a few days in a temperate, European climate (Theresa Kamper, pers. comm.). This rotting would be accompanied by pungent smells, which would impact on effective hunting and raise the possibility of even small cuts becoming septic. Rawhide is clearly inappropriate for clothing, while soft, wet hides were surely barely tolerable, although they at least provide a malleable, if very short-lived, raw material. If untreated hides were used, there would therefore have been a rapidly repeated cycle of hide acquisition, use and discard, one which would have extended into the winter months when clothing is most likely to have been needed, and which seems unfeasible.

So what would be required to treat a hide? MacDonald (2018) summarised the core elements of the process: stretching, cleaning (scraping), softening (*e.g.* kneading) and treating (*e.g.* greasing with ochre and/or fat). Rifkin (2011, 139) summarised five specific methods by which a raw hide can be converted into leather: tanning (treatment with tannins and tannic acid derived from plants); tawing (treatment with mineral powders); chamoising (treatment with animal oils and fats from the brains and around the kidneys); smoke tanning; and combinations of the above processes. Rifkin (2011) also demonstrated the use of red ochre as a hide preservation treatment, increasing resistance to putrification and dessication, but the presence of red ochre on European archaeological sites is currently not known prior to the early Middle Palaeolithic (Roebroeks *et al.* 2012). Softening of hides by working them in the mouth is also known ethnographically and has been suggested for Neanderthals (Clement *et al.* 2012): it might also be reflected in the high wear rates on the teeth samples at the Sima de los Huesos (Bermúdez de Castro *et al.* 2003a). Striations and other traces (*e.g.* enamel flakes and polished enamel) on the anterior teeth from both the Sima de los Huesos and Boxgrove have been interpreted as evidence for the working of materials, potentially including sinews, nerves, skins and vegetable fibre strips (Lozano *et al.* 2008; Hillson *et al.* 2010). However it would seem that such behaviour was not universal: the tooth fractures, attrition and dental lesions (suggesting periodontal disease) in Sima de los Huesos cranium 5 (adult male, 35+) have been interpreted as a product of masticatory processes, not the use of the teeth as a tool in the manner of the Inuit and, possibly, Neanderthals (Gracia-Téllez *et al.* 2013).

An outline hide treatment process derived from Rifkin (2011) does not necessarily require any tools or materials beyond the reach of Lower Palaeolithic hominins, when the chamoising or smoke tanning options are used (Table 6.7). However, what *is* implied is a controlled space, in which hides can be worked (scraped, treated and stretched), dried and/or smoked, and favourable weather conditions – to enable open-air drying (the typical humidity levels in caves would make 'indoor' drying unfeasible). The other key question is whether such multi-stage processes were within the cognitive reach of Lower Palaeolithic hominins and, if so, how the treatments were discovered? Questions regarding the sources of technical innovations have similarly been asked with reference to Neanderthal birch bark pitch production (Kozowyk *et al.* 2017). However, those associations of birch bark and fire, and the potential for accidental pitch production, is a much more likely day-to-day scenario. Although the softening effects of soaking untreated 'rawhides', whether through rainfall or immersion in ponds or streams, are very likely to have been known, the accidental application of fats to hides, or the accidental smoking of a hide, are more difficult to envisage. One possible source of insight might be the laying out of brains and other fatty organs/deposits on a freshly removed and scraped hide, but this is obviously speculative.

As for cognitive capacity, the evidence from both handaxes and other organic tools, with reference to their production processes as known from finished forms, rough-outs and débitage in the archaeological record, and modern experimental insights, strongly supports the conclusion that at least some of Europe's Lower Palaeolithic hominins could undertake multi-stage technological processes (Table 6.8). Moreover, these processes utilised the properties of multiple types of materials (although there is as yet no confirmed evidence of composite tools; Rots *et al.* 2015). Therefore, the processing demands of hides may not have been beyond the cognitive abilities of Lower Palaeolithic hominins. However, the time and space demands do suggest that such work would have required a 'safe' space and relatively good weather and been more likely to occur when good quality hides were most widely available: in other words, autumn.

Table 6.7: Potential tanning methods for hides (after Rifkin 2011)

Task	Lower Palaeolithic feasibility & tools/ requirements?
Skinning	Yes – biface, flake
Remove excess flesh & fat from inner hide surface (wet scraping)	Yes – scraper
Remove skin residue (& any tanning ingredients) from inner hide surface	Yes – scraper
Tanning ingredients (see below) applied by hand to clean, damp hides	Yes
Hides stretched by hand	Yes – involving multiple individuals
Hides dried	Yes – but requires a safe venue
Tanning methods — Vegetable tanning: • Bark samples air-dried and ground to fine particles • Ground bark soaked in water to produce tannin extract	No? Could hammerstones be used as grindstones? Water containers?
Mineral tawing: • Minerals (e.g. iron salts) ground into power & applied to hides	No? Could hammerstones be used as grindstones?
Chamoising: • **Brain fats and fatty deposits from kidneys & below skin** • **Fats softened through sun exposure prior to application to hides**	**Yes**
Smoke tanning: • **Smoke-rich fire, with temperature control to optimise smoking & reduce charring** • **Suspend hide on frame**	**Yes – pyrotechnology & organic artefacts**

Possible tanning methods for the Lower Palaeolithic (based on materials and technologies most likely to be available and/or suggested by the available evidence) are highlighted in bold

Table 6.8: Multi-stage and multi-material Lower Palaeolithic processes

	Artefact	
	Handaxe	*Spear*
Stages	• Select lithic raw material (e.g. core) & hammer[s] (e.g. hard and/or soft-stone hammerstones & antler soft hammer[1])	• Select lithic raw material (e.g. core), hammer[s] (e.g. hammerstone) & wood raw material:
	° Journey to (& from?) raw material source(s) ° Working of organic materials (e.g. antler) prior to use as percussor ° Nodule selection/blank production	° Journey to (& from?) raw material source(s)
		• Knap a stone tool
	• Primary handaxe knapping:	• Shape a wooden tool (produce spear):
	° Roughing out	° Cut-down tree ° Remove side branches
	• Secondary handaxe knapping (iterative process):	° Smooth bases of branches ° Strip off bark ° Rework spear's form & surface ° Trim the tip
	° Platform preparation ° Thinning ° Finishing	
	• Butcher an animal (or other task)	• Hunt an animal
	• Meet basic needs (e.g. hunger)	• Meet basic needs (e.g. hunger)
Materials	Stone (core/blank) Stone (hammerstone) Antler/bone/wood	Stone (core/blank) Stone (hammerstone) Wood

Sources: Wenban-Smith 1989; 1999; Haidle 2009; Stout *et al.* 2014). [1]Assumes use of soft-hammer in handaxe manufacturing, as evident on many of Europe's Lower Palaeolithic handaxe sites.

An ability to plan? A lithic perspective ...

The previous discussions of food storage (whether internal or external) and the possible preparation of winter clothing to some extent implies, rather than demonstrates, an ability to plan and anticipate future needs, fundamentally because of the low (or absent) archaeological visibility of some of those behaviours. So can we see material evidence for planning and anticipation in other aspects of the European Lower Palaeolithic record? At the broader scale it is only occasionally evident in the examples of longer-distance raw material acquisitions (Chap. 5 & Table 5.9). A harsh critic of Lower Palaeolithic hominins' cognitive abilities might view those lithic transfer patterns as evidence that while skill is most certainly present (as anyone who has tried to replicate a Boxgrove-style handaxe or, perhaps less likely, hunt a horse with a wooden spear can testify) there seems to be little will, or need, to habitually move raw materials more than a few kilometres from one place to another. Is this reflecting an inability to anticipate future needs and plan ahead? Perhaps this is so, and perhaps it can be linked to the smaller cognitive capacities of *H. heidelbergensis* (*c.* 1250 cc) and *H. antecessor* (*c.* 1000 cc), relative to *H. sapiens.* And yet, can this really be squared with the evidence, both direct and inferred, for technological expertise, in a suite of realms (organic and inorganic), foraging skills, residential and logistical mobility, and knowledge (of plants, animals and landscapes) which are strongly suggested by both direct evidence and the requirements of seasonal Pleistocene Europe?

To test these differing views of planning ability, what level of variation is evident in the most widespread component of the Lower Palaeolithic record, the lithic technological signatures of individual site assemblages and regional traditions? Does it reveal or refute planning and/or an ability to anticipate future needs?

An alternative view of the inherently local patterns in the site and assemblage-level lithic transfer data (see also Chap. 5) is that they reflect the practical energy demands of stone transport and situational needs, not overarching cognitive capabilities. In short, stone tools were made as they were needed *because they could be*, with effective functional edges produced on whatever the locally available materials were. In some cases this would permit the production of large cutting tools on large, fresh nodules, while elsewhere it would lead to the making of small, conditioned[7] handaxes on water-rolled cobbles or nodules (Fig. 6.15), while elsewhere again it would result in small flake and core tools made on small pebbles (*e.g.* as at Vértesszőlős, Hungary; Dobosi 2003, fig. 4). The local lithic transfer patterns would suggest that only if no tool-making materials were available were artefacts or raw materials imported from elsewhere (see also White 1998). This making *in the moment* generates some, although

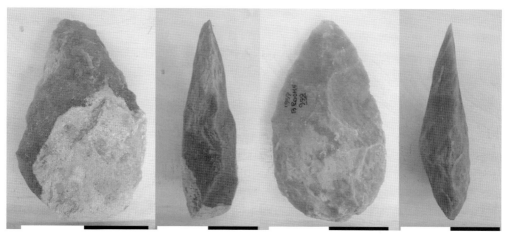

Figure 6.15: 'Conditioned' (left: Broom, artefact no. 879; image © Jennifer Chambers) and 'unconditioned' handaxes (right: Broom, artefact no. 999; image © Jennifer Chambers). Scales: 40 mm intervals. Digital handaxe models from a variety of European sites can be found at: https://sketchfab.com/P. Garcia-Medrano/models (© Paula García Medrano).

not all, of the variability seen in the archaeological record, and emphasises situational flexibility, rather than an inherent inability to plan.

Moreover, this strong reliance on local lithic material does not mean that no artefacts moved and that when hominins shifted to new foraging and hunting grounds and favourable natural shelters they carried nothing with them but their very youngest infants. It is instead likely that other types of materials, harder to obtain and produce and infuriatingly hard for archaeologists to detect, such as shelter coverings, digging sticks and spears, containers, and portable foodstuffs may have been the primary 'luggage' in the Lower Palaeolithic. Thus, the limited evidence for stone tool mobility does not demonstrate an inability to think ahead,[8] but rather an appreciation that other items were more important. This might be especially true in the context of the colder months and autumnal relocations,[9] if there was a need to enhance the insulation of natural shelters, but it would probably apply throughout the year in the case of organic food-getting tools such as digging sticks. Moreover, it is important to re-emphasise that such moves are residential, rather than short, food-getting trips and other types of local resource provision (*i.e.* logistical mobility). In the latter instances I suspect that stone tools, if needed, *were* transported. This is suggested by the evidence from the Schöningen 'spear horizon' site (13 II-4), where the flaking debris is predominantly from the end of the knapping sequence – suggesting that the tools, predominantly retouched flakes, were mostly brought onto the site in their finished form (Serangeli and Conard 2015). It is notable that the fist-sized flint nodules and the frost-fractured flakes used at Schöningen for tool-making were often much larger than any naturally occurring stones in the fine-grained sediments of the archaeological site. In short, the hominins may have known where they were heading, or at the very least carried pre-made tools with them in anticipation of carrying out tasks at a location poor in useable stone – such as somewhere along the lake shore. A similar argument has been made with regards to the spruce spears in locality 13 II-4, with the palaeoenvironmental evidence suggesting the artefacts, or at least the raw materials, were introduced to the site from elsewhere (Bigga *et al.* 2015).

Such behavioural flexibility and adaptability is also evident in the increasingly rich evidence for dynamic artefact life histories: *e.g.* previously-made bifaces and cores introduced into, and just-made artefacts removed from, specific locations (Tuffreau *et al.* 1997; Hallos 2005, fig. 6; Pope and Roberts 2005; Lhomme 2007), or the extensively re-sharpened scrapers in the Hoxne Upper Industry (Ashton 2016). Yet one of the most notable aspects of Lower Palaeolithic technological behaviour is the apparently limited evidence for contextually-driven variability in tool typology: in other words, hominins do not seem to have been using different tools for different tasks in different settings, and this could be seen as an inability to anticipate future needs. Corbey *et al.* (2016) highlighted this point more broadly in their observation that handaxes fail to track environmental variation (see also Chap. 5: Box N). But to what extent is this lack of contextual variation a genuine pattern? This question is difficult to answer because large cutting tools (LCTs), principally bifacial handaxes, dominate the overall artefact record west of the Rhine, with a greater representation of cleavers in the Iberian Peninsula (Santonja and Villa 2006). But this is at least partly a by-product of selective 19th and early 20th century collecting from re-worked Pleistocene river deposits exposed within sand and gravel quarries (see also Chap. 2: Box F). It is therefore necessary to turn to the artefacts from primary context, excavated sites. LCTs occur at both kill-butchery sites (*e.g.* Boxgrove and Soucy: Roberts and Parfitt 1999; Lhomme 2007) and 'domestic' sites (defined here as sites with hearths, *e.g.* Beeches Pit and Foxhall Road; White and Plunkett 2004; Gowlett 2006; Preece *et al.* 2006). This probably reflects the role of LCTs as a heavy-duty butchery knife, a conclusion supported by use-wear evidence (*e.g.* Keeley 1980; 1993; Mitchell 1995; Solodenko *et al.* 2015), cut-marks (Bello *et al.* 2009; Yravedra *et al.* 2010), and the likely occurrence of butchery tasks at both kill and domestic sites. This may in part explain why some primary context sites, such as Boxgrove, Soucy and Porto Maior (Méndez-Quintas *et al.* 2018), are so rich in them, as the immediate availability of raw materials to produce those large cutting tools,[10] *combined* with the butchery activities (or other tasks in the case of Porto Maior) undertaken at these sites, created a specific scenario in which handaxe and other LCT manufacturing was favoured. However, it is also true that the numbers of such artefacts can potentially be over-interpreted, with insufficient regard to the geochronological context. In the case of Boxgrove, 459 handaxes and roughouts have been recovered from the Q1/B waterhole (García-Medrano *et al.* 2019), and are associated with a landsurface that may have existed for *c.* 75–100 years (unit 4c and the equivalent units; Roberts and Parfitt 1999; Roberts and Pope 2009). The number of handaxes discarded around the waterhole therefore average out at around 5/ year. This is a formulaic calculation, but it certainly suggests that handaxes were not necessarily being made on a daily basis, and possibly that they were not used in every butchery activity (although it is possible that handaxes were retained and re-used over several separate events).

The possibility that handaxes were not quite as important as the record sometimes suggests is also borne out on a wider European scale. First, there are butchery sites at which LCTs, and the characteristic debris from their manufacture, do not occur. Some of these sites occur in parts of Europe where LCTs were otherwise commonplace (*e.g.* the elephant horizon at Southfleet Road in southeast Britain; Wenban-Smith 2013). Other butchery sites occur in regions where LCTs were relatively rare throughout the Lower Palaeolithic (*e.g.* Schöningen in northern Germany; Haidle and Pawlik 2010; Serangeli and Conard 2015), although examples, including backed knives, are known (Brühl 2003; McNabb 2007). Finally, butchery sites also appear during those periods prior to the widespread appearance of LCTs in Europe (*e.g.* Gran Dolina TD-6.2; Carbonell *et al.* 1999). It is therefore clear that LCTs were by no means necessary for butchery, of either small or large animals (although the majority of LCT use-wear evidence does relate to butchery, and experimental work has confirmed that they can perform the task very efficiently; Machin *et al.* 2007).

Table 6.9: Lower Palaeolithic tool use-wear and residue evidence

Tool	Contact material (use-wear/residue)	Sites	Reference
LCTs (i.e. handaxe, cleaver, pick)	Meat/fat & bone	Aridos 2; Boxgrove; Hoxne	Keeley 1993; Mitchell 1995; Yravedra *et al.* 2010
	Plant	La Noira	B. Hardy *et al.* 2018
	Hard materials (*e.g.* wood, bone)	Porto Maior	Méndez-Quintas *et al.* 2018
Flake tools (retouched and unretouched)	Hide	Hoxne; Schöningen	Keeley 1993; Rots *et al.* 2015
	Meat, fat &/or bone	La Noira[1]; Revadim Quarry; Schöningen	Solodenko *et al.* 2015; Rots *et al.* 2015; B. Hardy *et al.* 2018; Venditti *et al.* 2019
	Plant	Hoxne; La Noira[1]; Schöningen	Keeley 1993; Rots *et al.* 2015; B. Hardy *et al.* 2018
	Wood	Hoxne; La Noira[1]; Schöningen	Keeley 1993; Rots *et al.* 2015; B. Hardy *et al.* 2018

[1]La Noira flake tools include unretouched flakes, becs, denticulates, notches, retouched points & scrapers (B. Hardy *et al.* 2018, table 4).

Moreover, increasing numbers of use-wear studies have highlighted that flake tools were used in a highly flexible manner (Table 6.9).

While use-wear examples are still relatively rare, the available studies highlight the subtle and changing relationships between artefact form, the uses to which they were put, and how they were held and applied. Analysis at La Noira suggests *ad hoc*, flexible tool use, with artefact 'types' used flexibly and variably in accordance with their edge properties (B. Hardy *et al.* 2018). The impression is that while 'types' may have been made according to social and technological traditions (see also Chap. 5: Box N) they were used very flexibly. Similarly, at the late Lower Palaeolithic site of Revadim in Israel, on the borders of Europe, use-wear suggests the use of a biface to work a medium-hard material, probably hide, with a transversal movement (perhaps scraping?), while a scraper and flakes smaller than 2–3 cm were probably used to cut soft and medium materials, particularly animal tissues, with predominantly longitudinal movements and precise actions (Solodenko *et al.* 2015; Venditti *et al.* 2019; Fig. 6.16). This is not surprising, since skeletal morphology from the Sima de los Huesos indicates a powerful precision hand grip for hominins of the Middle Pleistocene period (Arsuaga *et al.* 2015; although the hand evidence for *H. antecessor* is more limited, the phalanges' dimensions are similar to modern humans: Lorenzo *et al.* 1999). Moreover, materials have different structures (*e.g.* wood grain direction), and the artefact record is characterised by considerable variability in overall dimensions (*e.g.* length, width, thickness) and in the dimensions and position of the working 'edge': all of which will impact on an artefact's prehensile, or grasping, properties. Such appreciations of task needs and artefact forms are perhaps evident at Schöningen 13 II-4: the tool blanks (*i.e.* the flakes or natural pieces of flint upon which the tools are made) are very diverse, but the scrapers are carefully made, with a degree of standardisation evident in the retouched edges. As Serangeli and Conard

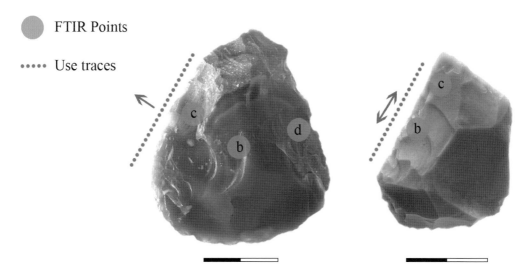

FTIR Points

Use traces

Figure 6.16: Working edge locations (based on use-wear and residue evidence) and use-motions of a biface (left) and scraper (right) from Revadim (Solodenko et al. 2015, figs 4 & 7).

(2015) note, this latter characteristic probably relates to the resharpening practices that were demanded by the limited raw material supply and the intensive nature of the carcass processing tasks, but nonetheless the care afforded to those resharpening tasks would appear to signal a concern for the 'end product' and an appreciation of what was needed for the job at hand.

So there does appear to be flexibility, variability and adaptability in tool production and functionality, and this is also evident in regional-scale raw material patterns. However, these patterns are by no means straightforward. Regional-scale divisions are often drawn in northern Europe between the raw materials available to the east and west of the Rhine, with more limited flint sources available in modern-day Germany and Poland (Fig. 6.17), where they were replaced by other, coarser materials, often occurring in smaller sizes (*e.g.* limestone, quartzite and quartz). A resulting, raw material-driven, regional-scale pattern of artefact variability is true to some extent, with small artefact examples evident at classic central European sites including Schöningen, Bilzingsleben, Korolevo and Vértesszőlős, and the proposal of a Lower Palaeolithic Microlithic Tradition in this region (Haidle and Pawlik 2010). What is also evident however is the diverse use of this small, supposedly 'limiting' material, as can be seen in the range of tool forms at Bilzingsleben (Brühl 2003, figs 3 & 5–7). Moreover, such variations in raw materials also occur to the west of the Rhine, and small, non-flint materials are again associated with non- or handaxe-rare assemblages, such as those at Arago, Isernia la Pineta, Monte Poggiolo and Saint-Colomban (Santonja and Villa 2006; McNabb 2007). But such simple correlations are not always possible – while raw material types are essentially stable at Arago (layers D–G) for example, the proportions of handaxes vary over time. Moreover the earlier European Lower Palaeolithic industries, essentially those prior to *c.* 700 kya and La Noira, show no

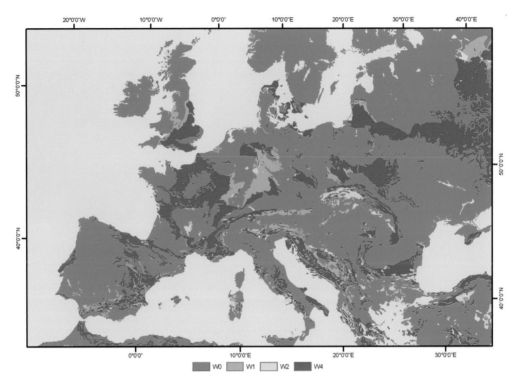

Figure 6.17: Distribution of tool-quality-yielding bedrock geological formations across Europe. Colours indicate absence (dark green) or increasing quality (light green–red) of bedrock potential at each location (e.g. Cretaceous bedrock classified as W4; Duke and Steele 2010, fig. 6).

persistent evidence of LCT technology, with a fundamental reliance on core and flake technologies at sites in both southern (*e.g.* the Orce Basin, Gran Dolina TD-6, Barranc de la Boella, Vallparadís, Monte Poggiolo, Pont de Lavaud) and, more occasionally, northern Europe (*e.g.* Happisburgh III, Pakefield), although occasional LCTs have been claimed, both convincingly (*e.g.* Barranc de la Boella; Mosquera *et al.* 2016) and with varying levels of debate as to their status (*e.g.* Solana del Zamborino and Cueva Negra del Estrecho; Jiménez-Arenas *et al.* 2011). As with the later non-LCT sites, these Early Pleistocene assemblages were clearly functionally effective (Carbonell and Rodríguez 2006), whatever the exact nature of the hominins' survival strategies within those highly competitive environments.

These larger and smaller-scale patterns in space and time again highlight flexibility and adaptability in tool-making and using, and perhaps suggest that the handaxes and other LCTs, so often seen as Palaeolithic 'Swiss army knives' (*e.g.* Brumm 2014; Finkel and Barkai 2018), were sometimes a little less ubiquitous and necessary in day-to-day hominin life.

This sense of a non-LCT world (sometimes) is also supported by the occasional multi-layered sites or groups of sites which reveal larger-scale fluctuations between artefact assemblages with and without handaxes, such as Arago, Notarchirico, and the sites of southeast Britain during MIS 11 (Piperno *et al.* 1998; de Lumley *et al.* 2004; Santonja and Villa 2006; Ashton 2016), and offer a different perspective on artefact variability in the European 'Acheulean era' (effectively the Middle Pleistocene after *c.* 700–600 kya). Ashton *et al.* (2016) have recently re-interpreted the shift in the British early MIS 11 record from Clactonian (core and flake) to Acheulean (handaxe) industries as reflecting changing hominin groups (although probably not species) with different tool-making habits, and population dynamics between southern Britain and the north-west of the continental 'mainland', although various other interpretations of the Clactonian have been offered over the years (see White 2000; McNabb 2007 for comprehensive reviews). The British record is based on large-scale patterns across multiple sites, correlated through pollen assemblages (Ashton *et al.* 2016, fig 3), and since the multiple sites are likely to represent multiple times of the year, it suggests that British early MIS 11c (core and flake) and later MIS 11c–11a (handaxe) tool-kits did not vary on a seasonal basis.

Might this explanation also apply to Arago and Notarchirico? Possibly, but the changing character of the Arago evidence (Table 6.10) also raises the possibility that the different technologies are the products not of different groups with different tool-making traditions, but of different durations of hominin behaviour, with different archaeological layers characterised by short-term visits and sustained occupations, with differing seasonal associations (McNabb 2007).

Notarchirico offers another different perspective, as the fluctuating appearances and disappearances of bifaces may be associated with changing local conditions (McNabb 2007). This might suggest changes in the types of raw materials that were easily available (Table 6.11), although this has not been supported by Santagata (2016). Certainly there is not a clear association, suggesting again that a combination of factors were likely at play. What is noteworthy is that Piperno *et al.* (1998) rejected the notion of two cultural traditions (see also Santonja and Villa 2006), suggesting a flexible approach to tool-making, one that emerged variably from a single underlying technical tradition.

A further factor is the day-to-day, situational contexts of tool-making. In short, when effective sharp edges are needed to butcher an encountered carcass before competitors arrive the production of a highly symmetrical handaxe (even if the local raw material permits it) would seem to be an unnecessary, and probably unwise, luxury. The technology at a number of short-lived megafauna sites, such as Aridos 1 and 2 (Yravedra *et al.* 2010) and Southfleet Road (Wenban-Smith *et al.* 2006; Wenban-Smith 2013) looks interesting in this regard – and are excellent demonstrations that both handaxes (Aridos) and core and flake technologies (Southfleet Road) make functional tool-kits for butchering megafauna. By

Table 6.10: Occupation histories and technological characteristics at Arago Cave, by layer (after McNabb 2007, table 8.2; see also de Lumley et al. 2004)

Unit	Nature of occupation	Bifaces present?
E	Long-term	Yes
F	Seasonal, spring to summer	Rare & atypical
F/G	Short-term	No
G	1 year or more	Yes, in small numbers

Table 6.11: Depositional characteristics, occupation histories and technological characteristics at Notarchirico, by layer (after McNabb 2007, table 8.4; see also Piperno et al. 1998)

Layer	Depositional environment	Bifaces present?
A1	Unstable conditions, bank collapse & solifluction	Yes
B	River eroding laterally & reworking its channel deposits	Yes
C	Unstable conditions, bank collapse & solifluction	No
D	River eroding laterally & reworking its channel deposits	Yes
E & E1	River eroding laterally	No
F	River eroding laterally & reworking its channel deposits	Yes

contrast, the quality, and similarity, of the handaxe-making at Foxhall Road (White and Plunkett 2004; Hopkinson and White 2005, fig. 2.1) looks equally intriguing in light of the suggested associated hearth. The impression that is given is of a relaxed, social setting, in which time could be devoted to careful tool-making.

One important implication of the above discussions is that handaxes were perhaps less important to Lower Palaeolithic hominins than they are to 19th–21st century archaeologists, at least in terms of day-to-day, task-specific needs and survival. This may seem to be verging on heresy! The handaxe is probably the most recognisable symbol, at least in terms of artefacts, for the human Stone Age – it is the symbol for the UK's Royal Academy of Engineering (https://www.raeng.org.uk/). Its curatable, portable nature may also have been an important attribute within increasingly mobile and logistical landscape-use strategies in the later Middle Pleistocene (Lewis *et al.* 2019). And yet the evidence from various sites, including the Schöningen 13 II-4 'spear site', the Southfleet Road 'Ebbsfleet elephant' site and the Swanscombe Lower Loam 'knapping floor', suggests that complex and skilled butchery of large animals was sometimes carried out with sharp flakes and flake tools, without recourse to handaxes (Wenban-Smith 2013; Serangeli and Conard 2015), while use-wear suggests that non-LCTs could be used to work a wide range of materials (Rots *et al.* 2015). At larger regional and chronological scales too, handaxes and other LCTs barely feature in the Lower Palaeolithic records of Germany and Central Europe (Serangeli and Conard 2015), although other types of bifacial tools are present (McNabb 2007), and are almost completely absent across Europe prior to *c.* 600–700 kya (Moncel 2010). This is not to say that handaxes and LCTs are unimportant – both to us as archaeologists (as a means of understanding technological skill, tool use, and learning and social environments: *e.g.* Gamble 1998a; Kohn and Mithen 1999; Hodgson 2009; Spikins 2012; White and Foulds 2018) and to Lower Palaeolithic hominins (as one means of extracting resources from the environment, and perhaps in social life as well) – but it is to argue that the conflation of the Lower Palaeolithic and the handaxe may blind us to the nuanced, flexible and dynamic behaviours of this period. Related concerns have been highlighted with regards to the definition of the Acheulean, emphasising the diversity of assemblages labelled as 'Acheulean' and questioning whether 'Acheulean' traits (both technological, *e.g.* bifacial flaking of discoidal cores and complex débitage systems, and behavioural, *e.g.* landscape mobility and functional distinctions between sites) can be identified in 'non-Acheulean' sites (Mosquera *et al.* 2013; 2016; Moncel *et al.* 2015; Ollé *et al.* 2016; Rocca *et al.* 2016).

So overall, does the lithic artefact record for the European Lower Palaeolithic support evidence for longer-term planning and anticipation of needs? There are glimpses, most obviously the insights into artefact life history and mobility (*e.g.* being introduced onto or removed from, specific sites), although this evidence typically lacks a clear temporal or spatial dimension. Beyond this however I think the blunt answer is not obviously, although there is considerable evidence for flexibility and adaptability.

Autumn: season of mists and mellow fruitfulness?

The focus of autumn living thus seems to have been centred around the exploitation of resources, primarily foodstuffs but probably also including high quality animal

Figure 6.18: An autumn strategy.

hides (Fig. 6.18). As with spring and summer, there may have again been important opportunities, and possibly a need, for younger children to contribute significantly to foraging, through the collection of late-year fruits and nuts. However, there was probably also a tendency to look ahead, towards local winter survival, in the form of storage, perhaps both internally and externally, local relocations, and possibly the preparation of 'clothing'. These behaviours are difficult to demonstrate, but the needs seem likely (see also Chap. 3), and the meeting of them implies a degree of anticipation and awareness, a detailed understanding of animal behaviours, and 'safe' spaces.

Yet autumn may also have been a socially and emotionally demanding period, reflecting the involvement of many or even all hominins in potentially dangerous hunting of aggregating animals. This would have been made perilous both by the timing of ungulate breeding seasons, and the likely heightened alertness and aggression of the prey, and by the importance for hominins of not missing out on this critical, seasonal opportunity.

Notes

1. Pleistocene wild boar behaviour might also have been a valuable cue to the presence of various plant foods, including 'invisible' buried resources such as roots and tubers (Genov 1981).
2. Cohen *et al.* (2012) defined the coastal zone as including saline habitats, freshwater ecotones of inland coastal plains, the lower reaches of river valleys, hill-slopes bounding the coastal plains, and lower valleys, and emphasised that Atlantic conditions can reach several tens of kilometres inland (and potentially as much as 100–200 km or more).
3. The leg of Przewalski's horse has been described as shorter and thicker than that of the domestic horse, resulting in a shorter stride, although the species is still described as being well adapted for speed and endurance (Sasaki *et al.* 1999).
4. However other residential moves, perhaps especially those of the summer and early autumn when plant foods were abundant, may also have been dictated by the exhaustion of plant resources within the core foraging radius (Grove 2009).
5. However, the use of clo units are only relatively useful for discussing pre-Holocene clothing, since they are derived from modern day, woven fabric, clothing, which has different thermal properties to animal hides and furs. Clo units also apply to wind-free conditions (Gilligan 2010).
6. Beavers' castoreum secretion has analgesic and anti-inflammatory properties, as does salicin in willow bark, the presence of which in the beaver diet explains the origins and the benefits of castoreum (Pincock 2005).
7. In some cases the shape of the original blank 'conditioned' human action (*e.g.* pointed handaxes made on thick, narrow, elongated blanks), whereas in others there was little or no evidence of blank conditioning (*e.g.* ovates produced on flat and/or large nodules; White 1998; see also Ashton and McNabb 1994).
8. It might even indicate that hominins *were* thinking ahead (*i.e.* they rarely transported stone because they knew that other suitable materials were typically available in the places they were travelling to).
9. It is possible that stone tools were more heavily curated during generally harsher conditions, reflecting the greater energetic costs and risks of mobility (and perhaps also increased intensity of tool use). This is difficult to demonstrate, but the heavily retouched scrapers in the late MIS 11 Hoxne Upper Industry are intriguing in that regard (Ashton 2016).
10. Although at some of the Soucy localities (*e.g.* Soucy 5: level 1) bifaces were not produced on-site but rather introduced from elsewhere (Lhomme 2007).

Chapter 7

A year in a supremely skilled life?

The hominin year: a seasonal perspective

Chapters 3–6 offer a seasonally-structured perspective on Lower Palaeolithic life in Europe (Fig. 7.1). While the specifics would have varied geographically, chronologically and across the fluctuations of Early and Middle Pleistocene climate cycles, the challenges and opportunities of the seasons (see Fig. 1.9) suggest the general presence of shared, 'benchmark' moments in the hominin year: early spring relocations in response to animal movements, late spring/early summer births, late summer/early autumn targeting of animal aggregations and abundant plant foods, and winters characterised by local living and a reliance on scarce, and possibly stored, foodstuffs. Focusing on the hominin strategies associated with those moments significantly enhances our understanding of Lower Palaeolithic Europe with regards to the adaptability of early *Homo*, their ecological position within the wider Pleistocene world, and the overall settlement record.

Adapting to a seasonal Europe?

High levels of environmental diversity across time and space have been argued to favour behavioural plasticity and adaptability in many organisms, including hominins (Potts 2012). Climatic and habitat variability is clearly evident in Lower Palaeolithic Europe, in terms of its annual seasonality, overprinted by glacial–interglacial and sub-stage variations (Chap. 2). In such environments adaptive versatility reduces the risk of extinction, or the need to move in response to climate change (Fig. 7.2).

Yet the European Lower Palaeolithic record suggests that hominins were never sufficiently versatile to be constantly present throughout Europe: initially there was only a small-scale presence during the Early Pleistocene, predominantly around the

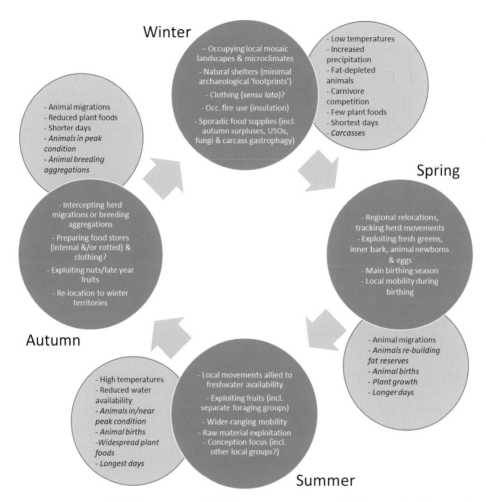

Figure 7.1: A year in the life in Lower Palaeolithic Europe. Seasonal challenges (in plain text) and opportunities (in italics) are summarised in the grey circles (see also Fig. 1.9).

northern rim of the Mediterranean (below *c.* 45°N), associated with *H. antecessor* (and possibly other species) and core and flake technologies; subsequently an expansion during the early Middle Pleistocene, with occupations extending into north-western Europe, associated with *H. heidelbergensis* (*sensu lato*) and a mixture of Acheulean and core and flake technologies; and finally a widespread site and artefact record in the later Middle Pleistocene, but with a stronger presence to the west of the Rhine in northern Europe and punctuated by regular northern, and possibly Europe-wide, extirpations (Carbonell *et al.* 1996; Dennell and Roebroeks 1996; Carbonell and Rodríguez 2006; Moncel 2010; Mosquera *et al.* 2013; Moncel *et al.* 2015; Hosfield and Cole 2018).

This overall settlement record seems to suggest a 'glass ceiling' to Lower Palaeolithic adaptations in Europe, and hominin populations that were 'dancing to the rhythms of

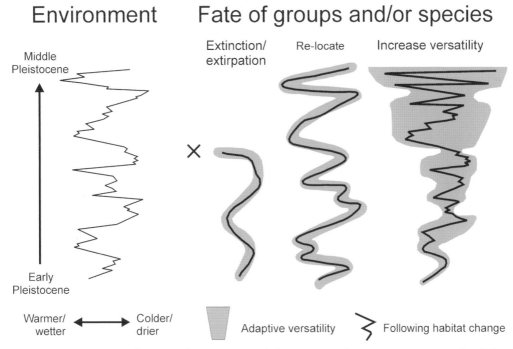

Figure 7.2: Hominin evolution and environmental dynamics in the European Lower Palaeolithic (redrawn and modified after Potts 2012, fig. 6). The European record suggests that while hominins' adaptive versatility increased over time, it was cylically 'squeezed' during periods of harsh climatic conditions.

the Pleistocene' (Gamble 1999, 125). Viewed at the large scale the earliest Europeans can therefore potentially be characterised as having constrained or 'banded' versatility, typified by extinctions or movements in the face of environmental change (Fig. 7.2). They might be described as a narrowly adapted bipedal social carnivore.

But this view ignores the fact that as well as cyclical extirpations there *were* changes in the European distribution of hominins across the duration of the Lower Palaeolithic: from 'Homo mediterraneous' to 'Homo europa'. While the fully glacial north remained beyond them, their warm stage range expanded significantly in scope and size after c. 600 kya: in Potts' terms they most certainly became more versatile. This can be seen through changes over time in various Lower Palaeolithic behaviours, which enabled hominins to cope more successfully with the particular seasonal challenges of Early and Middle Pleistocene Europe. Specifically, transformations in foraging, insulation, planning ability, social life, and site and landscape use were all important in meeting the seasonal challenges of Europe, and overcoming them at an increasingly large scale. Critically these transformations are all detectable, to a greater and lesser extent, in the archaeological and fossil records.

Securing food in a seasonal world

While southern winters were relatively mild in the Early Pleistocene the seasonal cycle is still likely to have resulted in hominins facing periods of resource stress (*i.e.* reduced plant food availability and animals in poorer condition in winter/early spring), while the levels of carnivore competition would probably have reduced opportunities for winter carcass access and/or for internal storage through autumn gorging. These seasonal difficulties in acquiring sufficient food may therefore have been an important factor behind the temporary and small-scale character of the earliest hominin dispersals into southern Europe, originating from western Asia.

From a seasonal perspective, the possible shift towards increasingly stable, although still small-scale, hominin occupations of southern Europe towards the end of the Early Pleistocene and in the early Middle Pleistocene implies an ability to secure reliable food supplies throughout the year. Detecting a clear dietary shift in the European Lower Palaeolithic is difficult, not least because of the paucity of direct foraging evidence for this period. However while there appears to be a similar range of butchery tasks and prey/carcasses throughout the Lower Palaeolithic (*e.g.* horse, bison, deer and occasionally larger animals such as rhinoceros and elephant), the review of subsistence at Gran Dolina (TD-6.2 and TDW4) and the Sima del Elefante by Huguet *et al.* (2013) concluded that early animal resource use prior to *c.* 900 kya was opportunistic and indiscriminate, while carcass access is likely to have been even more unreliable for the very earliest Europeans. This offers an interesting contrast to the much younger sites of Schöningen and Gran Dolina TD-10.1 and TD-10.2 which show evidence for possible specialisation in the late Middle Pleistocene in the respective dominance of horse, red deer and bison within the three assemblages (Fig. 7.3) and the first direct evidence for hunting weapons. While these faunal profiles may be a reflection of the specific circumstances of the settings of those sites, it is perhaps also an early expression of dietary behavioural shifts that led towards the hunting specialisation characteristic of a number of Middle Palaeolithic Neanderthal sites (White *et al.* 2016). Overall, these changing assemblage profiles are suggestive of more reliable animal food provision by the late Middle Pleistocene. This is significant given the seasonal dietary shortages of winter and early spring and the specific dietary needs of late spring (the 3rd trimester and childbirth), and the wider context of increasingly sustained and widespread occupations by large-brained *Homo* in a period of more marked seasonality.

Interestingly, TD-6.2 offers an 'interim' picture. While the nature of the site's butchery evidence and the occasional 'over-printing' of stone tool cut-marks with carnivore tooth marks (but not *vice versa*; Saladié *et al.* 2011) suggests that the hominins could secure primary access to a kill, the prey focus is clearly more mixed than at the later sites (Fig. 7.3). To some extent this is also true at the early Middle Pleistocene sites of Boxgrove (Parfitt and Roberts 1999; Smith 2013) and Isernia la Pineta (Hohenstein *et al.* 2009): these are all occupations which fall prior to, or at the very beginning of, the more sustained occupations of the later Middle Pleistocene.

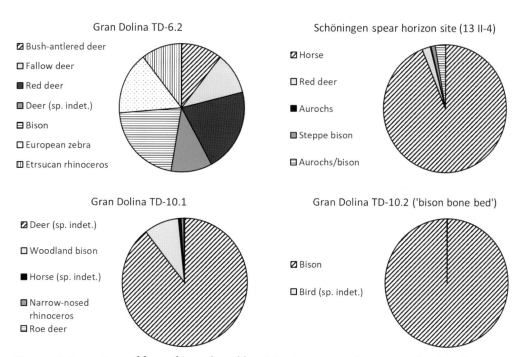

Figure 7.3: *Proportions of fauna from selected hominin sites. Gran Dolina TD-6.2 data based on MNI, as hominin modifications divided by size class (Saladié et al. 2011, table 2; n=19); Schöningen 13 II-4 data based on all hominin modifications (van Kolfschoten et al. 2015a, table 1; n=834); Gran Dolina TD-10.1 data based on cut-marks (Rodríguez-Hidalgo et al. 2015, table 5; n=444); Gran Dolina TD-10.2 ('bison bone bed') data based on cut-marks (Rodriguez-Hidalgo et al. 2017; n=1,020).*

By contrast the nature of hominins' engagement with the very largest fauna is uncertain throughout the period. Sites from Fuente Nueva-3 onwards have produced evidence for rhinoceros and elephant butchery, from Greece to Britain (*e.g.* Villa 1990; Parfitt and Roberts 1999; Yravedra *et al.* 2010; Saladié *et al.* 2011; Saccà 2012; Espigares *et al.* 2013; Wenban-Smith 2013; Mosquera *et al.* 2015; Rodríguez-Hidalgo *et al.* 2015; Konidaris *et al.* 2018), although in all cases there is uncertainty as to the cause of death. Discussion of Neanderthal healthcare (Spikins *et al.* 2019) highlights the potential risks associated with the hunting of larger animals while the consequences to small hominin groups arising from the death of one or more experienced adults might favour a scavenging-type approach where these particular animals were concerned. Irrespective of the access method, access to fats might be a key motivating factor behind the repeated associations of stone tools and remains (Ben-Dor *et al.* 2011), especially in winter and early spring, as might the sheer scale of the resource. If not hunted (*cf.* Wenban-Smith 2013) then regular monitoring of likely carcass accumulation points is implied (although the relative success rates of such monitoring remains unknown). What is clear however, from many sites, is that the exploitation of large, thick-skinned carcasses was not beyond the butchery expertise, technology or physical strength of Lower Palaeolithic hominins.

Alongside animal foods is the question of plant foods. While direct evidence remains limited, there are clearly glimpses and the palaeoecological indicators suggest a diverse range of accessible foods that could be consumed raw, in many cases with low processing costs. Combined with nutritional (the protein ceiling), ecological (the temporal and spatial patchiness of European resources) and life history models (early weaning and extended childhoods), and perhaps especially with likely hunting success rates, this suggests that plant foods were always a significant component of Lower Palaeolithic survival across Europe. Critically, it is clear that plant foods were available in each of the seasons, although with variations in types and abundance.

Adults and children in a seasonal world

These food acquisition strategies were intrinsically linked to the demands and character of hominin life history at various stages: modern dietary studies indicate that diverse diets lower infant mortality and increase life expectancy (Hockett and Haws 2003). Successful reproduction for Lower Palaeolithic European *Homo* clearly demanded reliable, high-quality foraging and food provision in a seasonal environment, with all of the cognitive demands that this entails. Growth and development, using an essentially modern human model, was likely a major challenge for European *Homo*, and may have initially restricted the ability of Lower Palaeolithic hominins to colonise at least the more northerly parts of Europe. There is an indication of those challenges in the dental data from the Sima de los Huesos: while rates of enamel hypoplasia are lower than in Neanderthal samples, they are concentrated in years 0–6, with the 3rd year of life being the most stressful. This would appear to coincide with, and may be related to, the onset of weaning and its particular dietary demands and metabolic stresses (Cunha *et al.* 2004). What is currently less clear from the data available is whether and to what extent such stress patterns (and the associated behaviours) varied over the course of the European Lower Palaeolithic.

The nutritional ecology approach also suggests that population increases may result from dietary changes that lead to improved health (*e.g.* reduced mortality) rather than the other way around, and this offers an interesting perspective on the pre-/post-*c*. 600 kya shift in the scale of the Lower Palaeolithic record. The expanded hominin distributions of the later Middle Pleistocene may therefore also have been associated with the increasing use of separate foraging groups, and the early participation of children in food-getting (and learning how to do it). This offers a partial solution to the problems of seasonally variable food availability, and also highlights the importance of reliable food sharing strategies, extended childhoods, and alloparenting.

Controlling temperatures in a seasonal world

An ability to tolerate colder temperatures, probably through a combination of physiological and behavioural traits, would have been required both in warm stage winters and, in association with cold stage occupations in southern Europe, throughout the year. Fire use appears to be a later emergence, perhaps only from the later Middle Pleistocene onwards, although this may reflect taphonomic bias. If genuine, this late appearance may well be due to the cognitive and practical demands and costs of both fire production and its maintenance (Twomey 2013; Henry *et al.* 2018) and may also indicate that the needs for, and uses of, fire varied markedly between different areas of Europe (*e.g.* the Mediterranean, the Atlantic northwest, and central–eastern Europe). If the chronological pattern is correct, fire may only have been critical to a sustained northern expansion. Moreover, the importance of fire may still have been highly seasonal and focused towards heat and light (*i.e.* a winter technology) rather than cooking, perhaps explaining its apparent absence on Early Pleistocene sites, both open-air and caves, in southern Europe. This is suggested by a consideration of the costs of fires, especially, but not exclusively, in terms of fuel-getting, set against the possibly limited benefits of cooked food, the potential existence of alternative methods for externally 'pre-digesting' food and the environmental contexts. Lower Palaeolithic hominins may therefore have been episodic, rather than obligate, fire-users (see also Henry *et al.* 2018).

By contrast, skinning evidence has a much longer antiquity in Lower Palaeolithic Europe, extending back to at least *c.* 900 kya at Gran Dolina TD-6.2 (Saladié *et al.* 2011; Huguet *et al.* 2013). Skinning of an animal does not necessarily mean the production of clothing or other items for keeping warm but sustained occupations in southern Europe had to meet the challenges of seasonally cool and arid environments and the chillier and more open habitats associated with cold-stage intervals. It is also likely that hide technologies emerged and developed in a punctuated manner, as with fire (Gowlett 2016): a greater reliance on these technologies is perhaps indicated by the mono-species faunal exploitation evidence from the later Middle Pleistocene. 'Style' and the degree of clothing use may also have varied seasonally, perhaps especially in the warmer summer climates of the Mediterranean and the continental interior. Nonetheless I suggest that European hominins may have worn (or slept under) mobile insulation long before they sat around fires.

Cold-climate tolerances may have increased significantly after MIS 12 in the later Middle Pleistocene, when the central European record starts to expand (Haidle and Pawlik 2010; Szymanek and Julien 2018). The palaeoenvironmental evidence suggests that winter challenges may not have been an insurmountable problem during this time: the sites show a preference for the early and late portions of warm stages (not the thermal maximums), and are associated with open-forest or forest-steppe environments. Moreover, there is occasional evidence for occupations in cool/cold and open conditions, such as at Korolevo VI and Kärlich H (Szymanek and Julien 2018). The chronology of the first secure European evidence for controlled fire use, from MIS 11 onwards, is intriguing in that regard, raising the possibility that reliable winter pyrotechnology was critical for more substantial central European occupations, and that once present it also opened up more 'challenging' environments to hominins.

Spaces and places in a seasonal world

Finally, while occupation histories may still have been discontinuous there is perhaps evidence for more organised use of landscapes in the latter half of the European Lower Palaeolithic. From an on-site perspective it is reflected not only in occasional task-specific locations (*e.g.* the Charco Hondo flint 'quarries') but also from longer distance raw material provisioning, the glimpses of artefact mobility, on both LCT and non-LCT sites (*e.g.* Soucy and Schöningen) and, by extension, from the sense of a 'wider and deeper' approach to planning, anticipation and 'landscape management' (Moncel *et al.* 2015; Lewis *et al.* 2019). This change may have been necessitated by the increasing seasonality of the Middle Pleistocene (*e.g.* summer aridity in the south and winter cold in the north), with its implications for resource availability and access. Those pressures, and the associated behaviours, may also have selected for increasingly sophisticated learning opportunities and organisational abilities (*e.g.* extended childhood/adolescence and linguistic communication).

Continuity in a seasonal world

Yet other aspects of behaviour show relatively little change. The evidence from Arago cave, Gran Dolina and Menez-Dregan, and selected open-air sites (*e.g.* Boxgrove and Soucy), suggests that some places in the landscape were re-used and it is notable that this behaviour again stretches back to *c.* 900 kya at TD-6.2. Such re-uses were presumably testament to the properties or affordances of these places (*e.g.* shelter, access to key resources), while their detectability reflects distinctive, and often exceptional, preservation conditions, but are these comparable to the long rockshelter occupations of the Middle and Upper Palaeolithic? The temporal evidence is admittedly limited but with the possible exception of Menez-Dregan and Arago the impression is of intermittent occupations of sites and local areas for a few years or decades, characterised by small bands (and very low-density regional populations), frequent residential mobility (albeit varying across the seasons), limited investment in place, and consequently occupational discontinuity. This seems more akin to Ashton's (2018) notion of habitual local landscapes, than to the persistent places concept of the later Palaeolithic (*e.g.* Pope *et al.* 2018).

This sense of discontinuity is also evident in artefact records, such as the British handaxe/core and flake fluctuations in MIS 11, and possibly at other sites with fluctuating industries (*e.g.* Notarchirico during MIS 16). While this discontinuity may, at least in part, have been driven by stage and sub-stage-level environmental fluctuations, it perhaps also reflects a fundamentally locally-focused but spatially shifting life, with hominin bands periodically relocating their small core foraging areas in response to resource availability and/or changing predatory threats. Small core areas are suggested in part by lithic data, and are appropriate responses to the respective demands of winter and early spring survival. There is limited evidence from on-site records (*e.g.* measures of occupation intensity) that this residential/mobility aspect of Lower Palaeolithic life fundamentally changes, although the more 'efficient' hunting abilities (*e.g.* increasingly mono-specific profiles) suggested by late Lower Palaeolithic sites (Schöningen and Gran Dolina TD-10) might enhance the resilience of local communities.

Coping with seasonal Europe: biology and technology

Some of the specific behavioural changes reviewed above, and the shifting intensity in European occupation signatures before and after *c.* 600 kya, may be linked to the flexible nature of humans. Wells (2012) emphasised that humans have a limited genetic commitment to specific conditions and niches – we inhabit a wide range of environments but have limited genetic variability in the present, although this is also related to *H. sapiens'* relatively recent evolution in Africa. This flexibility may be driven by environmental variability and uncertainty, as the frequency of environmental stochasticity (*i.e.* unpredictability) increases relative to the length of life – humans are relatively long-lived and Pleistocene Europe was both variable and uncertain, at seasonal and longer timescales. Such conditions do not favour systematic adaptations but nor do they favour developmental plasticity, which reflects genotype/environment interactions in early life and tends to be relatively irreversible. Rather it may well be better to increase the breadth of one's environmental tolerances – *i.e.* phenotypic plasticity. In particular, Wells stressed the notion of energy capital and the ability to extend the gap between gaining energy and 'spending' it. Two types of energy capital were suggested, of which social capital (*i.e.* you do not always have to generate all of your own energy and can sometimes rely on others) is of particular interest here, especially since social capital makes you potentially vulnerable (*e.g.* to a lack of reciprocity). It is possible that the occupation intensity changes of the later Lower Palaeolithic might be a reflection of Wells' enhanced phenotypic flexibility, with an emphasis on new models of reproduction (*i.e.* 'cooperative breeding', with significant provisioning), multiple and differently focused, foraging-groups within each hominin band, and perhaps also shifting models of energy storage (food of course, but potentially also other socially co-operative mechanisms of 'banking' energy, such as clothing, shelter and fire).

So in light of the seasonal challenges of Europe, how important was technological buffering, *e.g.* fire and clothing, and how did it interact with biological adaptations such as metabolic heat production? Extricating the cultural from the physical is difficult but it is clear that the niche created by big-brained and big-bodied *Homo* in Lower Palaeolithic Europe was at least partially reliant on organic and inorganic technologies and their associated behaviours. Those technologies extended beyond the basics of immediate food getting, *e.g.* cutting tools and digging sticks, and enhanced hominin tolerances by effectively increasing their energy budgets, *e.g.* through the gains provided by clothing and shelter. This buffering was enabled by encephalisation but was also necessary for the procurement of the energetic resources required by it (Galway-Witham *et al.* 2019). The archaeological evidence from Europe suggests that such buffering was markedly enhanced in the later Middle Pleistocene, associated with *H. heidelbergensis*, and perhaps also the emergence of Neanderthals, and with the widespread and sustained presence of hominins across northern Europe during warm stages. But, as argued throughout this book, I also think that cognitive buffering, the ability to anticipate and plan, was equally important in the seasonal environments of the north.

How different were *H. heidelbergensis* and *H. antecessor*? A seasonal perspective

There are clear differences between the geographical distribution and scale of the occupation records of *H. heidelbergensis* and *H. antecessor*. While the total number of fossils is relatively small the appearance of *H. heidelbergensis* broadly maps onto the transformation in the artefact record that occurs in the early Middle Pleistocene and which indicates a sustained hominin expansion to the north of the Alps during warm stages. Intra-annual fluctuations in climatic conditions and food resources have been highlighted in this book and key aspects of those, particularly winter temperatures and the length of the growing season, vary markedly on a latitudinal basis and, to some extent, longitudinally. Successful hominin expansions above 45°N were therefore dependent on overcoming the more marked seasonal challenges of both northern Europe and the Middle Pleistocene, *i.e.* a fully four-season year, colder winters and shorter growing seasons. A seasonal approach therefore offers valuable new perspectives on possible differences between *H. heidelbergensis* and *H. antecessor*.

Solutions to those marked seasonal challenges would have focused primarily on insulation and food supply, with associated social transformations, although biological adaptations (*e.g.* enhanced vasoconstriction or non-shivering thermogenesis) may also have played a secondary role. Reliable plant and animal food provision throughout the year would also have been critical to satisfy the growth and development needs of large-brained *H. heidelbergensis*. This view is in-keeping with Stiner and Kuhn (2006, fig. 8), who argued for significant shifts at *c.* 500 kya in three dimensions of the hominin ecological niche: ungulate age/sex selectivity (shifting from possible 'non-selective hunting' to 'prime-age-biased ungulate hunting'); prey type evenness (possible 'increased use of large game' shifting to 'nearly exclusive focus on highly ranked prey'), and energy retention efficiency (from 'fire as a heat source' to possibly 'greater hide working'). The first two are clearly supported by the evidence from Gran Dolina and Schöningen, although the evidence for the third niche is generally more ambiguous. Step-changes in these, and other, characteristics are likely to be an important part of the distinction between *Homo antecessor* and *Homo heidelbergensis*.

Meeting the demands of insulation and food supply are likely to have required the ability to anticipate the future, such as pending winter shortages or the need for cultural insulation, and respond accordingly. This anticipation, while drawing on environmental cues, would be grounded in accumulated knowledge and experience of previous years. The increasing segregation and aggregation of resources, in both space and time, would also have selected for intricate strategies of mobility and sub-divided roles in food getting, with further implications for life history models and learning environments.

Thus culture, both material and non-material, was therefore critical for *H. heidelbergensis* to more consistently buffer the environments of seasonal Europe. In that sense, the crossing of the mid-latitude rubicon is suggested to have been a key milestone, although by no means the only one, in the transition of *Homo* from a bipedal, tool-using, social carnivore comparable to wolves and hyenas, to a recognisably human forager.

While aspects of these behaviours may also have been present in *H. antecessor* and other early *Homo* populations in the Mediterranean zone during the Early Pleistocene, it is likely that there would have been little mutual recognition or understanding between *H. antecessor* and *H. heidelbergensis*. The European Lower Palaeolithic therefore oversees a transition from a fundamentally African hominin, albeit one that episodically extended its range around the Mediterranean rim, to a European specialist, firmly on the road towards the Neanderthals.

A seasonal world: hominins on the edge?

A recurring theme of this book has been the seasonal challenges faced by European hominins during the Lower Palaeolithic. Stresses, both climatic and dietary, would make small hominin groups vulnerable to minor fluctuations in the passage of the seasons, and while the low resolution of the record tends to lead archaeologists to study long-term demographic trends, those patterns emerge from the experiences, decisions and fates of individuals (French 2016). A late arriving spring, for example, might have had serious immediate implications for health and mortality, especially for those with the most demanding dietary requirements such as pregnant/lactating females and weaned infants, principally due to its impacts on plant and animal food availability. Numerous examples can be found among the animals of the Białowieża Primeval Forest today: bison reproductive success is significantly influenced by May temperatures and masting,[1] while their survival rates are impacted by harsh winters (*i.e.* snow and low temperatures; Mysterud *et al.* 2007). The reverse could theoretically also apply, potentially reducing mortality and increasing group sizes. However, for hominins this is complicated by their long childhood, and possible multi-generational legacies (*e.g.* the DOHaD hypothesis), which thus requires a succession of 'good' years. As the examples from the BPF have shown, high frequency cycles of both 'good' and 'bad' years are much more common. While the specific processes and impacts would have varied for hominins, the general message of seasonally and annually fluctuating resources and weather conditions is key.

But is this seasonally-based picture of hominins on the edge supported by larger-scale evidence? This is a more difficult question to answer, since estimating demographic trends and settlement histories is difficult for most archaeological periods (*e.g.* Hassan 1982; Chamberlain 2006). It is perhaps particularly difficult for the Lower Palaeolithic (French 2016), where time-depth, hominin behaviour and taphonomy combine to remove data-sets such as site density, site size, radiocarbon chronologies and, increasingly, genetics, all which have been used for the late Middle and Upper Palaeolithic and Holocene prehistory (*e.g.* Bocquet-Appel *et al.* 2005; Gamble *et al.* 2005; Bocquet-Appel and Degioanni 2013; French and Collins 2015; French 2016). However, there are nonetheless approaches to which we can turn. Ashton and Lewis (2002) have exploited the tendency of Pleistocene rivers to rework and concentrate artefacts in commercially valuable sand and gravel deposits to build artefact density models for the River Thames in the UK, subsequently extended to the Solent River landscape (Ashton and Hosfield 2010; Ashton *et al.* 2011). While all such 'accumulations'-based approaches incorporate a wide range of potential biasing factors (French 2016), these specific studies suggested a decline in warm stage populations between MIS 13 and MIS 7. This pattern may relate both to changes in the timing and durations of Britain's palaeogeographic connections to the continent (see Figs 6.2 & 6.3), but also possibly to changing habitat preference across the Lower/Middle Palaeolithic transition. But the British picture, and that of northern Europe generally, is complicated by the glacial cycles of the Middle Pleistocene: hominins were absent from the north during these coldest phases, with landscapes either covered by ice sheets, or characterised by harsh periglacial conditions.

This sense of unstable populations, and of fluctuating demography and settlement histories, is also evident in other samples of lithic assemblages. White (2015) has demonstrated chronological patterning in the modal handaxe types from British assemblages between MIS 15/13 and 9, a pattern that seems best explained by the repopulating of Britain, after each glacial 'abandonment', by different hominin lineages. More tentatively, Lewis *et al.* (2019) have suggested three assemblage groups in Britain during MIS 13 (and possibly slightly earlier), which may reflect different hominin incursions at different times (see also Davis and Ashton 2019). On a larger scale, Moncel *et al.* (2016) have highlighted the diversity in both bifacial and non-bifacial assemblages, in particular those between *c.* 700 and 500 kya. While Moncel *et al.* acknowledge that this may in part reflect raw material variations, and perhaps also different tasks and activities, they also highlight episodic dispersals as a possible factor. Dennell *et al.* (2011) have similarly argued for frequent local extinctions of 'sink' populations in northern Europe during the Middle Pleistocene, a point also made by White (2015) for Britain. Seasonal challenges may have meant that for hominins warm stage Europe was, at best, a zone of disjunct distribution (ZDD), if not a zone of periodic extinction (ZPE; after Roebroeks 2006, fig. 2). The distributions of those zones changed over time, with southern Europe forming a ZDD/ZPE during the Early Pleistocene, while northern Europe, and perhaps sometimes all of Europe, formed a ZDD/ZPE during the Middle Pleistocene (Fig. 7.4). Such patterns also fit with the highly varied morphology of *H. heidelbergensis sensu lato* (Chap. 2) and the mosaic/

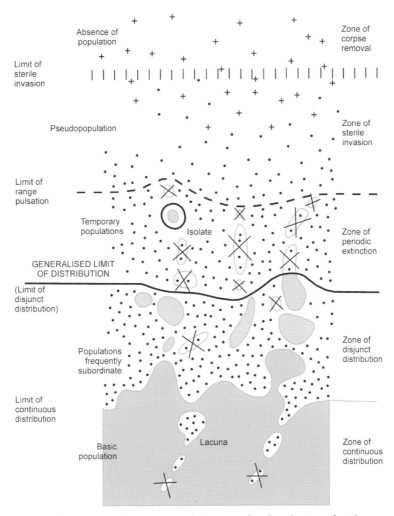

Figure 7.4: Limits to geographic ranges (Roebroeks 2006, fig. 2).

accretion model proposed for the development of *H. neanderthalensis* in the later Middle Pleistocene (Hublin 2009; Martinón-Torres *et al.* 2012).

A consequence of fragmented populations and small, isolated groups might be significant levels of long-term in-breeding. This has been recently suggested for Neanderthals (Sánchez-Quinto and Lalueza-Fox 2015), and high-levels of endogamy (reproduction within a specific social group) and consanguinity (being descended from the same ancestor) have also been suggested as possible explanations for the numbers of congenital abnormalities in the El Sidrón Neanderthals (Ríos *et al.* 2019), and more broadly for the levels of developmental abnormalities in Pleistocene populations (Trinkaus 2018).[2]

Thus, both seasonal and MIS-scale approaches make a case for persistently small-scale regional Lower Palaeolithic hominin populations in Europe throughout the Middle and Late Pleistocene, a pattern underpinned by seasonal and annual fluctuations in resources, weather conditions, climatic regimes, and hominin survival. This demographic pattern may also be linked to the ultimate 'extinction fate'[3] of European hominins, in the form of Neanderthals, which has recently been linked to population size and structure (*e.g.* Vaesen *et al.* 2019). The strategies of European Lower Palaeolithic hominins are therefore perhaps early indicators of adaptations which were frequently effective *enough*, but ultimately not sustainable.

Benefits of a seasonal perspective

This book's seasonal perspective has highlighted the fluctuating challenges presented by Early and Middle Pleistocene Europe, but also allowed possible solutions to be explored. Winter survival (Chap. 3) probably relied upon reduced mobility (especially in the coldest landscapes), the sustained re-use of residential sites and the stockpiling (internally and/or externally) of food resources during autumn through cooperative foraging (Chap. 6). At other times of the year, group sub-divisions and differently-focused foraging activities may have been key to survival, collectively involving both children and adults in the critical tasks of provisioning and childcare (Chaps 4 & 5). Limitations in some of these behaviours (*e.g.* a reliance on scavenging or opportunistic hunting; a partial ability to anticipate future opportunities and threats) may explain the sporadic and geographically limited nature of Early Pleistocene occupations.

The archaeological record, in places, offers key insights that support aspects of these scenarios. Some of this evidence is scarce indeed, such as the isotopic evidence for relatively local Lower Palaeolithic lives or the direct indicators of fire, shelters and clothing, but other categories, such as the evidence for animal exploitation, are increasingly rich.

This is not to suggest however that European survival was at all straightforward in the Lower Palaeolithic. Hominins would have had to cope with challenging conditions, from mid-winter cold and short days to summer aridity, and the vagaries of resources. Threats from predators and tensions during encounters with other hominin groups would be part of the dramatic back-drop of life. Survival was a challenge, with small groups particularly vulnerable to the deaths of individuals.

Yet the European archaeological record as it stands is suggestive that after *c.* 600–500 kya hominins were able to buffer those adverse effects somewhat more successfully. In part this probably reflects major environmental changes, both climatic and faunal. But it was also based on emerging behavioural 'innovations' such as fire and shelter (Mania and Mania 2005; Preece *et al.* 2006; Roebroeks and Villa 2011), specialised foraging, landscape management and social flexibility which, together, helped hominins to survive the challenges of a seasonal Europe: in short, an emerging 'human' culture. It is perhaps not only in anatomical terms that Palaeolithic Europe after *c.* 450 kya began to look like an increasingly Neanderthal world.

This book has argued that the seasons of Early and Middle Pleistocene Europe are knowable and that, in combination with the archaeological and palaeoenvironmental evidence, a seasonal framework helps to explain Europe's earliest hominin occupation evidence. This is due to the relationships between seasonal challenges, hominin behaviour, and the long-term patterns in the archaeological record. I therefore finish by offering six predictions for the European Lower Palaeolithic that can be tested against ongoing and future research:

1. Dental wear studies and isotopic analysis will demonstrate that at least some of the key ungulate species were essentially all year-round local residents, with low annual mobility. This would reduce demands on hominin mobility with regards to hunting and/or scavenging, as prey species would be locally available across all four seasons;
2. Mono-specific predation events will remain limited to the later Middle Pleistocene. This is because such strategies, especially in autumn and spring, were critical to reliable, demographic stability, in light of winter food shortages and weather conditions, and widespread and sustained hominin presence across Europe is clearly bias towards the later Middle Pleistocene;
3. Dental calculus analysis and other methods and data (*e.g.* pounding tool residues) will expand the range of plant foods for which there is direct evidence for consumption, with examples from across the entire Lower Palaeolithic. This is due to the hominin need for a balanced diet, the diversity of available plant food resources in Pleistocene Europe and the potential for the involvement of children in plant foraging and provisioning, especially during the longer days of late spring–summer;
4. Further life history studies will confirm a longer, rather than shorter, middle childhood and adolescence, although these stages will not necessarily be equivalent to those of modern humans. This reflects the key roles of foraging and alloparenting, and the associated learning, required by seasonal Europe and concentrated childbirth. I also predict a possible increase in the length of the sub-adult phase from *H. antecessor* (Early Pleistocene) to *H. heidelbergensis sensu lato* (Middle Pleistocene), given the expanded learning of knowledge, skills and experiences required by pan-European, later Middle Pleistocene strategies (see also prediction 3);
5. Further examples of planning and anticipation, *e.g.* raw material provisioning from task-specific 'quarry' sources and/or other seasonal opportunities, will be identified, with a possible bias towards the later Middle Pleistocene. This reflects the importance of an increasingly 'task-structured' year, *e.g.* targeting and storing key food resources during periods of peak availability, in widespread and sustained European occupations;
6. Evidence for fire use will remain limited to the Middle Pleistocene, and probably to the north of the Alps. This is based on the costs of the technology and the existence of other ways of providing cultural insulation and the 'pre-digested' benefits of cooked foods.

As introduced in Chapter 1, this book has seen the earliest Europeans as 'dextrous and skilled' rather than 'clumsy hybrids'. Most importantly, the particular seasonal challenges of Pleistocene Europe highlight the need for those skills and dexterity to be seen holistically: both manual and cognitive, allied together to successfully exploit the changing seasons, as humans started to truly live beyond the here and now.

Notes

1. The highly variable annual production of the fruits of forest trees and shrubs (*e.g.* acorns and other nuts).
2. Although the evidence for developmental abnormalities in Lower Palaeolithic hominins is actually limited in comparison to later Pleistocene species, this is likely to be a product of palaeontological preservation and research focus, rather than a significant difference.
3. The use of the 'extinction fate' phrase acknowledges the presence of a small component of the Neanderthal genome in all non-African populations of modern humans, as a result of inter-breeding (*e.g.* Sankararaman *et al.* 2014).

Appendix A: Key European Lower Palaeolithic sites

Site	Country	Chronology (MIS)[1]	Selected references[2]
Abbeville (Carrière Carpentier; Somme Formation VII)	France	15/13	Voinchet et al. 2015
Ambrona	Spain	11–9?	Falguères et al. 2006
Arago (levels P–F)	France	14–11	Falguères et al. 2015
Aridos (1 & 2)	Spain	11–9	Villa 1990
Aroeira	Portugal	11	Daura et al. 2017
Barnham	UK	11	Ashton et al. 1998
Barranc de la Boella	Spain	25–20	Mosquera et al. 2015
Barranco León	Spain	49–43	Gibert et al. 1998
Beeches Pit	UK	11	Preece et al. 2006
Bilzingsleben	Germany	11	Mania and Mania 2005
Boxgrove (Slindon Formation)	UK	13	Roberts and Parfitt 1999
Brooksby	UK	13	Coope 2006
Cagny la Garenne (I & II; Somme Formation V)	France	12	Moncel et al. 2015
Castel di Guido	Italy	9	Saccà 2012
Ceprano (calvarium)	Italy	11	Manzi et al. 2010
Charco Hondo	Spain	c. 9–6?	del Cueto et al. 2016
Clacton	UK	11	Singer et al. 1973
Dealul Guran	Romania	11	Iovita et al. 2012
Elveden	UK	11	Ashton et al. 2005
Fuente Nueva	Spain	49–43	Gibert et al. 1998
Gran Dolina (TD-6)	Spain	21	Falguères et al. 1999
Gran Dolina (TD-10)	Spain	11–9	Rodríguez-Hidalgo et al. 2015
Happisburgh I	UK	13	Lewis et al. 2019
Happisburgh III	UK	25 or 21	Parfitt et al. 2010
High Lodge	UK	13	Ashton et al. 1992
Hoxne	UK	11	Singer et al. 1993 ; Ashton et al. 2008a
Isernia la Pineta	Italy	15	Peretto et al. 2015
Kärlich A	Germany	23	Bosinski 1995
Kärlich H	Germany	12	Bosinski 1995
Kärlich "Seeufer"	Germany	11	Bosinski 1995
Kent's Cavern (Breccia)	UK	13?	Cook and Jacobi 1998
Korolevo VI	Bulgaria	14	Koulakovska et al. 2010
Kozarnika (layers 13–11a)	Bulgaria	c. 55–45	Sirakov et al. 2010
La Celle	France	11	Limondin-Lozouet et al. 2010
Le Vallonnet	France	36	Michel et al. 2017
Lézignan-la-Cèbe (Bois-de-Riquet)	France	c. 40–33	Bourguignon et al. 2016

(Continued)

Key European Lower Palaeolithic sites (Continued)

Site	Country	Chronology (MIS)[1]	Selected references[2]
Lunery-Rosières (Units 3 & 2)	France	32 & 24	Despriée *et al.* 2011
Marathousa 1	Greece	13–11	Konidaris *et al.* 2018
Mauer	Germany	15	Wagner *et al.* 2011
Medzhibozh (locality 1)	Ukraine	15/13–11	Stepanchuk and Moigne 2016
Miesenheim I	Germany	13	Turner 1999
Menez-Dregan	France	13–11 (9?)	Monnier *et al.* 2016
Monte Poggiolo	Italy	*c.* 31/30 or 21	Falguères 2003; Muttoni *et al.* 2011
Notarchirico	Italy	16	Pereira *et al.* 2015
Pakefield	UK	19 or 17	Parfitt *et al.* 2005
Pirro Nord	Italy	*c.* 55–40	Arzarello *et al.* 2015
Polledrara	Italy	9	Santucci *et al.* 2016
Pont-de-Lavaud	France	*c.* 33–32	Despriée *et al.* 2011
Račiněves	Czech Republic	15/13 or 11?	Fridrich and Sýkorová 2003; Bridgland *et al.* 2006
Rodafnidia	Greece (Lesvos)	19–6	Galanidou *et al.* 2016
Rusko	Poland	11	Burdukiewicz 2003
Saint-Acheul (Somme Formations V & VI)	France	12–11	Bahain *et al.* 2007
Saint-Colomban	France	14–10 (13–11?)	Ravon 2017
Schöningen	Germany	9	Conard *et al.* 2015
Sima del Elefante	Spain	37	Carbonell *et al.* 2008
Sima de los Huesos	Spain	12	Arnold *et al.* 2014
Soucy	France	9	Lhomme 2007
Southfleet Road	UK	11	Wenban-Smith 2013
Steinheim	Germany	11	van Asperen 2013
Stránská skála I	Czech Republic	17–16	Valoch 2013
Swanscombe	UK	11	Conway *et al.* 1996
Terra Amata	France	11?	de Lumley 1969; Villa 1982; Yokoyama *et al.* 1985
Torralba	Spain	11–8?	Villa *et al.* 2005
Trzebnica	Poland	15 or 13?	Burdukiewicz 2003
Vallparadís	Spain	*c.* 27–15	Martínez *et al.* 2013b
Vértesszőlős	Hungary	13	Bridgland *et al.* 2006
Waverley Wood	UK	13	Shotton *et al.* 1993

[1]Chronology refers to key archaeological deposits, not necessarily the full range of Pleistocene deposits represented at the site. MIS numbers used for consistency, but the source publications use a mixture of MIS and absolute ages: see the listed references for the details of individual site chronologies, including error margins on absolute ages. [2]References refer to chronology and other information (*e.g.* hominin behaviour and/or palaeoenvironmental data) where possible.

Appendix B: Common names for key plant genus

Genus/family (Latin)	Common name
Abies	Fir
Alnus	Alder
Artemisia	Genus within *Asteraceae* (species include mugwort and wormwood)
Asteraceae	Daisy or Aster
Betula	Birch
Carpinus	Hornbeam
Carya	Hickory
Chenopodiaceae	Amaranth family (annual or perennial herbs, and shrubs)
Corylus	Hazel
Cyperaceae	Sedges
Fagus	Beech
Fraxinus	Ash
Larix	Larch
Ostrya	Hop-hornbeam
Picea	Spruce
Pinus	Pine
Poaceae	Grasses
Pterocarya	Wingnut
Quercus	Oak
Salix	Willow
Tilia	Lime
Tsuga	Hemlock
Ulmus	Elm

—

Appendix C: Common names for key mammal species

Genus/family (Latin)	Common name
Acinonyx pardinensis	Giant cheetah
Bison priscus	Steppe bison
Bison schoetensacki	Woodland bison
Bison sp.	Bison
Bos primigenius	Aurochs
Canis lupus	Wolf
Capreolus capreolus	Roe deer
Castor fiber	Beaver
Cervus elaphus	Red deer
Crocuta crocuta	Spotted hyena
Cuon alpinus	Dhole
Dama dama	Fallow deer
Dama dama clactoniana	Clacton fallow deer
Elephantidae	Elephant
Equus ferus	Horse
Equus hydruntinus	Ass
Felix sylvestris	Wild cat
Hemitragus bonali	Bonal tahr
Hippopotamus antiquus	Hippopotamus
Homotherium latidens	Scimitar-toothed cat
Lynx pardinus	Iberian lynx
Lynx pardinus spelaeus	Mediterranean cave lynx
Mammuthus trogontherii	Steppe mammoth
Megaloceros sp.	Giant deer
Megantereon cultridens	Sabre-toothed cat
Meles meles	Badger
Mustela erminea	Stoat
Mustela lutreola	Mink
Mustela nivalis	Weasel
Mustela putorius	European polecat
Pachycrocuta brevirostris	Giant, short-faced hyena
Palaeoloxodon antiquus	Straight-tusked elephant
Panthera gombaszoegensis	European jaguar
Panthera leo	Lion
Panthera leo spelaea	Cave lion

(Continued)

Common names for key mammal species (Continued)

Genus/family (Latin)	Common name
Panthera pardus	Leopard
Stephanorhinus etruscus	Etruscan rhinoceros
Stephanorhinus hemitoechus	Narrow-nosed rhinoceros
Stephanorhinus hundsheimensis	Rhinoceros
Stephanorhinus kirchbergensis	Forest or Merck's rhinoceros
Sus scrofa	Wild boar
Ursus arctos	Brown bear
Ursus deningeri	Deninger's bear
Vulpes vulpes	Red fox

References

Ackroyd, P., 2008. *Thames: sacred river*. Vintage, London.

Adolphs, R., 2013. The Biology of Fear. *Current Biology* 23, R79–R93.

Adovasio, J.M., Soffer, O. and Klima, B., 1996. Upper Palaeolithic fibre technology: interlaced woven finds from Pavlov I, Czech Republic, *c*. 26,000 years ago. *Antiquity* 70, 526–534.

Agam, A. and Barkai, R., 2016. Not the brain alone: the nutritional potential of elephant heads in Paleolithic sites. *Quaternary International* 406, 218–226.

Agustí, J., Blain, H.-A., Cuenca-Bescós, G. and Bailon, S., 2009. Climate forcing of first hominid dispersal in western Europe. *Journal of Human Evolution* 57, 815–821.

Agustí, J., Blain, H.-A., Lozano-Fernández, I., Piñero, P., Oms, O., Furió, M., Blanco, A., López-García, J.M. and Sala, R., 2015. Chronological and environmental context of the first hominin dispersal into western Europe: the case of Barranco León (Guadix-Baza Basin, SE Spain). *Journal of Human Evolution* 87, 87–94.

Aiello, L.C., 2007. Notes on the implications of the Expensive Tissue Hypothesis for human biological and social evolution, in: Roebroeks, W. (ed.), *Guts and Brains: an integrative approach to the hominin record*. Leiden University Press, Leiden, 17–28.

Aiello, L.C. and Key, C., 2002. Energetic consequences of being a *Homo erectus* female. *American Journal of Human Biology* 14, 551–565.

Aiello, L.C. and Wheeler, P., 1995. The expensive-tissue hypothesis: the brain and the digestive system in human and primate evolution. *Current Anthropology* 36, 199–221.

Aiello, L.C. and Wheeler, P., 2003. Neanderthal Thermoregulation and the Glacial Climate, in: Van Andel, T.H., Davies, W. (eds), *Neanderthals and Modern Humans in the European Landscape During the Last Glaciation*. McDonald Institute for Archaeological Research, Cambridge, 147–166.

Aleksiuk, M., 1970. The function of the tail as a fat storage depot in the beaver (*Castor canadensis*). *Journal of Mammalogy* 51, 145–148.

Allington-Jones, L., 2015. The Clacton spear: The last one hundred years. *Archaeological Journal* 172, 273–296.

Allué, E., Cáceres, I., Expósito, I., Canals, A., Rodríguez, A., Rosell, J., Bermúdez de Castro, J.M. and Carbonell, E., 2015. *Celtis* remains from the Lower Pleistocene of Gran Dolina, Atapuerca (Burgos, Spain). *Journal of Archaeological Science* 53, 570–577.

Alperson-Afil, N., 2008. Continual fire-making by hominins at Gesher Benot Ya'aqov, Israel. *Quaternary Science Reviews* 27, 1733–1739.

Andrews, K., 2017. Chimpanzee mind reading: don't stop believing. *Philosophy Compass* 12(1), p.e12394.

Andrews, P. and Ghaleb, B., 1999. Taphonomy of the Westbury Cave bone assemblages, in: Andrews, P., Cook, J., Currant, A. and Stringer, C.B. (eds), *Westbury Cave: The Natural History Museum Excavations, 1976-1984*. Western Academic & Specialist Press, Bristol, 87–126.

Antoine, P., Auguste, P., Bahain, J.-J., Chaussé, C., Falguères, C., Ghaleb, B., Limondin-Lozouet, N., Locht, J.-L. and Voinchet, P., 2010. Chronostratigraphy and palaeoenvironments of Acheulean occupations in northern France (Somme, Seine and Yonne valleys). *Quaternary International* 223–224, 456–461.

Antoine, P., Moncel, M.-H., Locht, J.-L., Limondin-Lozouet, N., Auguste, P., Stoetzel, E., Dabkowski, J., Voinchet, P., Bahain, J.-J. and Falgueres, C., 2015. Dating the earliest human occupation of western Europe: new evidence from the fluvial terrace system of the Somme basin (northern France). *Quaternary International* 370, 77–99.

Aranguren, B., Revedin, A., Amico, N., Cavulli, F., Giachi, G., Grimaldi, S., Macchioni, N. and Santaniello, F., 2018. Wooden tools and fire technology in the early Neanderthal site of Poggetti Vecchi (Italy). *Proceedings of the National Academy of Sciences* 115, 2054–2059.

Arbib, M.A., 2012. Evolutionary parallels between language and tool use, in: Scott-Phillips, T.C., Tamariz, M., Cartmill, E.A. and Hurford, J.R. (eds), *The Evolution Of Language*. World Scientific, Singapore, 3–10.

Arnold, L.J., Demuro, M., Parés, J.M., Arsuaga, J.L., Aranburu, A., Bermúdez de Castro, J.M. and Carbonell, E., 2014. Luminescence dating and palaeomagnetic age constraint on hominins from Sima de los Huesos, Atapuerca, Spain. *Journal of Human Evolution* 67, 85–107.

Arribas, A. and Palmqvist, P., 1999. On the ecological connection between sabre-tooths and hominids: faunal dispersal events in the Lower Pleistocene and a review of the evidence for the first human arrival in Europe. *Journal of Archaeological Science* 26, 571–585.

Arsuaga, J.-L., Martínez, I. and Gracia, A., 2001. Analyse phylogénétique des Hominidés de la Sierra de Atapuerca (Sima de los Huesos et Gran Dolina TD-6): l'évidence crânienne. *L'Anthropologie* 105, 161–178.

Arsuaga, J.-L., Lorenzo, C., Carretero, J.-M., Gracia, A., Martínez, I., García, N., Bermúdez de Castro, J.M. and Carbonell, E., 1999. A complete human pelvis from the Middle Pleistocene of Spain. *Nature* 399, 255–258.

Arsuaga, J.-L., Martínez, I., Gracia, A., Carretero, J.-M. and Carbonell, E., 1993. Three new human skulls from the Sima de los Huesos Middle Pleistocene site in Sierra de Atapuerca, Spain. *Nature* 362, 534–537.

Arsuaga, J.L., Carretero, J.-M., Lorenzo, C., Gómez-Olivencia, A., Pablos, A., Rodríguez, L., García-González, R., Bonmatí, A., Quam, R.M., Pantoja-Pérez, A., Martínez, I., Aranburu, A., Gracia-Téllez, A., Poza-Rey, E., Sala, N., García, N., Alcázar de Velasco, A., Cuenca-Bescós, G., Bermúdez de Castro, J.M. and Carbonell, E., 2015. Postcranial morphology of the Middle Pleistocene humans from Sima de los Huesos, Spain. *Proceedings of the National Academy of Sciences* 112, 11524–11529.

Arsuaga, J.L., Carretero, J.M., Lorenzo, C., Gracia, A., Martínez, I., Bermúdez de Castro, J.M. and Carbonell, E., 1997a. Size variation in Middle Pleistocene humans. *Science* 277, 1086–1088.

Arsuaga, J.L., Martínez, I., Gracia, A. and Lorenzo, C., 1997b. The Sima de los Huesos crania (Sierra de Atapuerca, Spain). A comparative study. *Journal of Human Evolution* 33, 219–281.

Arsuaga, J.L., Martínez, I., Arnold, L., Aranburu, A., Gracia-Téllez, A., Sharp, W., Quam, R., Falguères, C., Pantoja-Pérez, A. and Bischoff, J., 2014. Neandertal roots: cranial and chronological evidence from Sima de los Huesos. *Science* 344, 1358–1363.

Arzarello, M. and Peretto, C., 2010. Out of Africa: the first evidence of Italian peninsula occupation. *Quaternary International* 223, 65–70.

Arzarello, M., Peretto, C. and Moncel, M.-H., 2015. The Pirro Nord site (Apricena, Fg, Southern Italy) in the context of the first European peopling: Convergences and divergences. *Quaternary International* 389, 255–263.

Arzarello, M., Pavia, G., Peretto, C., Petronio, C. and Sardella, R., 2012. Evidence of an Early Pleistocene hominin presence at Pirro Nord (Apricena, Foggia, southern Italy): P13 site. *Quaternary International* 267, 56–61.

Arzarello, M., Marcolini, F., Pavia, G., Pavia, M., Petronio, C., Petrucci, M., Rook, L. and Sardella, R., 2007. Evidence of earliest human occurrence in Europe: the site of Pirro Nord (southern Italy). *Naturwissenschaften* 94, 107–112.

Ashton, N.M., 2015. Ecological niches, technological developments and physical adaptations of early humans in Europe: the handaxe-*heidelbergensis* hypothesis, in: Coward, F., Hosfield, R., Pope, M. and Wenban-Smith, F.F. (eds), *Settlement, Society and Cognition in Human Evolution: Landscapes in Mind*. Cambridge University Press, Cambridge, pp. 138–153.

Ashton, N.M., 2016. The human occupation of Britain during the Hoxnian Interglacial. *Quaternary International* 409, 41–53.

Ashton, N.M., 2017. *Early Humans*. William Collins, London.

Ashton, N.M., 2018. Landscapes of habit and persistent places during MIS 11 in Europe: a return journey from Britain, in: Pope *et al.* (eds) 2018, 164–186.

Ashton, N.M. and Hosfield, R., 2010. Mapping the human record in the British early Palaeolithic: evidence from the Solent River system. *Journal of Quaternary Science* 25, 737–753.

Ashton, N.M. and Lewis, S.G., 2002. Deserted Britain: declining populations in the British late Middle Pleistocene. *Antiquity* 76, 388–396.

Ashton, N.M. and Lewis, S.G., 2012. The environmental contexts of early human occupation of northwest Europe: the British Lower Palaeolithic record. *Quaternary International* 271, 50–64.

Ashton, N.M. and McNabb, J., 1994. Bifaces in perspective, in: Ashton, N.M. and David, N. (eds), *Stories in Stone*. Lithic Studies Society, London, 182–191.

Ashton, N.M. and McNabb, J., 1996. The flint industries from the Waechter excavations, in: Conway *et al.* (eds) 2006, 201–236.

Ashton, N.M., Lewis, S.G. and Hosfield, R., 2011. Mapping the human record: population change in Britain during the Early Palaeolithic, in: Ashton, N.M., Lewis, S.G. and Stringer, C.B. (eds), *The Ancient Human Occupation of Britain*. Elsevier, Amsterdam, 39–51.

Ashton, N.M., Lewis, S.G. and Parfitt, S.A. (eds), 1998. *Excavations at the Lower Palaeolithic site at East Farm, Barnham, Suffolk 1989-94*. British Museum Press Occasional Paper 125, London.

Ashton, N.M., Cook, J., Lewis, S.G. and Rose, J., 1992. *High Lodge: Excavations by G. de G. Sieveking 1962-1968 and J. Cook 1988*. British Museum Press, London.

Ashton, N.M., Lewis, S.G., Parfitt, S.A. and White, M.J., 2006. Riparian landscapes and human habitat preferences during the Hoxnian (MIS 11) Interglacial. *Journal of Quaternary Science* 21, 497–505.

Ashton, N.M., Lewis, S.G., Parfitt, S.A., Davis, R.J. and Stringer, C., 2016. Handaxe and non-handaxe assemblages during Marine Isotope Stage 11 in northern Europe: recent investigations at Barnham, Suffolk, UK. *Journal of Quaternary Science* 31, 837–843.

Ashton, N.M., Lewis, S.G., Parfitt, S.A., Penkman, K.E. and Coope, G.R., 2008a. New evidence for complex climate change in MIS 11 from Hoxne, Suffolk, UK. *Quaternary Science Reviews* 27, 652–668.

Ashton, N.M., Parfitt, S.A., Lewis, S.G., Coope, G.R., Larkin, N., 2008b. Happisburgh Site 1 (TG388307), in: Candy, I., Lee, J.R. and Harrison, A.M. (eds), *The Quaternary of Northern East Anglia Field Guide*. Quaternary Research Association, London, 151–156.

Ashton, N.M., Lewis, S.G., De Groote, I., Duffy, S.M., Bates, M., Bates, R., Hoare, P., Lewis, M., Parfitt, S.A. and Peglar, S., 2014. Hominin footprints from Early Pleistocene deposits at Happisburgh, UK. *PLOS ONE* 9, e88329.

Ashton, N.M., Lewis, S.G., Parfitt, S.A., Candy, I., Keen, D.H., Kemp, R., Penkman, K.E.H., Thomas, G.N., Whittaker, J.E. and White, M.J., 2005. Excavations at the Lower Palaeolithic Site at Elveden, Suffolk, UK. *Proceedings of the Prehistoric Society* 71, 1–61.

Aureli, D., Contardi, A., Giaccio, B., Jicha, B., Lemorini, C., Madonna, S., Magri, D., Marano, F., Milli, S. and Modesti, V., 2015. *Palaeoloxodon* and human interaction: depositional setting, chronology and archaeology at the Middle Pleistocene Ficoncella site (Tarquinia, Italy). *PLOS ONE* 10, e0124498.

Austin, L.A., Bergman, C.A., Roberts, M.B. and Wilhelmsen, K.H., 1999. Archaeology of excavated areas, in: Roberts and Parfitt (eds) 1999, 312–378.

Bahain, J.-J., Falguères, C., Laurent, M., Voinchet, P., Dolo, J.-M., Antoine, P. and Tuffreau, A., 2007. ESR chronology of the Somme river terrace system and first human settlements in northern France. *Quaternary Geochronology* 2, 356–362.

Bailey, G.N., 1981. Concepts of resource exploitation: continuity and discontinuity in palaeoeconomy. *World Archaeology* 13, 1–15.

Bailey, G.N. and Davidson, I., 1983. Site exploitation territories and topography: two case studies from Palaeolithic Spain. *Journal of Archaeological Science* 10, 87–115.

Baldwin, H. and Smithson, P., 1979. Wind chill in upland Britain. *Weather* 34, 294–308.

Ballatore, M. and Breda, M., 2013. *Stephanorhinus hundsheimensis* (Rhinocerontidae, Mammalia) teeth from the early Middle Pleistocene of Isernia La Pineta (Molise, Italy) and comparison with coeval British material. *Quaternary International* 302, 169–183.

Barham, L., 2013. *From Hand to Handle: the first industrial revolution.* Oxford University Press, Oxford.

Barkai, R., Gopher, A. and LaPorta, P.C., 2002. Palaeolithic landscape of extraction: flint surface quarries and workshops at Mt Pua, Israel. *Antiquity* 76, 672–680.

Barkai, R., Gopher, A. and LaPorta, P.C., 2006. Middle Pleistocene landscape of extraction: quarry and workshop complexes in northern Israel, in: Goren-Inbar, N. and Sharon, G. (eds), *Axe Age: Acheulian tool-making from quarry to discard.* Equinox Publishing, London, 7–44.

Barkai, R., Rosell, J., Blasco, R. and Gopher, A., 2017. Fire for a reason: barbecue at Middle Pleistocene Qesem Cave, Israel. *Current Anthropology* 58, S314–S328.

Barker, D.J., 2012. Developmental origins of chronic disease. *Public Health* 126, 185–189.

Barron, E., van Andel, T.H. and Pollard, D., 2003. Glacial environments II: reconstructing the climate of Europe in the last glaciation, in: Van Andel and Davies (eds) 2003, 57–78.

Barsky, D., Vergès, J.-M., Sala, R., Menéndez, L. and Toro-Moyano, I., 2015. Limestone percussion tools from the late Early Pleistocene sites of Barranco León and Fuente Nueva 3 (Orce, Spain). *Philosophical Transactions of the Royal Society B: Biological Sciences* 370(1682), p.20140352.

Bartoš, L., 1985. Social activity and the antler cycle in red deer stags. *Biology of Deer Production* 22, 269–272.

Bartoš, L., Fričová, B., Bartošová-Víchová, J., Panama, J., Šustr, P. and Šmídová, E., 2007. Estimation of the probability of fighting in fallow deer (*Dama dama*) during the rut. *Aggressive Behavior* 33, 7–13.

Bassinot, F.C., Labeyrie, L.D., Vincent, E., Quidelleur, X., Shackleton, N.J. and Lancelot, Y., 1994. The astronomical theory of climate and the age of the Brunhes–Matuyama magnetic reversal. *Earth and Planetary Science Letters* 126, 91–108.

Bekoff, M., 1974. Social play and play-soliciting by infant canids. *American Zoologist* 14, 323–340.

Belinsky, D.L. and Kuhnlein, H.V., 2000. Macronutrient, mineral, and fatty acid composition of Canada Goose (*Branta canadensis*): an important traditional food resource of the Eastern James Bay Cree of Quebec. *Journal of Food Composition and Analysis* 13, 101–115.

Bellai, D., 1995. Techniques d'exploitation du cheval à la Caune de l'Arago (Tautavel, Pyrénées-Orientales). *Paléo* 7, 139–155.

Bello, S.M., Parfitt, S.A. and Stringer, C.B., 2009. Quantitative micromorphological analyses of cut marks produced by ancient and modern handaxes. *Journal of Archaeological Science* 36, 1869–1880.

Beluhan, S. and Ranogajec, A., 2011. Chemical composition and non-volatile components of Croatian wild edible mushrooms. *Food Chemistry* 124, 1076–1082.

Ben-Dor, M., Gopher, A., Hershkovitz, I. and Barkai, R., 2011. Man the fat hunter: the demise of *Homo erectus* and the emergence of a new hominin lineage in the Middle Pleistocene (*ca.* 400 kyr) Levant. *PLOS ONE* 6, e28689.

Bennett, D. and Hoffmann, R.S., 1999. *Equus caballus* Linnaeus, 1758 Horse. *Mammalian Species* 628, 1–14.

Berger, T.D. and Trinkaus, E., 1995. Patterns of trauma among the Neandertals. *Journal of Archaeological Science* 22, 841–852.

Bermúdez de Castro, J.M. and Pérez, P.J., 1995. Enamel hypoplasia in the Middle Pleistocene hominids from Atapuerca (Spain). *American Journal of Physical Anthropology* 96, 301–314.

Bermúdez de Castro, J.M., Martinón-Torres, M., Sarmiento, S., Lozano, M., Arsuaga, J. and Carbonell, E., 2003a. Rates of anterior tooth wear in Middle Pleistocene hominins from Sima de los Huesos (Sierra de Atapuerca, Spain). *Proceedings of the National Academy of Sciences* 100, 11992–11996.

Bermúdez de Castro, J.M., Arsuaga, J.L., Carbonell, E., Rosas, A., Martínez, I. and Mosquera, M., 1997. A hominid from the Lower Pleistocene of Atapuerca: possible ancestor to Neanderthals and Modern Humans. *Science* 276, 1392–1395.

Bermúdez de Castro, J.M., Martinón-Torres, M., Lozano, M., Sarmiento, S. and Muela, A., 2004. Paleodemography of the Atapuerca-Sima de Los Huesos hominin sample: a revision and new approaches to the paleodemography of the European Middle Pleistocene population. *Journal of Anthropological Research* 60, 5–26.

Bermúdez de Castro, J.M., Martinón-Torres, M., Martín-Francés, L., Modesto-Mata, M., Martínez de Pinillos, M., García, C. and Carbonell, E., 2017. *Homo antecessor*: the state of the art eighteen years later. *Quaternary International* 433, 22–31.

Bermúdez de Castro, J.M., Ramirez Rozzi, F.V., Martinón-Torres, M., Sarmiento Perez, S. and Rosas, A., 2003b. Patterns of dental development in Lower and Middle Pleistocene hominins from Atapuerca (Spain), in: Thompson, J.L., Krovitz, G.E. and Nelson, A.J. (eds), *Patterns of Growth and Development in the Genus Homo*. Cambridge University Press, Cambridge, 246–270.

Bermúdez de Castro, J.M., Rosas, A., Carbonell, E., Nicolás, M.E., Rodríguez, J. and Arsuaga, J.L., 1999. A modern human pattern of dental development in Lower Pleistocene hominids from Atapuerca-TD6 (Spain). *Proceedings of the National Academy of Sciences* 96, 4210–4213.

Bigga, G., 2014. *Die Pflanzen von Schöningen: Botanische Makroreste aus den Mittelpleistozänen Ablagerungen und das Nutzungspotential einer Interglazialen Paläoflora*. Unpublished PhD thesis, Eberhard Karls Universität Tübingen.

Bigga, G., Schoch, W.H. and Urban, B., 2015. Paleoenvironment and possibilities of plant exploitation in the Middle Pleistocene of Schöningen (Germany). Insights from botanical macro-remains and pollen. *Journal of Human Evolution* 89, 92–104.

Binford, L.R., 1978. *Nunamiut Ethnoarchaeology*. Academic Press, New York.

Binford, L.R., 1980. Willow smoke and dogs' tails: hunter-gatherer settlement systems and archaeological site formation. *American Antiquity* 45, 1–17.

Binford, L.R., 1981. *Bones: Ancient Men and Modern Myths*. Academic Press, New York.

Binford, L.R., 1982. The archaeology of place. *Journal of Anthropological Archaeology* 1, 5–31.

Binford, L.R., 1991. When the going gets tough, the tough get going: Nunamiut local groups, camping patterns and economic organization, in: Gamble and Boismier (eds) 1991, 25–137.

Binford, L.R., 2001. *Constructing Frames of Reference: An Analytical Method for Archaeological Theory Building Using Ethnographic and Environmental Data Sets*. University of California Press, Berkeley.

Birch, S.E.P., Miracle, P.T., Stevens, R.E. and O'Connell, T.C., 2016. Late Pleistocene/Early Holocene migratory behavior of ungulates using isotopic analysis of tooth enamel and its effects on forager mobility. *PLOS ONE* 11, e0155714.

Bird, D.W. and Bliege Bird, R., 2000. The ethnoarchaeology of juvenile foragers: shellfishing strategies among Meriam children. *Journal of Anthropological Archaeology* 19, 461–476.

Bird, D.W. and Bliege Bird, R., 2005. Martu children's hunting strategies in the Western Desert, Australia, in: Hewlett, B.S. and Lamb, M.E. (eds), *Hunter-gatherer Childhoods: evolutionary, developmental, and cultural perspectives*. Routledge, New York, 129–146.

Bird, R., 1999. Cooperation and conflict: The behavioral ecology of the sexual division of labor. *Evolutionary Anthropology: issues, news, and reviews* 8, 65–75.

Birdsell, J.B., 1953. Some environmental and cultural factors influencing the structuring of Australian Aboriginal populations. *American Naturalist* 87, 171–207.

Blain, H.-A., Bailon, S., Cuenca-Bescos, G., Arsuaga, J.L., Bermúdez de Castro, J.M. and Carbonell, E., 2009. Long-term climate record inferred from Early–Middle Pleistocene amphibian and squamate reptile assemblages at the Gran Dolina Cave, Atapuerca, Spain. *Journal of Human Evolution* 56, 55–65.

Blain, H.-A., Bailon, S., Cuenca-Bescós, G., Bennàsar, M., Rofes, J., López-García, J.M., Huguet, R., Arsuaga, J.L., Bermúdez de Castro, J.M. and Carbonell, E., 2010. Climate and environment of the earliest West European hominins inferred from amphibian and squamate reptile assemblages: Sima del Elefante Lower Red Unit, Atapuerca, Spain. *Quaternary Science Reviews* 29, 3034–3044.

Blain, H.-A., Cuenca-Bescós, G., Burjachs, F., López-García, J.M., Lozano-Fernández, I. and Rosell, J., 2013. Early Pleistocene palaeoenvironments at the time of the *Homo antecessor* settlement in the Gran Dolina cave (Atapuerca, Spain). *Journal of Quaternary Science* 28, 311–319.

Blain, H.-A., Cuenca-Bescós, G., Lozano-Fernández, I., López-García, J.M., Ollé, A., Rosell, J. and Rodríguez, J., 2012. Investigating the Mid-Brunhes Event in the Spanish terrestrial sequence. *Geology* 40, 1051–1054.

Blain, H.-A., Lozano-Fernández, I., Agustí, J., Bailon, S., Menéndez Granda, L., Espígares Ortiz, M.P., Ros-Montoya, S., Jiménez Arenas, J.M., Toro-Moyano, I., Martínez-Navarro, B. and Sala, R., 2016. Refining upon the climatic background of the Early Pleistocene hominid settlement in western Europe: Barranco León and Fuente Nueva-3 (Guadix-Baza Basin, SE Spain). *Quaternary Science Reviews* 144, 132–144.

Blain, H.-A., Santonja, M., Pérez-González, A., Panera, J. and Rubio-Jara, S., 2014. Climate and environments during Marine Isotope Stage 11 in the central Iberian Peninsula: the herpetofaunal assemblage from the Acheulean site of Áridos-1, Madrid. *Quaternary Science Reviews* 94, 7–21.

Blasco, R., Blain, H.-A., Rosell, J., Díez Fernández-Lomana, J.C., Huguet Pàmies, R., Rodríguez, J., Arsuaga, J.L., Bermúdez de Castro, J.M. and Carbonell, E., 2011. Earliest evidence for human consumption of tortoises in the European Early Pleistocene from Sima del Elefante, Sierra de Atapuerca, Spain. *Journal of Human Evolution* 61, 503–509.

Blasco, R., Finlayson, C., Rosell, J., Marco, A.S., Finlayson, S., Finlayson, G., Negro, J.J., Pacheco, F.G. and Vidal, J.R., 2014. The earliest pigeon fanciers. *Scientific Reports* 4, 5971.

Blasco, R., Rosell, J., Arsuaga, J.L., Bermúdez de Castro, J.M. and Carbonell, E., 2010. The hunted hunter: the capture of a lion (*Panthera leo fossilis*) at the Gran Dolina site, Sierra de Atapuerca, Spain. *Journal of Archaeological Science* 37, 2051–2060.

Blasco, R., Rosell, J., Fernández Peris, J., Arsuaga, J.L., Bermúdez de Castro, J.M. and Carbonell, E., 2013. Environmental availability, behavioural diversity and diet: a zooarchaeological approach from the TD10-1 sub-level of Gran Dolina (Sierra de Atapuerca, Burgos, Spain) and Bolomor Cave (Valencia, Spain). *Quaternary Science Reviews* 70, 124–144.

Bliege Bird, R. and Bird, D.W., 2008. Why women hunt: risk and contemporary foraging in a Western Desert aboriginal community. *Current Anthropology* 49, 655–693.

Bliege Bird, R., Codding, B.F. and Bird, D.W., 2009. What explains differences in men's and women's production? *Human Nature* 20, 105–129.

Blumenschine, R.J., 1991. Hominid carnivory and foraging strategies, and the socio-economic function of early archaeological sites. *Philosophical Transactions of the Royal Society B: Biological Sciences* 334, 211–221.

Blumenschine, R.J., 1995. Percussion marks, tooth marks, and experimental determinations of the timing of hominid and carnivore access to long bones at FLK Zinjanthropus, Olduvai Gorge, Tanzania. *Journal of Human Evolution* 29, 21–51.

Blumenschine, R.J. and Cavallo, J.A., 1992. Scavenging and human evolution. *Scientific American* 267, 90–97.

Blundell, L., 2020. *A critical evaluation of the Lower–Middle Palaeolithic archaeological record of the Chalk uplands of Northwest Europe.* Unpublished PhD thesis, University College London (Institute of Archaeology), London.

Boardman, J., Ligneau, L., de Roo, A. and Vandaele, K., 1994. Flooding of property by runoff from agricultural land in northwestern Europe, in: Morisawa, M. (ed.), *Geomorphology and Natural Hazards*. Elsevier, Amsterdam, 183–196.

Bobiec, A., 2002. Białowieża Primeval Forest: The largest area of natural deciduous lowland forest in Europe. *International Journal of Wilderness* 8, 33–37.

Bocherens, H., Drucker, D.G., Billiou, D., Patou-Mathis, M. and Vandermeersch, B., 2005. Isotopic evidence for diet and subsistence pattern of the Saint-Césaire I Neanderthal: review and use of a multi-source mixing model. *Journal of Human Evolution* 49, 71–87.

Bocquet-Appel, J.-P. and Degioanni, A., 2013. Neanderthal demographic estimates. *Current Anthropology* 54, S202–S213.

Bocquet-Appel, J.-P., Demars, P.-Y., Noiret, L. and Dobrowsky, D., 2005. Estimates of Upper Palaeolithic meta-population size in Europe from archaeological data. *Journal of Archaeological Science* 32, 1656–1668.

Boesch, C., 1991. Teaching among wild chimpanzees. *Animal Behaviour* 41, 530–532.

Boesch, H. and Boesch, C., 1984. Possible causes of sex differences in the use of natural hammers by wild chimpanzees. *Journal of Human Evolution* 13, 415–440.

Bogin, B. and Smith, B.H., 1996. Evolution of the human life cycle. *American Journal of Human Biology* 8, 703–716.

Boismier, W.A., 1991. Site formation among subarctic peoples: an Ethnohistorical Approach, in: Gamble and Boismier (eds) 1991, 189–214.

Bonmatí, A., Gómez-Olivencia, A., Arsuaga, J.-L., Carretero, J.M., Gracia, A., Martínez, I., Lorenzo, C., Bérmudez de Castro, J.M. and Carbonell, E., 2010. Middle Pleistocene lower back and pelvis from an aged human individual from the Sima de los Huesos site, Spain. *Proceedings of the National Academy of Sciences* 107, 18386–18391.

Bosinski, G., 1995. The earliest occupation of Europe: western Central Europe, in: Roebroeks and van Kolfschoten (eds) 1995, 103–128.

Bourguignon, L., Crochet, J.-Y., Capdevila, R., Ivorra, J., Antoine, P.-O., Agustí, J., Barsky, D., Blain, H.-A., Boulbes, N., Bruxelles, L., Claude, J., Cochard, D., Filoux, A., Firmat, C., Lozano-Fernández, I., Magniez, P., Pelletier, M., Rios-Garaizar, J., Testu, A., Valensi, P. and De Weyer, L., 2016. Bois-de-Riquet (Lézignan-la-Cèbe, Hérault): a late Early Pleistocene archeological occurrence in southern France. *Quaternary International* 393, 24–40.

Bradshaw, R. and Mitchell, F.J., 1999. The palaeoecological approach to reconstructing former grazing–vegetation interactions. *Forest Ecology and Management* 120, 3–12.

Bradshaw, R.H., Hannon, G.E. and Lister, A.M., 2003. A long-term perspective on ungulate–vegetation interactions. *Forest Ecology and Management* 181, 267–280.

Bramble, D.M. and Lieberman, D.E., 2004. Endurance running and the evolution of *Homo*. *Nature* 432, 345–352.

Bratlund, B., 1996. Hunting strategies in the Late Glacial of Northern Europe: a survey of the faunal evidence. *Journal of World Prehistory* 10, 1–48.

Braun, D.R., Harris, J.W.K., Levin, N.E., McCoy, J.T., Herries, A.I.R., Bamford, M.K., Bishop, L.C., Richmond, B.G. and Kibunjiai, M., 2010. Early hominin diet included diverse terrestrial and aquatic animals 1.95 Ma in East Turkana, Kenya. *Proceedings of the National Academy of Sciences* 107, 10002–10007.

Bridgland, D.R., 2000. River terrace systems in north-west Europe: an archive of environmental change, uplift and early human occupation. *Quaternary Science Reviews* 19(13), 1293–1303.

Bridgland, D.R., Antoine, P., Limondin-Lozouet, N., Santisteban, J.I., Westaway, R. and White, M.J., 2006. The Palaeolithic occupation of Europe as revealed by evidence from the rivers: data from IGCP 449. *Journal of Quaternary Science* 21, 437–455.

Britton, K., Grimes, V., Dau, J. and Richards, M.P., 2009. Reconstructing faunal migrations using intra-tooth sampling and strontium and oxygen isotope analyses: a case study of modern caribou (*Rangifer tarandus granti*). *Journal of Archaeological Science* 36, 1163–1172.

Britton, K., Grimes, V., Niven, L., Steele, T.E., McPherron, S., Soressi, M., Kelly, T.E., Jaubert, J., Hublin, J.-J. and Richards, M.P., 2011. Strontium isotope evidence for migration in late Pleistocene *Rangifer*: implications for Neanderthal hunting strategies at the Middle Palaeolithic site of Jonzac, France. *Journal of Human Evolution* 61, 176–185.

Bro-Jørgensen, J., 2013. Evolution of sprint speed in African savannah herbivores in relation to predation. *Evolution* 67, 3371–3376.

Brugal, J.P., 1999. Middle Palaeolithic subsistence on large bovids: La Borde and Coudoulous (Lot), in: Gaudzinski, S. and Turner, E. (eds), *The Role of Early Humans in the Accumulation of European Lower and Middle Palaeolithic Bone Assemblages*. RGZM, Bonn, 263–266.

Brühl, E., 2003. The small flint tool industry from Bilzingsleben-Steinrinne, in: Burdukiewicz, J.M. and Ronen, A. (eds), *Lower Palaeolithic Small Tools in Europe and the Levant*. British Archaeological Report S1115, Oxford, 49–63.

Brumm, A., 2014. Handaxes and biface technology, in: Smith, C. (ed.), *Encyclopedia of Global Archaeology*. Springer, New York, 3202–3208.

Bruner, E., 2018. Human paleoneurology and the evolution of the parietal cortex. *Brain, Behavior and Evolution* 91, 136–147.

Bruner, E. and Manzi, G., 2005. CT-based description and phyletic evaluation of the archaic human calvarium from Ceprano, Italy. *The Anatomical Record Part A: Discoveries in Molecular, Cellular, and Evolutionary Biology* 285, 643–657.

Buck, L.T. and Stringer, C.B., 2014a. Having the stomach for it: a contribution to Neanderthal diets? *Quaternary Science Reviews* 96, 161–167.

Buck, L.T. and Stringer, C.B., 2014b. *Homo heidelbergensis*. *Current Biology* 24, R214–R215.

Buck, L.T. and Stringer, C.B., 2015. A rich locality in South Kensington: the fossil hominin collection of the Natural History Museum, London. *Geological Journal* 50, 321–337.

Buck, L.T., Berbesque, J.C., Wood, B.M. and Stringer, C.B., 2016. Tropical forager gastrophagy and its implications for extinct hominin diets. *Journal of Archaeological Science: Reports* 5, 672–679.

Buck, L.T., De Groote, I., Hamada, Y. and Stock, J.T., 2018. Humans preserve non-human primate pattern of climatic adaptation. *Quaternary Science Reviews* 192, 149–166.

Bunn, H.T. and Ezzo, J.A., 1993. Hunting and scavenging by Plio-Pleistocene hominids: nutritional constraints, archaeological patterns, and behavioural implications. *Journal of Archaeological Science* 20, 365–398.

Burch, E.S., 1972. The caribou/wild reindeer as a human resource. *American Antiquity* 37, 339–368.

Burdukiewicz, J.M., 2003. Lower Palaeolithic sites with small artefacts in Poland, in: Burdukiewicz, J.M. and Ronen, A. (eds), *Lower Palaeolithic Small Tools in Europe and the Levant*. British Archaeological Report S1115, Oxford, 65–92.

Butterworth, P.J., Ellis, P.R., Wollstonecroft, M., Hardy, K. and Kubiak-Martens, L., 2016. Why protein is not enough: the roles of plants and plant processing in delivering the dietary requirements of modern and early *Homo*, in: Hardy, K. and Kubiak-Martens, L. (eds), *Wild Harvest: plants in the hominin and pre-agrarian human worlds*. Studying Scientific Archaeology 2, Oxford, 31–54.

Butzer, K.W., 1981. Cave sediments, Upper Pleistocene stratigraphy and Mousterian facies in Cantabrian Spain. *Journal of Archaeological Science* 8, 133–183.

Butzer, K.W., 2008. Challenges for a cross-disciplinary geoarchaeology: the intersection between environmental history and geomorphology. *Geomorphology* 101, 402–411.

Bynoe, R., 2018. The submerged archaeology of the North Sea: enhancing the Lower Palaeolithic record of northwest Europe. *Quaternary Science Reviews* 191, 1–14.

Bynoe, R., Dix, J.K. and Sturt, F., 2016. Of mammoths and other monsters: historic approaches to the submerged Palaeolithic. *Antiquity* 90, 857–875.

Candy, I. and Alonso-Garcia, M., 2018. A 1 Ma sea surface temperature record from the North Atlantic and its implications for the early human occupation of Britain. *Quaternary Research* 90, 406–417.

Candy, I. and Mcclymont, E.L., 2013. Interglacial intensity in the North Atlantic over the last 800,000 years: investigating the complexity of the mid-Brunhes Event. *Journal of Quaternary Science* 28, 343–348.

Candy, I., Rose, J. and Lee, J., 2006. A seasonally 'dry' interglacial climate in eastern England during the early Middle Pleistocene: palaeopedological and stable isotopic evidence from Pakefield, UK. *Boreas* 35, 255–265.

Candy, I., Schreve, D. and White, T.S., 2015. MIS 13–12 in Britain and the North Atlantic: understanding the palaeoclimatic context of the earliest Acheulean. *Journal of Quaternary Science* 30, 593–609.

Candy, I., Schreve, D.C., Sherriff, J. and Tye, G.J., 2014. Marine Isotope Stage 11: palaeoclimates, palaeoenvironments and its role as an analogue for the current interglacial. *Earth-Science Reviews* 128, 18–51.

Capasso, L., Michetti, E. and D'Anastasio, R., 2008. A *Homo erectus* hyoid bone: possible implications for the origin of the human capability for speech. *Collegium Antropologicum* 32(4), 1007–1011.

Capasso, L., D'Anastasio, R., Mancini, L., Tuniz, C. and Frayer, D.W., 2016. New evaluation of the Castel di Guido 'hyoid'. *Journal of Anthropological Sciences* 94, 231–235.

Carbone, C., Du Toit, J. and Gordon, I., 1997. Feeding success in African wild dogs: does kleptoparasitism by spotted hyenas influence hunting group size? *Journal of Animal Ecology* 66, 318–326.

Carbonell, E. and Rodríguez, X.P., 2006. The first human settlement of Mediterranean Europe. *Comptes Rendus Palevol* 5, 291–298.

Carbonell, E., Mosquera, M., Rodríguez, X.P. and Sala, R., 1996. The first human settlement of Europe. *Journal of Anthropological Research* 52, 107–114.

Carbonell, E., Bermúdez de Castro, J.M., Arsuaga, J.L., Diez, J.C., Rosas, A., Cuenca-Bescós, G., Sala, R., Mosquera, M. and Rodríguez, X., 1995. Lower Pleistocene hominids and artifacts from Atapuerca-TD6 (Spain). *Science* 269, 826–830.

Carbonell, E., Bermúdez de Castro, J.M., Arsuaga, J.L., Allue, E., Bastir, M., Benito, A., Cáceres, I., Canals, T., Díez, J.C., van der Made, J., Mosquera, M., Ollé, A., Pérez-González, A., Rodríguez, J., Rodríguez, X.P., Rosas, A., Rosell, J., Sala, R., Vallverdú, J. and Vergés, J.M., 2005. An Early Pleistocene hominin mandible from Atapuerca-TD6, Spain. *Proceedings of the National Academy of Sciences* 102, 5674–5678.

Carbonell, E., Bermúdez de Castro, J.M., Pares, J.M., Perez-Gonzalez, A., Cuenca-Bescos, G., Olle, A., Mosquera, M., Huguet, R., van der Made, J., Rosas, A., Sala, R., Vallverdu, J., Garcia, N., Granger, D.E., Martinón-Torres, M., Rodriguez, X.P., Stock, G.M., Vergès, J.M., Allue, E., Burjachs, F., Caceres, I., Canals, A., Benito, A., Diez, C., Lozano, M., Mateos, A., Navazo, M., Rodriguez, J., Rosell, J. and Arsuaga, J.L., 2008. The first hominin of Europe. *Nature* 452, 465–469.

Carbonell, E., Cáceres, I., Lozano, M., Saladié, P., Rosell, J., Lorenzo, C., Vallverdú, J., Huguet, R., Canals, A. and Bermúdez de Castro, J.M., 2010. Cultural cannibalism as a paleoeconomic system in the European Lower Pleistocene: The case of level TD6 of Gran Dolina (Sierra de Atapuerca, Burgos, Spain). *Current Anthropology* 51, 539–549.

Carbonell, E., García-Antón, M., Mallol, C., Mosquera, M., Ollé, A., Rodriguez, X.P., Sahnouni, M., Sala, R. and Vergès, J.M., 1999. The TD6 level lithic industry from Gran Dolina, Atapuerca (Burgos, Spain): production and use. *Journal of Human Evolution* 37, 653–693.

Carranza, J. and de Reyna, L.A., 1987. Spatial organization of female groups in red deer (*Cervus elaphus* L.). *Behavioural Processes* 14, 125–135.

Carranza, J., Alvarez, F. and Redondo, T., 1990. Territoriality as a mating strategy in red deer. *Animal Behaviour* 40, 79–88.

Carretero, J.M., Lorenzo, C. and Arsuaga, J.L., 1999. Axial and appendicular skeleton of *Homo antecessor*. *Journal of Human Evolution* 37, 459–499.

Cassidy, K.A., MacNulty, D.R., Stahler, D.R., Smith, D.W. and Mech, L.D., 2015. Group composition effects on aggressive interpack interactions of gray wolves in Yellowstone National Park. *Behavioral Ecology* 26, 1352–1360.

Chamberlain, A.T., 2006. *Demography in Archaeology*. Cambridge University Press, Cambridge.

Chen, J., Weng, Q., Chao, J., Hu, D. and Taya, K., 2008. Reproduction and development of the released Przewalski's Horses (*Equus przewalskii*) in Xinjiang, China. *Journal of Equine Science* 19, 1–7.

Chen, Z., Auler, A.S., Bakalowicz, M., Drew, D., Griger, F., Hartmann, J., Jiang, G., Moosdorf, N., Richts, A., Stevanovic, Z., Veni, G. and Goldscheider, N., 2017. The World Karst Aquifer Mapping project: concept, mapping procedure and map of Europe. *Hydrogeology Journal* 25, 771–785.

Christian, H.J., Blakeslee, R.J., Boccippio, D.J., Boeck, W.L., Buechler, D.E., Driscoll, K.T., Goodman, S.J., Hall, J.M., Koshak, W.J. and Mach, D.M., 2003. Global frequency and distribution of lightning as observed from space by the Optical Transient Detector. *Journal of Geophysical Research: Atmospheres* 108, 4.1–4.15.

Chu, W., 2009. A functional approach to Paleolithic open-air habitation structures. *World Archaeology* 41, 348–362.

Churchill, S.E., 1993. Weapon technology, prey size selection, and hunting methods in modern hunter-gatherers: implications for hunting in the Palaeolithic and Mesolithic, in: Peterkin, G.L., Bricker, H.M. and Mellars, P. (eds), *Hunting and Animal Exploitation in the Later Palaeolithic and Mesolithic of Eurasia*. American Anthropological Association Archaeological Paper 4, Washington, DC, 11–24.

Churchill, S.E. and Rhodes, J.A., 2009. The evolution of the human capacity for "killing at a distance": the human fossil evidence for the evolution of projectile weaponry, in: Hublin, J.-J. and Richards, M. (eds), *The Evolution of Hominin Diets: integrating approaches to the study of Palaeolithic subsistence*. Springer, New York, 201–210.

Churchill, S.E., Walker, C.S. and Schwartz, A.M., 2016. Home-range size in large-bodied carnivores as a model for predicting Neandertal territory size. *Evolutionary Anthropology: Issues, News, and Reviews* 25, 117–123.

Clark, J.D. and Harris, J.W.K., 1985. Fire and its roles in early hominid lifeways. *African Archaeological Review* 3, 3–27.

Clement, A.F., Hillson, S.W. and Aiello, L.C., 2012. Tooth wear, Neanderthal facial morphology and the anterior dental loading hypothesis. *Journal of Human Evolution* 62, 367–376.

Clutton-Brock, T.H., Guinness, F.E. and Albon, S.D., 1982. *Red Deer: behaviour and ecology of two sexes*. Edinburgh University Press, Edinburgh.

Cochard, D., Brugal, J.-P., Morin, E. and Meignen, L., 2012. Evidence of small fast game exploitation in the Middle Paleolithic of Les Canalettes Aveyron, France. *Quaternary International* 264, 32–51.

Cohen, K.M., MacDonald, K., Joordens, J.C., Roebroeks, W. and Gibbard, P.L., 2012. The earliest occupation of north-west Europe: a coastal perspective. *Quaternary International* 271, 70–83.

Cole, J., 2015. Handaxe symmetry in the Lower and Middle Palaeolithic: implications for the Acheulean Gaze in: Coward, F., Hosfield, R., Pope, M. and Wenban-Smith, F.F. (eds), *Settlement, Society and Cognition in Human Evolution: landscapes in mind*. Cambridge University Press, Cambridge, 234–257.

Cole, J., 2017. Assessing the calorific significance of episodes of human cannibalism in the Palaeolithic. *Scientific Reports* 7, 44707.

Coles, B., 2006. *Beavers in Britain's Past*. Oxbow Books, Oxford.

Collard, M., Tarle, L., Sandgathe, D. and Allan, A., 2016. Faunal evidence for a difference in clothing use between Neanderthals and early modern humans in Europe. *Journal of Anthropological Archaeology* 44, 235–246.

Collins, W.A., Madsen, S.D. and Susman-Stillman, A., 2002. Parenting during middle childhood, in: Bornstein, M.H. (ed.), *Handbook of Parenting Volume I: children & parenting*. Lawrence Erlbaum Associates, New Jersey, 73–101.

Combourieu-Nebout, N., Bertini, A., Russo-Ermolli, E., Peyron, O., Klotz, S., Montade, V., Fauquette, S., Allen, J., Fusco, F. and Goring, S., 2015. Climate changes in the central Mediterranean and Italian vegetation dynamics since the Pliocene. *Review of Palaeobotany and Palynology* 218, 127–147.

Conard, N.J., Serangeli, J., Böhner, U., Starkovich, B.M., Miller, C.E., Urban, B. and van Kolfschoten, T., 2015. Excavations at Schöningen and paradigm shifts in human evolution. *Journal of Human Evolution* 89, 1–17.

Conway, B., McNabb, J. and Ashton, N.M. (eds), 1996. *Excavations at Barnfield Pit, Swanscombe 1968-72*. British Museum Press Occasional Paper 94, London.

Cook, J. and Jacobi, R., 1998. Observations on the artefacts from the breccia at Kent's Cavern, in: Ashton, N.M., Healy, F. and Pettitt, P. (eds), *Stone Age Archaeology: essays in honour of John Wymer*. Oxbow Books, Oxford, 77–89.

Coolidge, F.L. and Wynn, T., 2005. Working memory, its executive functions, and the emergence of modern thinking. *Cambridge Archaeological Journal* 15, 5–26.

Coolidge, F.L. and Wynn, T., 2016. An introduction to cognitive archaeology. *Current Directions in Psychological Science* 25, 386–392.

Coope, G.R., 1993. Late-Glacial (Anglian) and Late-Temperate (Hoxnian) Coleoptera, in: Singer *et al.* (eds) 1993, 156–162.

Coope, G.R., 2006. Insect faunas associated with Palaeolithic industries from five sites of pre-Anglian age in central England. *Quaternary Science Reviews* 25, 1738–1754.

Cooper, T. and Symonds, J., 2014. Gravel extraction: history of the aggregates extraction in the Trent valley, in: Bridgland, D.R., Howard, A.J., White, M.J. and White, T.S. (eds), *Quaternary of the Trent*. Oxbow Books, Oxford, 236–242.

Copeland, L., 2016. Food carbohydrates from plants, in: Hardy, K. and Kubiak Martens, L. (eds), *Wild Harvest: plants in the hominin and pre-agrarian human worlds*. Studying Scientific Archaeology 2, Oxford, 19–30.

Corbet, G.B. and Harris, S., 1991. *The Handbook of British Mammals* (3rd edn). Blackwell Scientific Publications, Oxford.

Corbey, R., Jagich, A., Vaesen, K. and Collard, M., 2016. The Acheulean handaxe: more like a bird's song than a Beatles' tune. *Evolutionary Anthropology: issues, news, and reviews* 25, 6–19.

Cordain, L., Miller, J.B., Eaton, S.B., Mann, N., Holt, S.H. and Speth, J.D., 2000. Plant-animal subsistence ratios and macronutrient energy estimations in worldwide hunter-gatherer diets. *The American Journal of Clinical Nutrition* 71, 682–692.

Coxworth, J.E., Kim, P.S., McQueen, J.S. and Hawkes, K., 2015. Grandmothering life histories and human pair bonding. *Proceedings of the National Academy of Sciences* 112, 11806–11811.

Creel, D., 1991. Bison hides in late prehistoric exchange in the southern plains. *American Antiquity* 56, 40–49.

Crittenden, A.N., 2011. The importance of honey consumption in human evolution. *Food and Foodways* 19, 257–273.

Crittenden, A.N., Conklin-Brittain, N.L., Zes, D.A., Schoeninger, M.J. and Marlowe, F.W., 2013. Juvenile foraging among the Hadza: implications for human life history. *Evolution and Human Behavior* 34, 299–304.

Cunha, E., Rozzi, F.R., Bermúdez de Castro, J.M., Martinón-Torres, M., Wasterlain, S.N. and Sarmiento, S., 2004. Enamel hypoplasias and physiological stress in the Sima de los Huesos Middle Pleistocene hominins. *American Journal of Physical Anthropology* 125, 220–231.

Dabkowski, J., Limondin-Lozouet, N., Antoine, P., Andrews, J., Marca-Bell, A. and Robert, V., 2012. Climatic variations in MIS 11 recorded by stable isotopes and trace elements in a French tufa (La Celle, Seine Valley). *Journal of Quaternary Science* 27, 790–799.

Daleszczyk, K., 2004. Mother–calf relationships and maternal investment in European bison *Bison bonasus*. *Acta Theriologica* 49, 555–566.

Daleszczyk, K., Krasińska, M., Krasiński, Z.A. and Bunevich, A.N., 2007. Habitat structure, climatic factors, and habitat use by European bison (*Bison bonasus*) in Polish and Belarusian parts of the Białowieża Forest, Poland. *Canadian Journal of Zoology* 85, 261–272.

Daura, J., Sanz, M., Arsuaga, J.L., Hoffmann, D.L., Quam, R.M., Ortega, M.C., Santos, E., Gómez, S., Rubio, A., Villaescusa, L., Souto, P., Mauricio, J., Rodrigues, F., Ferreira, A., Godinho, P., Trinkaus, E. and Zilhão, J., 2017. New Middle Pleistocene hominin cranium from Gruta da Aroeira (Portugal). *Proceedings of the National Academy of Sciences* 114, 3397–3402.

Dávid-Barrett, T. and Dunbar, R.I.M., 2016. Bipedality and hair loss in human evolution revisited: the impact of altitude and activity scheduling. *Journal of Human Evolution* 94, 72–82.

Davies, W., Valdes, P., Ross, C. and Van Andel, T.H., 2003. The human presence in Europe during the Last Glacial period III: site clusters, regional climates and resource attractions, in: van Andel and Davies (eds) 2003, 191–220.

Davis, R. and Ashton, N., 2019. Landscapes, environments and societies: The development of culture in Lower Palaeolithic Europe. *Journal of Anthropological Archaeology* 56, 101107.

Davis, R.J., Lewis, S.G., Ashton, N.M., Parfitt, S.A., Hatch, M.T. and Hoare, P.G., 2017. The early Palaeolithic archaeology of the Breckland: current understanding and directions for future research. *Journal of Breckland Studies* 1, 28–44.

Dean, C., Leakey, M.G., Reid, D., Schrenk, F., Schwartz, G.T., Stringer, C. and Walker, A., 2001. Growth processes in teeth distinguish modern humans from *Homo erectus* and earlier hominins. *Nature* 414, 628–631.

Demuro, M., Arnold, L.J., Aranburu, A., Sala, N. and Arsuaga, J.-L., 2019. New bracketing luminescence ages constrain the Sima de los Huesos hominin fossils (Atapuerca, Spain) to MIS 12. *Journal of Human Evolution* 131, 76–95.

Dennell, R.W., 2003. Dispersal and colonisation, long and short chronologies: how continuous is the Early Pleistocene record for hominids outside East Africa? *Journal of Human Evolution* 45, 421–440.

Dennell, R.W., 1979. Prehistoric diet and nutrition: some food for thought. *World Archaeology* 11, 121–135.

Dennell, R.W. and Roebroeks, W., 1996. The earliest colonization of Europe: the short chronology revisited. *Antiquity* 70, 535–542.

Dennell, R.W., Martinón-Torres, M. and Bermúdez de Castro, J.M., 2010. Out of Asia: the initial colonisation of Europe in the Early and Middle Pleistocene. *Quaternary International* 223–224, 439.

Dennell, R.W., Martinón-Torres, M. and Bermúdez de Castro, J.M., 2011. Hominin variability, climatic instability and population demography in Middle Pleistocene Europe. *Quaternary Science Reviews* 30, 1511–1524.

Despriée, J., Voinchet, P., Tissoux, H., Bahain, J.-J., Falguères, C., Courcimault, G., Dépont, J., Moncel, M.-H., Robin, S., Arzarello, M., Sala, R., Marquer, L., Messager, E., Puaud, S. and Abdessadok, S., 2011. Lower and Middle Pleistocene human settlements recorded in fluvial deposits of the middle Loire River Basin, Centre Region, France. *Quaternary Science Reviews* 30, 1474–1485.

Despriée, J., Voinchet, P., Tissoux, H., Moncel, M.-H., Arzarello, M., Robin, S., Bahain, J.-J., Falgueres, C., Courcimault, G. and Depont, J., 2010. Lower and Middle Pleistocene human settlements in the middle Loire River basin, Centre region, France. *Quaternary International* 223, 345–359.

de Lumley, H., 1969. A Paleolithic camp at Nice. *Scientific American* 220, 42–51.

de Lumley, H., 2006. Il y a 400,000 ans: la domestication du feu, un formidable moteur d'hominisation. *Comptes Rendus Palevol* 5, 149–154.

de Lumley, H., Grégoire, S., Barsky, D., Batalla, G., Bailon, S., Belda, V., Briki, D., Byrne, L., Desclaux, E., El Guenouni, K., Fournier, A., Kacimi, S., Lacombat, F., de Lumley, M.-A., Moigne, A.-M., Moutoussamy, J., Paunescu, C., Perrenoud, C., Pois, V., Quiles, J., Rivals, F., Roger, T. and Testu, A., 2004. Habitat et mode de vie des chasseurs paléolithiques de la Caune de l'Arago (600,000–400,000 ans). *L'Anthropologie* 108, 159–184.

de Lumley, M.-A., 2015. L'homme de Tautavel. Un *Homo erectus* européen évolué. *Homo erectus tautavelensis. L'Anthropologie* 119, 303–348.

de Waal, F., 1997. Are we in anthropodenial? *Discover* 18, 50–53.

del Cueto, S.B., Preysler, J.B., Pérez-González, A., Torres, C., Pérez, I.R., de Miguel, J.V., 2016. Acheulian flint quarries in the Madrid Tertiary basin, central Iberian Peninsula: First data obtained from geoarchaeological studies. *Quaternary International* 411, 329–348.

Dibble, H.L., Abodolahzadeh, A., Aldeias, V., Goldberg, P., McPherron, S.P. and Sandgathe, D.M., 2017. How did hominins adapt to ice age Europe without fire? *Current Anthropology* 58, S278–S287.

Divišová, M. and Šída, P., 2015. Plant use in the Mesolithic period: archaeobotanical data from the Czech Republic in a European context – a review. *Interdisciplinaria Archaeologica* VI, 95–106.

Dobosi, V.T., 2003. Changing environment – unchanged culture at Vértesszőlős, Hungary, in: Burdukiewicz, J.M. and Ronen, A. (eds), *Lower Palaeolithic Small Tools in Europe and the Levant*. British Archaeological Report S1115, Oxford, 101–111.

Domínguez-Rodrigo, M., 2002. Hunting and scavenging by early humans: the state of the debate. *Journal of World Prehistory* 16, 1–54.

Doronichev, V.B., 2008. The Lower Paleolithic in Eastern Europe and the Caucasus: a reappraisal of the data and new approaches. *PaleoAnthropology* 2008, 107–157.

do Amaral, L.Q., 1996. Loss of body hair, bipedality and thermoregulation. Comments on recent papers in the *Journal of Human Evolution*. *Journal of Human Evolution* 30, 357–366.

Duke, C. and Steele, J., 2010. Geology and lithic procurement in Upper Palaeolithic Europe: a weights-of-evidence based GIS model of lithic resource potential. *Journal of Archaeological Science* 37, 813–824.

Dunbar, R.I., 1998. The social brain hypothesis. *Evolutionary Anthropology: Issues, news, and reviews* 6, 178–190.

Dunbar, R.I., 2003. The social brain: mind, language, and society in evolutionary perspective. *Annual Review of Anthropology* 32, 163–181.

Dunbar, R.I., 2007. Why hominins had big brains, in: Roebroeks, W. (ed.), *Guts and Brains: an integrative approach to the hominin record*. Leiden University Press, Leiden, 91–105.

Dunbar, R.I., 2009. The social brain hypothesis and its implications for social evolution. *Annals of Human Biology* 36, 562–572.

Dunbar, R.I. and Gowlett, J.A.J., 2014. Fireside chat: the impact of fire on hominin socioecology, in: Dunbar, R.I., Gamble, C.S. and Gowlett, J.A.J. (eds), *Lucy to Language: The Benchmark Papers*. Oxford University Press, Oxford, 277–296.

Dunbar, R.I. and Shultz, S., 2007. Evolution in the social brain. *Science* 317, 1344–1347.

Eaton, S.B. and Konner, M., 1985. Paleolithic nutrition: a consideration of its nature and current implications. *New England Journal of Medicine* 312, 283–289.

Eaton, S.B., Eaton III, S.B. and Konner, M.J., 1997. Paleolithic nutrition revisited: A twelve-year retrospective on its nature and implications. *European Journal Of Clinical Nutrition* 51, 207–216.

Edmonds, M., 2012. *Stone Tools and Society*. Routledge, London.

Espigares, M.P., Martínez-Navarro, B., Palmqvist, P., Ros-Montoya, S., Toro, I., Agustí, J. and Sala, R., 2013. *Homo* vs. *Pachycrocuta*: earliest evidence of competition for an elephant carcass between scavengers at Fuente Nueva-3 (Orce, Spain). *Quaternary International* 295, 113–125.

European Commission, 2004. *Mountain Areas in Europe: analysis of mountain areas in EU member states, acceding and other European countries – final report*. Nordic Centre for Spatial Development (European Commission contract No 2002.CE.16.0.AT.136; https://ec.europa.eu/regional_policy/sources/docgener/studies/pdf/montagne/mount1.pdf).

Fa, J.E., Stewart, J.R., Lloveras, L. and Vargas, J.M., 2013. Rabbits and hominin survival in Iberia. *Journal of Human Evolution* 64, 233–241.

Falguères, C., 2003. ESR dating and the human evolution: contribution to the chronology of the earliest humans in Europe. *Quaternary Science Reviews* 22, 1345–1351.

Falguères, C., Bahain, J.-J., Pérez-González, A., Mercier, N., Santonja, M. and Dolo, J.-M., 2006. The Lower Acheulian site of Ambrona, Soria (Spain): ages derived from a combined ESR/U-series model. *Journal of Archaeological Science* 33, 149–157.

Falguères, C., Bahain, J.-J., Yokoyama, Y., Arsuaga, J.L., Bermúdez de Castro, J.M., Carbonell, E., Bischoff, J.L. and Dolo, J.-M., 1999. Earliest humans in Europe: the age of TD6 Gran Dolina, Atapuerca, Spain. *Journal of Human Evolution* 37, 343–352.

Falguères, C., Shao, Q., Han, F., Bahain, J.J., Richard, M., Perrenoud, C., Moigne, A.M. and de Lumley, H., 2015. New ESR and U-series dating at Caune de l'Arago, France: A key-site for European Middle Pleistocene. *Quaternary Geochronology* 30, 547–553.

Falguères, C., Yokoyama, Y., Shen, G., Bischoff, J.L., Ku, T.-L. and de Lumley, H., 2004. New U-series dates at the Caune de l'Arago, France. *Journal of Archaeological Science* 31, 941–952.

Falk, D., 2012. Hominin paleoneurology: where are we now?, in: Hofman, M.A. and Falk, D. (eds), *Evolution of the Primate Brain*. Elsevier, Amsterdam, 255–272.

Féblot-Augustins, J., 1999. Raw material transport patterns and settlement systems in the European Lower and Middle Palaeolithic: continuity, change and variability, in: Roebroeks, W. and Gamble, C.S. (eds), *The Middle Palaeolithic Occupation of Europe*. University of Leiden, Leiden, 194–214.

Fediuk, K., Hidiroglou, N., Madere, R. and Kuhnlein, H.V., 2002. Vitamin C in Inuit traditional food and women's diets. *Journal of Food Composition and Analysis* 15, 221–235.

Fern, K., 1995–2010. *Plants for a Future: The PfaF Database* (https://pfaf.org/user/Default.aspx)

Fernández Peris, J., González, V.B., Blasco, R., Cuartero, F., Fluck, H., Sañudo, P., and Verdasco, C., 2012. The earliest evidence of hearths in southern Europe: the case of Bolomor Cave (Valencia, Spain). *Quaternary International* 247, 267–277.

Ferreira, I.C., Barros, L. and Abreu, R., 2009. Antioxidants in wild mushrooms. *Current Medicinal Chemistry* 16, 1543–1560.

Finkel, M. and Barkai, R., 2018. The Acheulean handaxe technological persistence: a case of preferred cultural conservatism? *Proceedings of the Prehistoric Society* 84, 1–19.

Finlay, N., 2015. Kid-knapped knowledge: changing perspectives on the child in lithic studies. *Childhood in the Past* 8, 104–112.

Finlayson, C., Brown, K., Blasco, R., Rosell, J., Negro, J.J., Bortolotti, G.R., Finlayson, G., Marco, A.S., Pacheco, F.G. and Vidal, J.R., 2012. Birds of a feather: Neanderthal exploitation of raptors and corvids. *PLOS ONE* 7, e45927.

Fiore, I., Gala, M. and Tagliacozzo, A., 2004. Ecology and subsistence strategies in the eastern Italian Alps during the Middle Palaeolithic. *International Journal of Osteoarchaeology* 14, 273–286.

Fisher, D.C., 1995. Experiments on subaqueous meat-caching. *Current Research in the Pleistocene* 12, 77–80.

Fisher, J.W.J. and Strickland, H.C., 1991. Dwellings and fireplaces: keys to Efe Pygmy campsite structure, in: Gamble and Boismier (eds) 1991, 215–236.

Fluck, H., 2007. Initial observations from experiments into the possible use of fire with stone tools in the manufacture of the Clacton point. *Lithics: The Journal of the Lithic Studies Society* 28, 15–19.

Fluck, H. and McNabb, J., 2007. Raw material exploitation at the Middle Pleistocene site of Vértesszőlős, Hungary. *Lithics: The Journal of the Lithic Studies Society* 28, 50–64.

Franciscus, R.G., 2009. When did the modern human pattern of childbirth arise? New insights from an old Neandertal pelvis. *Proceedings of the National Academy of Sciences* 106, 9125–9126.

Franks, J., 1960. Interglacial deposits at Trafalgar Square, London. *New Phytologist* 59, 145–152.

French, J.C., 2016. Demography and the Palaeolithic archaeological record. *Journal of Archaeological Method and Theory* 23, 150–199.

French, J.C. and Collins, C., 2015. Upper Palaeolithic population histories of Southwestern France: a comparison of the demographic signatures of ^{14}C date distributions and archaeological site counts. *Journal of Archaeological Science* 55, 122–134.

Fridrich, J. and Sýkorová, I., 2003. A new Lower Palaeolithic site with a small toolset at Račiněves (Central Bohemia), in: Burdukiewicz, J.M. and Ronen, A. (eds), *Lower Palaeolithic Small Tools in Europe and the Levant*. British Archaeological Report S1115, Oxford, 93–100.

Frings, R.M., Hillebrand, G., Gehres, N., Banhold, K., Schriever, S. and Hoffmann, T., 2019. From source to mouth: basin-scale morphodynamics of the Rhine River. *Earth-Science Reviews* 196, 102830.

Frison, G., 1989. Experimental use of Clovis weaponry and tools on African Elephants. *American Antiquity* 54(4), 766–784.

Froehle, A.W. and Churchill, S.E., 2009. Energetic competition between Neandertals and anatomically modern humans. *PaleoAnthropology* 2009, 96–116.

Galanidou, N., Cole, J., Iliopoulos, G. and McNabb, J., 2013. East meets west: the Middle Pleistocene site of Rodafnidia on Lesvos, Greece, *Antiquity* 87(336), http://www.antiquity.ac.uk/projgall/galanidou336/

Galanidou, N., Athanassas, C., Cole, J., Iliopoulos, G., Katerinopoulos, A., Magganas, A. and McNabb, J., 2016. The Acheulian Site at Rodafnidia, Lisvori, on Lesbos, Greece: 2010–2012, in: Harvati, K. and Roksandic, M. (eds), *Paleoanthropology of the Balkans and Anatolia: human evolution and its context*. Springer, Dordrecht, 119–138.

Galway-Witham, J. and Stringer, C.B., 2018. How did *Homo sapiens* evolve? *Science* 360, 1296–1298.

Galway-Witham, J., Cole, J. and Stringer, C.B., 2019. Aspects of human physical and behavioural evolution during the last 1 million years. *Journal of Quaternary Science* 34(6), 355–378.

Gamble, C.S., 1982. Interaction and alliance in Palaeolithic society. *Man* 17(1), 92–107.

Gamble, C.S., 1986. *The Palaeolithic Settlement of Europe*. Cambridge University Press, Cambridge.

Gamble, C.S., 1987. Man the shoveller: alternative models for Middle Pleistocene colonisation and occupation in northern latitudes, in: Soffer, O. (ed.), *The Pleistocene Old World: regional perspectives*. Plenum Press, New York, 81–98.

Gamble, C.S., 1991. An introduction to the living spaces of mobile peoples, in: Gamble and Boismier (eds) 1991, 1–23.

Gamble, C.S., 1996a. Hominid behaviour in the Middle Pleistocene: an English perspective, in: Gamble, C.S. and Lawson, A.J. (eds), *The English Palaeolithic Reviewed*. Wessex Archaeology, Salisbury, 61–71.

Gamble, C.S., 1996b. Making tracks: hominid networks and the evolution of the social landscape, in: Steele, J. and Shennan, S. (eds), *The Archaeology of Human Ancestry*. Routledge, London, 253–277.

Gamble, C.S., 1998a. Handaxes and Palaeolithic individuals, in: Ashton, N.M., Healy, F. and Pettitt, P. (eds), *Stone Age Archaeology: essays in honour of John Wymer*. Oxbow Books, Oxford, 105–109.

Gamble, C.S., 1998b. Palaeolithic society and the release from proximity: a network approach to intimate relations. *World Archaeology* 29, 426–449.

Gamble, C.S., 1999. *The Palaeolithic Societies of Europe*. Cambridge University Press, Cambridge.

Gamble, C.S. and Boismier, W.A. (eds), 1991. *Ethnoarchaeological Approaches to Mobile Campsites: hunter gatherer and pastoralist case studies*. Ethnoarchaeological Series 1, Ann Arbor, MI.

Gamble, C.S. and Marshall, G., 2001. The shape of handaxes, the structure of the Acheulian world, in: Milliken, S. and Cook, J. (eds), *A Very Remote Period Indeed: papers on the Palaeolithic presented to Derek Roe*. Oxbow Books, Oxford, 19–27.

Gamble, C.S., Gowlett, J.A.J. and Dunbar, R.I., 2014. *Thinking Big: how the evolution of social life shaped the human mind*. Thames & Hudson, London.

Gamble, C.S., Davies, W., Pettitt, P., Hazelwood, L. and Richards, M., 2005. The archaeological and genetic foundations of the European population during the Late Glacial: implications for 'agricultural thinking'. *Cambridge Archaeological Journal* 15, 193–223.

García-Medrano, P., Ollé, A., Ashton, N.M. and Roberts, M.B., 2019. The mental template in handaxe manufacture: new insights into Acheulean lithic technological behavior at Boxgrove, Sussex, UK. *Journal of Archaeological Method and Theory* 26, 396–422.

García, N. and Arsuaga, J.L., 2011. The Sima de los Huesos (Burgos, northern Spain): palaeoenvironment and habitats of *Homo heidelbergensis* during the Middle Pleistocene. *Quaternary Science Reviews* 30, 1413–1419.

García, N.G., Feranec, R., Arsuaga, J. and Bermúdez de Castro, J.M., Carbonell, E., 2009. Isotopic analysis of the ecology of herbivores and carnivores from the Middle Pleistocene deposits of the Sierra De Atapuerca, northern Spain. *Journal of Archaeological Science* 36, 1142–1151.

Garland, T., 1983. The relation between maximal running speed and body mass in terrestrial mammals. *Journal of Zoology* 199, 157–170.

Garriga, J.G., Martinez, K. and Yravedra, J., 2017. Hominin and carnivore interactions during the Early Pleistocene in western Europe. *L'Anthropologie* 121, 343–366.

Gaudzinski-Windheuser, S., Noack, E.S., Pop, E., Herbst, C., Pfleging, J., Buchli, J., Jacob, A., Enzmann, F., Kindler, L., Iovita, R., Street, M. and Roebroeks, W., 2018. Evidence for close-range hunting by last interglacial Neanderthals. *Nature Ecology & Evolution* 2, 1087–1092.

Gaudzinski, S., 1999. The faunal record of the Lower and Middle Palaeolithic of Europe: remarks on human interference, in: Roebroeks, W. and Gamble, C.S. (eds), *The Middle Palaeolithic Occupation of Europe*. University of Leiden, Leiden, 215–233.

Gaudzinski, S. and Roebroeks, W., 2000. Adults only. Reindeer hunting at the Middle Palaeolithic site Salzgitter Lebenstedt, northern Germany. *Journal of Human Evolution* 38, 497–521.

Geist, V., 1998. *Deer of the World: their evolution, behaviour, and ecology*. Stackpole Books, Mechanicsburg PA.

Geist, V., 2003. Of reindeer and man, modern and Neanderthal: A creation story founded on a historic perspective on how to conserve wildlife, woodland caribou in particular. *Rangifer* 23, 57–63.

Genov, P., 1981. Food composition of wild boar in north-eastern and western Poland. *Acta Theriologica* 26(10), 185–205.

Gibbard, P.L., 1988. The history of the great northwest European rivers during the past three million years. *Philosophical Transactions of the Royal Society B: Biological Sciences* 318, 559–602.

Gibbard, P.L. and Lewin, J., 2002. Climate and related controls on interglacial fluvial sedimentation in lowland Britain. *Sedimentary Geology* 151, 187–210.

Gibert, J., Gibert, L., Iglesias, A. and Maestro, E., 1998. Two 'Oldowan'assemblages in the Plio-Pleistocene deposits of the Orce region, southeast Spain. *Antiquity* 72, 17–25.

Gilby, I.C., Machanda, Z.P., O'Malley, R.C., Murray, C.M., Lonsdorf, E.V., Walker, K., Mjungu, D.C., Otali, E., Muller, M.N., Emery Thompson, M., Pusey, A.E. and Wrangham, R.W., 2017. Predation by female chimpanzees: toward an understanding of sex differences in meat acquisition in the last common ancestor of *Pan* and *Homo*. *Journal of Human Evolution* 110, 82–94.

Gilligan, I., 2010. The prehistoric development of clothing: archaeological implications of a thermal model. *Journal of Archaeological Method and Theory* 17, 15–80.

Gilligan, I., 2017. Clothing and hypothermia as limitations for midlatitude hominin settlement during the Pleistocene: a comment on Hosfield 2016. *Current Anthropology* 58, 534–535.

Godwin, H., 1975. *The History of the British Flora: a factual basis for phytogeography* (2nd edn). Cambridge University Press, Cambridge.

Golant, A., Nord, R.M., Paksima, N. and Posner, M.A., 2008. Cold exposure injuries to the extremities. *Journal of the American Academy of Orthopaedic Surgeons* 16, 704–715.

Gómez-Olivencia, A., Carretero, J.M., Lorenzo, C., Arsuaga, J.L., Bermúdez de Castro, J.M. and Carbonell, E., 2010. The costal skeleton of *Homo antecessor*: preliminary results. *Journal of Human Evolution* 59, 620–640.

Goodwin, D., 2003. Horse behaviour: evolution, domestication and feralisation, in: Waran, N. (ed), *The Welfare of Horses*. Kluwer Academic, New York, 1–18.

Goren-Inbar, N., Sharon, G., Melamed, Y. and Kislev, M., 2002. Nuts, nut cracking, and pitted stones at Gesher Benot Ya'aqov, Israel. *Proceedings of the National Academy of Sciences* 99, 2455–2460.

Goren-Inbar, N., Alperson, N., Kislev, M.E., Simchoni, O., Melamed, Y., Ben-Nun, A. and Werker, E., 2004. Evidence of hominin control of fire at Gesher Benot Ya'aqov, Israel. *Science* 304, 725–727.

Gosden, C., 2005. What do objects want? *Journal of Archaeological Method and Theory* 12, 193–211.

Gowland, R.L., 2015. Entangled lives: Implications of the developmental origins of health and disease hypothesis for bioarchaeology and the life course. *American Journal of Physical Anthropology* 158, 530–540.

Gowlett, J.A.J., 2005. Seeking the Palaeolithic individual in East Africa and Europe during the Lower–Middle Pleistocene, in: Gamble, C.S. and Porr, M. (eds), *The Hominid Individual in Context: archaeological investigations of Lower and Middle Palaeolithic landscapes, locales and artefacts*. Routledge, London, 50–67.

Gowlett, J.A.J., 2006. The early settlement of northern Europe: fire history in the context of climate change and the social brain. *Comptes Rendus Palevol* 5, 299–310.

Gowlett, J.A.J., 2010. Firing up the social brain, in: Dunbar, R.I., Gamble, C.S. and Gowlett, J.A.J. (eds), *Social Brain, Distributed Mind*. Oxford University Press, Oxford, 345–370.

Gowlett, J.A.J., 2016. The discovery of fire by humans: a long and convoluted process. *Philosophical Transactions of the Royal Society B: Biological Sciences* 371, 20150164.

Grabowski, A., Farley, C.T. and Kram, R., 2005. Independent metabolic costs of supporting body weight and accelerating body mass during walking. *Journal of Applied Physiology* 98, 579–583.

Grace, J. and Easterbee, N., 1979. The natural shelter for red deer (*Cervus elaphus*) in a Scottish glen. *Journal of Applied Ecology* 16, 37–48.

Gracia-Téllez, A., Arsuaga, J.-L., Martínez, I., Martín-Francés, L., Martinón-Torres, M., Bermúdez de Castro, J.M., Bonmatí, A. and Lira, J., 2013. Orofacial pathology in *Homo heidelbergensis*: the case of Skull 5 from the Sima de los Huesos site (Atapuerca, Spain). *Quaternary International* 295, 83–93.

Gramly, R.M., 1977. Deerskins and hunting territories: Competition for a scarce resource of the northeastern woodlands. *American Antiquity* 42, 601–605.

Graves-Brown, P., 1996. Their commonwealths are not as we supposed: sex, gender and material culture in human evolution, in: Steele, J. and Shennan, S. (eds), *The Archaeology of Human Ancestry: power, sex and tradition*. Routledge, Abingdon, 347–360.

Gray, P., 2009. Play as a foundation for hunter-gatherer social existence. *American Journal of Play* 1, 476–522.

Gronquist, R.M., 2013. *Furs of Alaska Mammals: a teachers guide*. Alaska Department of Fish & Game (Division of Wildlife Conservation), Juneau.

Grove, M., 2009. Hunter-gatherer movement patterns: causes and constraints. *Journal of Anthropological Archaeology* 28, 222–233.

Grove, M. Dunbar, R.I., 2015. Local objects, distant symbols: fission-fusion social systems and the evolution of human cognition, in: Coward, F., Hosfield, R., Pope, M. and Wenban-Smith, F.F. (eds), *Settlement, Society and Cognition in Human Evolution: Landscapes in Mind*. Cambridge University Press, Cambridge, 15–30.

Grove, M., Pearce, E. and Dunbar, R.I.M., 2012. Fission–fusion and the evolution of hominin social systems. *Journal of Human Evolution* 62, 191–200.

Grün, R., 1996. A re-analysis of electron spin resonance dating results associated with the Petralona hominid. *Journal of Human Evolution* 30, 227–241.

Guil-Guerrero, J.L., 2018. Comment on: 'Plant use in the Lower and Middle Palaeolithic: food, medicine and raw materials', by Karen Hardy. *Quaternary Science Reviews* 200, 406–408.

Güleç, E., White, T., Kuhn, S., Özer, I., Sağır, M., Yılmaz, H. and Howell, F.C., 2009. The Lower Pleistocene lithic assemblage from Dursunlu (Konya), central Anatolia, Turkey. *Antiquity* 83, 11–22.

Gupta, S., Collier, J.S., Garcia-Moreno, D., Oggioni, F., Trentesaux, A., Vanneste, K., De Batist, M., Camelbeeck, T., Potter, G., Van Vliet-Lanoë, B. and Arthur, J.C.R., 2017. Two-stage opening of the Dover Strait and the origin of island Britain. *Nature Communications* 8, 15101.

Gurven, M. and Hill, K., 2009. Why do men hunt? A reevaluation of "Man the Hunter" and the sexual division of labor. *Current Anthropology* 50, 51–74.

Gurven, M. and Hill, K., 2010. Moving beyond stereotypes of men's foraging goals: a reply to Hawkes, O'Connell, and Coxworth. *Current Anthropology* 51, 265–267.

Guthrie, R.D., 1990. *Frozen Fauna of the Mammoth Steppe*. University of Chicago Press, Chicago IL.

Haidle, M.N., 2009. How to think a simple spear, in: de Beaune, S.A., Coolidge, F.L. and Wynn, T. (eds), *Cognitive Archaeology And Human Evolution*. Cambridge University Press, Cambridge, 57–73.

Haidle, M.N. and Pawlik, A.F., 2010. The earliest settlement of Germany: Is there anything out there? *Quaternary International* 223, 143–153.

Haidt, A., Kamiński, T., Borowik, T. and Kowalczyk, R., 2018. Human and the beast – flight and aggressive responses of European bison to human disturbance. *PLOS ONE* 13, e0200635.

Hallos, J., 2005. '15 Minutes of Fame': exploring the temporal dimension of Middle Pleistocene lithic technology. *Journal of Human Evolution* 49, 155–179.

Haltenorth, T. and Diller, H., 1980. *A Field Guide to the Mammals of Africa, including Madagascar*. Collins, London.

Hammel, H., 1955. Thermal properties of fur. *American Journal of Physiology* 182, 369–376.

Hardaker, T. and MacRae, R.J., 2000. A lost river and some Palaeolithic surprises: new quartzite finds from Norfolk and Oxfordshire. *Lithics: The Newsletter of the Lithic Studies Society* 21, 52–59.

Hardy, B.L., 2010. Climatic variability and plant food distribution in Pleistocene Europe: implications for Neanderthal diet and subsistence. *Quaternary Science Reviews* 29, 662–679.

Hardy, B.L. and Moncel, M.-H., 2011. Neanderthal use of fish, mammals, birds, starchy plants and wood 125–250,000 years ago. *PLOS ONE* 6, e23768.

Hardy, B.L., Moncel, M.-H., Despriée, J., Courcimault, G. and Voinchet, P., 2018. Middle Pleistocene hominin behavior at the 700ka Acheulean site of la Noira (France). *Quaternary Science Reviews* 199, 60–82.

Hardy, K., 2018. Plant use in the Lower and Middle Palaeolithic: food, medicine and raw materials. *Quaternary Science Reviews* 191, 393–405.

Hardy, K., Buckley, S. and Copeland, L., 2018. Pleistocene dental calculus: Recovering information on Paleolithic food items, medicines, paleoenvironment and microbes. *Evolutionary Anthropology: Issues, News, and Reviews* 27, 234–246.

Hardy, K., Buckley, S. and Huffman, M., 2013. Neanderthal self-medication in context. *Antiquity* 87, 873–878.

Hardy, K., Brand-Miller, J., Brown, K.D., Thomas, M.G. and Copeland, L., 2015. The importance of dietary carbohydrate in human evolution. *Quarterly Review of Biology* 90, 251–268.

Hardy, K., Buckley, S., Collins, M.J., Estalrrich, A., Brothwell, D., Copeland, L., García-Tabernero, A., García-Vargas, S., de la Rasilla, M. and Lalueza-Fox, C., 2012. Neanderthal medics? Evidence for food, cooking, and medicinal plants entrapped in dental calculus. *Naturwissenschaften* 99, 617–626.

Hardy, K., Radini, A., Buckley, S., Blasco, R., Copeland, L., Burjachs, F., Girbal, J., Yll, R., Carbonell, E. and Bermúdez de Castro, J.M., 2017. Diet and environment 1.2 million years ago revealed through analysis of dental calculus from Europe's oldest hominin at Sima del Elefante, Spain. *Science of Nature* 104, https://doi.org/10.1007/s00114-00016-01420-x.

Hardy, K., Radini, A., Buckley, S., Sarig, R., Copeland, L., Gopher, A. and Barkai, R., 2016. Dental calculus reveals potential respiratory irritants and ingestion of essential plant-based nutrients at Lower Palaeolithic Qesem Cave Israel. *Quaternary International* 398, 129–135.

Harrington, F.H., 1987. Aggressive howling in wolves. *Animal Behaviour* 35, 7–12.

Harrington, F.H. and Mech, L.D., 1979. Wolf Howling and Its Role in Territory Maintenance. *Behaviour* 68, 207–249.

Harrington, F.H., Asa, C.S., Mech, L. and Boitani, L., 2003. Wolf communication, in: Mech, L. and Boitani, L. (eds), *Wolves: behavior, ecology, and conservation*. University of Chicago Press, Chicago IL, 66–103.

Harris, C.R., Ashton, N.M. and Lewis, S.G., 2019. From site to museum: A critical assessment of collection history on the formation and interpretation of the British Early Palaeolithic record. *Journal of Paleolithic Archaeology* 2, 1–25.

Harris, D.R., 2006. The interplay of ethnographic and archaeological knowledge in the study of past human subsistence in the tropics. *Journal of the Royal Anthropological Institute* 12, S63–S78.

Harrison, C.J.O. and Stewart, J.R., 1999. Arvifauna, in: Roberts and Parfitt (eds) 1999, 187–196.

Hart, J., 1956. Seasonal changes in insulation of the fur. *Canadian Journal of Zoology* 34, 53–57.

Harvati, K., Panagopoulou, E. and Runnels, C., 2009. The paleoanthropology of Greece. *Evolutionary Anthropology: issues, news, and reviews* 18, 131–143.

Hassan, F.A., 1982. Demographic archaeology, in: Schiffer, M. (ed.), *Advances in Archaeological Method and Theory*. Academic Press, London, 225–279.

Hatala, K.G., Roach, N.T., Ostrofsky, K.R., Wunderlich, R.E., Dingwall, H.L., Villmoare, B.A., Green, D.J., Harris, J.W.K., Braun, D.R. and Richmond, B.G., 2016. Footprints reveal direct evidence of group behavior and locomotion in *Homo erectus*. *Scientific Reports* 6, 28766.

Hawkes, K., 1996. Foraging differences between men and women: behavioural ecology of the sexual division of labour, in: Steele, J. and Shennan, S. (eds), *The Archaeology of Human Ancestry: power, sex and tradition*. Routledge, Abingdon, 279–305.

Hawkes, K., 2016. Genomic evidence for the evolution of human postmenopausal longevity. *Proceedings of the National Academy of Sciences* 113, 17–18.

Hawkes, K. and O'Connell, J.F., 2005. How old is human longevity? *Journal of Human Evolution* 49, 650–653.

Hawkes, K., O'Connell, J.F. and Coxworth, J.E., 2010. Family provisioning is not the only reason men hunt: a comment on Gurven and Hill. *Current Anthropology* 51, 259–264.

Hawkes, K., O'Connell, F. and Jones, N.B., 1995. Hadza children's foraging: juvenile dependency, social arrangements, and mobility among hunter-gatherers. *Current Anthropology* 36, 688–700.

Hawkes, K., O'Connell, J.F., Jones, N.B., Alvarez, H. and Charnov, E.L., 1998. Grandmothering, menopause, and the evolution of human life histories. *Proceedings of the National Academy of Sciences* 95, 1336–1339.

Haynes, G., 2006. Mammoth landscapes: good country for hunter-gatherers. *Quaternary International* 142, 20–29.

Head, M.J. and Gibbard, P.L., 2005. Early–Middle Pleistocene transitions: an overview and recommendation for the defining boundary, in: Head, M.J. and Gibbard, P.L. (eds), *Early-Middle Pleistocene Transitions: the land-ocean evidence.* Geological Society Special Publication 247, London, 1–18.

Heleno, S.A., Ferreira, R.C., Antonio, A.L., Queiroz, M.-J.R., Barros, L. and Ferreira, I.C., 2015. Nutritional value, bioactive compounds and antioxidant properties of three edible mushrooms from Poland. *Food Bioscience* 11, 48–55.

Helme, B., 2017. *Sticks and Stones: research into the use of Palaeolithic handaxes on wood via experimental methods.* Unpublished undergraduate dissertation, University of Reading

Henry, A.G., 2017. Neanderthal cooking and the costs of fire. *Current Anthropology* 58, S329–S336.

Henry, A.G., Brooks, A.S. and Piperno, D.R., 2011. Microfossils in calculus demonstrate consumption of plants and cooked foods in Neanderthal diets (Shanidar III, Iraq; Spy I and II, Belgium). *Proceedings of the National Academy of Sciences* 108, 486–491.

Henry, A.G., Büdel, T. and Bazin, P.-L., 2018. Towards an understanding of the costs of fire. *Quaternary International* 493, 96–105.

Hérisson, D., Coutard, S., Goval, E., Locht, J.-L., Antoine, P., Chantreau, Y. and Debenham, N., 2016. A new key-site for the end of Lower Palaeolithic and the onset of Middle Palaeolithic at Etricourt-Manancourt (Somme, France). *Quaternary International* 409, 73–91.

Herold, N., Yin, Q., Karami, M. and Berger, A., 2012. Modelling the climatic diversity of the warm interglacials. *Quaternary Science Reviews* 56, 126–141.

Herrero, J., Irizar, I., Laskurain, N.A., García-Serrano, A. and García-González, R., 2005. Fruits and roots: wild boar foods during the cold season in the southwestern Pyrenees. *Italian Journal of Zoology* 72, 49–52.

Hewlett, B.S. and Cavalli-Sforza, L.L., 1986. Cultural transmission among Aka pygmies. *American Anthropologist* 88, 922–934.

Hidiroglou, N., Peace, R.W., Jee, P., Leggee, D. and Kuhnlein, H., 2008. Levels of folate, pyridoxine, niacin and riboflavin in traditional foods of Canadian Arctic indigenous peoples. *Journal of Food Composition and Analysis* 21, 474–480.

Hijma, M.P., Cohen, K.M., Roebroeks, W., Westerhoff, W.E. and Busschers, F.S., 2012. Pleistocene Rhine–Thames landscapes: geological background for hominin occupation of the southern North Sea region. *Journal of Quaternary Science* 27, 17–39.

Hillson, S., Parfitt, S., Bello, S., Roberts, M. and Stringer, C., 2010. Two hominin incisor teeth from the Middle Pleistocene site of Boxgrove, Sussex, England. *Journal of Human Evolution* 59, 493–503.

Hinton, M., Oakley, K., Dines, H., King, W., Kennard, A., Hawkes, C., Warren, S.H., Cotton, M.A., Clark, W.L.G. and Morant, G., 1938. Report on the Swanscombe skull. *Journal of the Anthropological Institute of Great Britain and Ireland* 68, 17–98.

Hockett, B., 2012. The consequences of Middle Paleolithic diets on pregnant Neanderthal women. *Quaternary International* 264, 78–82.

Hockett, B. and Haws, J., 2003. Nutritional ecology and diachronic trends in Paleolithic diet and health. *Evolutionary Anthropology: Issues, News, and Reviews* 12, 211–216.

Hodgson, D., 2009. Evolution of the visual cortex and the emergence of symmetry in the Acheulean techno-complex. *Comptes Rendus Palevol* 8, 93–97.

Hohenstein, U.T., Di Nucci, A. and Moigne, A.-M., 2009. Mode de vie à Isernia La Pineta (Molise, Italie). Stratégie d'exploitation du Bison schoetensacki par les groupes humains au Paléolithique inférieur. *L'Anthropologie* 113, 96–110.

Holekamp, K.E., Smale, L., Berg, R. and Cooper, S.M., 1997. Hunting rates and hunting success in the spotted hyena (*Crocuta crocuta*). *Journal of Zoology* 242, 1–15.

Holliday, T.W., 1997. Postcranial evidence of cold adaptation in European Neandertals. *American Journal of Physical Anthropology* 104, 245–258.

Holliday, T.W., 1998. The ecological context of trapping among recent hunter-gatherers: implications for subsistence in terminal Pleistocene Europe. *Current Anthropology* 39, 711–719.

Holliday, T.W., 2012. Body size, body shape, and the circumscription of the genus *Homo*. *Current Anthropology* 53, S330–S345.

Holloway, P.S. and Alexander, G., 1990. Ethnobotany of the Fort Yukon Region, Alaska. *Economic Botany* 44, 214–225.

Holman, J.A., 1998. The herpetofauna. The interglacial mammalian fauna from Barnham, in: Ashton *et al.* (eds) 1998, 101–106.

Holman, J.A., 1999. Herpetofauna, in: Roberts and Parfitt (eds) 1999, 181–187.

Holmes, J.A., Atkinson, T., Fiona Darbyshire, D.P., Horne, D.J., Joordens, J., Roberts, M.B., Sinka, K.J. and Whittaker, J.E., 2010. Middle Pleistocene climate and hydrological environment at the Boxgrove hominin site (West Sussex, UK) from ostracod records. *Quaternary Science Reviews* 29, 1515–1527.

Hopkinson, T., Nowell, A. and White, M., 2013. Life histories, metapopulation ecology, and innovation in the Acheulian. *PaleoAnthropology* 2013, 61–76.

Hopkinson, T. and White, M.J., 2005. The Acheulean and the handaxe: structure and agency in the Palaeolithic, in: Gamble, C.S. and Porr, M. (eds), *The Hominid Individual in Context: archaeological investigations of Lower & Middle Palaeolithic landscapes, locales & artefacts*. Routledge, London, 13–28.

Hosfield, R., 2009. The unsung heroes. *Lithics: The Journal of the Lithic Studies Society* (Special Issue) 30, 185–200.

Hosfield, R., 2016. Walking in a winter wonderland? Strategies for Early and Middle Pleistocene survival in mid-latitude Europe. *Current Anthropology* 57, 653–683.

Hosfield, R. and Cole, J., 2018. Early hominins in north-west Europe: a punctuated long chronology? *Quaternary Science Reviews* 190, 148–160.

Hosfield, R., Cole, J., McNabb, J., 2018. Less of a bird's song than a hard rock ensemble. *Evolutionary Anthropology: Issues, News, and Reviews* 27, 9–20.

Howard, A.J. and Macklin, M.G., 1999. A generic geomorphological approach to archaeological interpretation and prospection in British river valleys: a guide for archaeologists investigating Holocene landscapes. *Antiquity* 73, 527–541.

Hublin, J.-J., 2009. The origin of Neandertals. *Proceedings of the National Academy of Sciences* 106, 16022–16027.

Hublin, J.-J. and Roebroeks, W., 2009. Ebb and flow or regional extinctions? On the character of Neandertal occupation of northern environments. *Comptes Rendus Palevol* 8, 503–509.

Hublin, J.-J., Neubauer, S. and Gunz, P., 2015. Brain ontogeny and life history in Pleistocene hominins. *Philosophical Transactions of the Royal Society B: Biological Sciences* 370, 20140062.

Huerta-Sánchez, E., Jin, X., Asan, Bianba, Z., Peter, B.M., Vinckenbosch, N., Liang, Y., Yi, X., He, M., Somel, M., Ni, P., Wang, B., Ou, X., Huasang, Luosang, J., Cuo, Z.X.P., Gao, G., Yin, Y., Wang, W., Zhang, X., Xu, X., Yang, H., Li, Y., Wang, J., Wang, J. and Nielsen, R., 2014. Altitude adaptation in Tibetans caused by introgression of Denisovan-like DNA. *Nature* 512(7513), 194–197.

Hughes, G.M., Teeling, E.C. and Higgins, D.G., 2014. Loss of olfactory receptor function in hominin evolution. *PLOS ONE* 9, e84714.

Huguet, R., Saladié, P., Cáceres, I., Díez, C., Rosell, J., Bennàsar, M., Blasco, R., Esteban-Nadal, M., Gabucio, M.J., Rodríguez-Hidalgo, A. and Carbonell, E., 2013. Successful subsistence strategies of the first humans in south-western Europe. *Quaternary International* 295, 168–182.

Huguet, R., Vallverdú, J., Rodríguez-Álvarez, X., Terradillos-Bernal, M., Bargalló, A., Lombera-Hermida, A., Menéndez, L., Modesto-Mata, M., Van der Made, J. and Soto, M., 2017. Level TE9c of Sima del Elefante (Sierra de Atapuerca, Spain): a comprehensive approach. *Quaternary International* 433, 278–295.

Ingold, T., 2013. *Making: anthropology, archaeology, art and architecture*. Routledge, Abingdon.

Iovita, R., Fitzsimmons, K.E., Dobos, A., Hambach, U., Hilgers, A. and Zander, A., 2012. Dealul Guran: evidence for Lower Palaeolithic (MIS 11) occupation of the Lower Danube loess steppe. *Antiquity* 86, 973–989.

Iovita, R., Tuvi-Arad, I., Moncel, M.-H., Despriee, J., Voinchet, P. and Bahain, J.-J., 2017. High hand-axe symmetry at the beginning of the European Acheulian: the data from la Noira (France) in context. *PLOS ONE* 12, e0177063.

Isaac, G.L., 1969. Studies of early culture in East Africa. *World Archaeology* 1, 1–28.

Isaac, G.L., 1981. Stone Age visiting cards: approaches to the study of early land use patterns, in: Hodder, I., Isaac, G.L. and Hammond, N. (eds), *Patterns of the Past: Studies in Honour of David Clarke*. Cambridge University Press, Cambridge, 131–155.

James, S.R., 1989. Hominid use of fire in the Lower and Middle Pleistocene: A review of the evidence. *Current Anthropology* 30, 1–26.

Jędrzejewski, W., Jędrzejewska, B., Okarma, H., Schmidt, K., Zub, K. and Musiani, M., 2000. Prey selection and predation by wolves in Białowieża Primeval Forest, Poland. *Journal of Mammalogy* 81, 197–212.

Jędrzejewski, W., Schmidt, K., Milkowski, L., Jędrzejewska, B. and Okarma, H., 1993. Foraging by lynx and its role in ungulate mortality: the local (Białowieża Forest) and the Palaearctic viewpoints. *Acta Theriologica* 38, 385–403.

Jędrzejewski, W., Schmidt, K., Theuerkauf, J., Jędrzejewska, B. and Okarma, H., 2001. Daily movements and territory use by radiocollared wolves (*Canis lupus*) in Białowieża Primeval Forest in Poland. *Canadian Journal of Zoology* 79, 1993–2004.

Jędrzejewski, W., Schmidt, K., Theuerkauf, J., Jędrzejewska, B., Selva, N., Zub, K. and Szymura, L., 2002. Kill rates and predation by wolves on ungulate populations in Białowieża Primeval Forest (Poland). *Ecology* 83, 1341–1356.

Jiménez-Arenas, J.M., Santonja, M., Botella, M. and Palmqvist, P., 2011. The oldest handaxes in Europe: fact or artefact? *Journal of Archaeological Science* 38, 3340–3349.

Joannes-Boyau, R., Adams, J.W., Austin, C., Arora, M., Moffat, I., Herries, A.I.R., Tonge, M.P., Benazzi, S., Evans, A.R., Kullmer, O., Wroe, S., Dosseto, A. and Fiorenza, L., 2019. Elemental signatures of *Australopithecus africanus* teeth reveal seasonal dietary stress. *Nature* 572, 112–115.

Jochim, M.A., 1981. *Strategies for Survival: cultural behaviour in an ecological context*. Academic Press, London.

Jones, N.B., Hawkes, K. and Draper, P., 1994. Foraging returns of !Kung adults and children: why didn't !Kung children forage? *Journal of Anthropological Research* 50, 217–248.

Jouzel, J., Masson-Delmotte, V., Cattani, O., Dreyfus, G., Falourd, S., Hoffmann, G., Minster, B., Nouet, J., Barnola, J.M., Chappellaz, J., Fischer, H., Gallet, J.C., Johnsen, S., Leuenberger, M., Loulergue, L., Luethi, D., Oerter, H., Parrenin, F., Raisbeck, G., Raynaud, D., Schilt, A., Schwander, J., Selmo, E., Souchez, R., Spahni, R., Stauffer, B., Steffensen, J.P., Stenni, B., Stocker, T.F., Tison, J.L., Werner, M. and Wolff, E.W., 2007. Orbital and millennial Antarctic climate variability over the past 800,000 years. *Science* 317, 793–796.

Julien, M.-A., Rivals, F., Serangeli, J., Bocherens, H. and Conard, N.J., 2015. A new approach for deciphering between single and multiple accumulation events using intra-tooth isotopic variations: application to the Middle Pleistocene bone bed of Schöningen 13 II-4. *Journal of Human Evolution* 89, 114–128.

Julien, M.-A., Bocherens, H., Burke, A., Drucker, D.G., Patou-Mathis, M., Krotova, O., and Péan, S., 2012. Were European steppe bison migratory? ^{18}O, ^{13}C and Sr intra-tooth isotopic variations applied to a palaeoethological reconstruction. *Quaternary International* 271, 106–119.

Kahlke, R.-D., 2014. The origin of Eurasian mammoth faunas (*Mammuthus-Coelodonta* faunal complex). *Quaternary Science Reviews* 96, 32–49.

Kahlke, R.-D. and Lacombat, F., 2008. The earliest immigration of woolly rhinoceros (*Coelodonta tologoijensis*, Rhinocerotidae, Mammalia) into Europe and its adaptive evolution in Palaearctic cold stage mammal faunas. *Quaternary Science Reviews* 27, 1951–1961.

Kahlke, R.-D., García, N., Kostopoulos, D.S., Lacombat, F., Lister, A.M., Mazza, P.P.A., Spassov, N. and Titov, V.V., 2011. Western Palaearctic palaeoenvironmental conditions during the Early and early Middle Pleistocene inferred from large mammal communities, and implications for hominin dispersal in Europe. *Quaternary Science Reviews* 30, 1368–1395.

Kalač, P., 2009. Chemical composition and nutritional value of European species of wild growing mushrooms: a review. *Food Chemistry* 113, 9–16.

Kamler, J.F., Jędrzejewski, W. and Jędrzejewska, B., 2008. Home ranges of red deer in a European old-growth forest. *American Midland Naturalist* 159, 75–82.

Kaplan, H.S., Gangestad, S.W., Gurven, M., Lancaster, J., Mueller, T. and Robson, A., 2007. The evolution of diet, brain and life history among primates and humans, in: Roebroeks, W. (ed), *Guts and Brains: an integrative approach to the hominin record*. Leiden University Press, Leiden, 47–90.

Karkanas, P., Shahack-Gross, R., Ayalon, A., Bar-Matthews, M., Barkai, R., Frumkin, A., Gopher, A. and Stiner, M.C., 2007. Evidence for habitual use of fire at the end of the Lower Paleolithic: site-formation processes at Qesem Cave, Israel. *Journal of Human Evolution* 53, 197–212.

Keeley, L., 1980. *Experimental Determination of Stone Tool Uses: a microwear analysis*. University of Chicago Press, Chicago IL.

Keeley, L.H., 1993. Microwear analysis of lithics, in: Singer *et al.* (eds) 1993, 129–138.

Kelly, R.L., 1983. Hunter-Gatherer mobility strategies. *Journal of Anthropological Research* 39, 277–306.

Kelly, R.L., 1995. *The Foraging Spectrum: diversity in hunter-gatherer lifeways*. Smithsonian Institution Press, Washington DC.

Kennedy, G.E., 2003. Palaeolithic grandmothers? Life history theory and early *Homo*. *Journal of the Royal Anthropological Institute* 9, 549–572.

Kershaw, G.P., Scott, P.A. and Welch, H.E., 1996. The shelter characteristics of traditional-styled Inuit snow houses. *Arctic* 49, 328–338.

King, C.M., 1983. *Mustela erminea*. *Mammalian Species* 195, 1–8.

Kleiber, M., 1961. *The Fire of Life. An Introduction to Animal Energetics*. Wiley, New York.

Klein, R.G., 1976. The mammalian fauna of the Klasies River mouth sites, southern Cape Province, South Africa. *South African Archaeological Bulletin* 31(123), 75–98.

Kleindienst, M.R., 1962. Components of the East African Acheulian assemblage: an analytic approach, in: Mortelmans, G. and Nenquin, J. (eds), *Actes du IVeme Congrès Panafricain de Préhistoire et de L'étude du Quaternaire*. Musée Royal de l'Afrique Centrale, Tervuren, 81–105.

Kleinen, T., Hildebrandt, S., Prange, M., Rachmayani, R., Müller, S., Bezrukova, E., Brovkin, V. and Tarasov, P.E., 2014. The climate and vegetation of Marine Isotope Stage 11 – model results and proxy-based reconstructions at global and regional scale. *Quaternary International* 348, 247–265.

Kohn, M. and Mithen, S.J., 1999. Handaxes: products of sexual selection? *Antiquity* 73, 518–526.

Kolen, J., De Loecker, D., Groenendijk, A.J. and de Warrimont, J.P., 1999. Middle Palaeolithic surface scatters: how informative? A case study from southern Limburg (the Netherlands), in: Roebroeks, W. and Gamble, C.S. (eds), *The Middle Palaeolithic Occupation of Europe*. Leiden University Press, Leiden, 177–191.

Konidaris, G.E., Athanassiou, A., Tourloukis, V., Thompson, N., Giusti, D., Panagopoulou, E. and Harvati, K., 2018. The skeleton of a straight-tusked elephant (*Palaeoloxodon antiquus*) and other large mammals from the Middle Pleistocene butchering locality Marathousa 1 (Megalopolis Basin, Greece): preliminary results. *Quaternary International* 497, 65–84.

Körner, C., 2007. The use of 'altitude' in ecological research. *Trends in Ecology & Evolution* 22, 569–574.

Koulakovska, L., Usik, V. and Haesaerts, P., 2010. Early Paleolithic of Korolevo site (Transcarpathia, Ukraine). *Quaternary International* 223–224, 116–130.

Kousis, I., Koutsodendris, A., Peyron, O., Leicher, N., Francke, A., Wagner, B., Giaccio, B., Knipping, M. and Pross, J., 2018. Centennial-scale vegetation dynamics and climate variability in SE Europe during Marine Isotope Stage 11 based on a pollen record from Lake Ohrid. *Quaternary Science Reviews* 190, 20–38.

Kowalczyk, R., Zalewski, A. and Jędrzejewska, B., 2004. Seasonal and spatial pattern of shelter use by badgers *Meles meles* in Białowieża Primeval Forest (Poland). *Acta Theriologica* 49, 75–92.

Kozowyk, P.R.B., Soressi, M., Pomstra, D. and Langejans, G.H.J., 2017. Experimental methods for the Palaeolithic dry distillation of birch bark: implications for the origin and development of Neandertal adhesive technology. *Scientific Reports* 7, 8033.

Kramer, K.L. and Otárola-Castillo, E., 2015. When mothers need others: The impact of hominin life history evolution on cooperative breeding. *Journal of Human Evolution* 84, 16–24.

Kramer, K.L. and Russell, A.F., 2015. Was monogamy a key step on the hominin road? Reevaluating the monogamy hypothesis in the evolution of cooperative breeding. *Evolutionary Anthropology: issues, news, and reviews* 24, 73–83.

Krech III, S., 2005. Birds and eskimos, in: King, J.C.H., Pauksztat, B. and Storrie, R. (eds), *Arctic Clothing*. British Museum Press, London, 62–68.

Kretzoi, M. and Vertes, L., 1965. Upper Biharian (Intermindel) pebble-industry occupation site in western Hungary. *Current Anthropology* 6, 74–87.

Kuhn, R.A., Ansorge, H., Godynicki, S. and Meyer, W., 2010. Hair density in the Eurasian otter *Lutra lutra* and the Sea otter *Enhydra lutris*. *Acta Theriologica* 55, 211–222.

Kuhn, S.L. and Stiner, M.C., 2006. What's a mother to do? The division of labor among Neandertals and modern humans in Eurasia. *Current Anthropology* 47, 953–981.

Kuhnlein, H.V., Receveur, O., Soueida, R. and Berti, P.R., 2008. Unique patterns of dietary adequacy in three cultures of Canadian Arctic indigenous peoples. *Public Health Nutrition* 11, 349–360.

Kuhnlein, H.V., Barthet, V., Farren, A., Falahi, E., Leggee, D., Receveur, O. and Berti, P., 2006. Vitamins A, D, and E in Canadian Arctic traditional food and adult diets. *Journal of Food Composition and Analysis* 19, 495–506.

Kuitems, M., van der Plicht, J., Drucker, D.G., van Kolfschoten, T., Palstra, S.W.L. and Bocherens, H., 2015. Carbon and nitrogen stable isotopes of well-preserved Middle Pleistocene bone collagen from Schöningen (Germany) and their paleoecological implications. *Journal of Human Evolution* 89, 105–113.

Kurtén, B., 1998. *On Evolution and Fossil Mammals*. Columbia University Press, New York.

Kurtén, B., 2017. *Pleistocene Mammals of Europe*. Routledge, London.

Lalueza-Fox, C., Rosas, A., Estalrrich, A., Gigli, E., Campos, P.F., García-Tabernero, A., García-Vargas, S., Sánchez-Quinto, F., Ramírez, O. and Civit, S., 2011. Genetic evidence for patrilocal mating behavior among Neandertal groups. *Proceedings of the National Academy of Sciences* 108, 250–253.

Lantis, M., 1950. The reindeer industry in Alaska. *Arctic* 3, 27–44.

Larivière, S. and Pasitschniak-Arts, M., 1996. *Vulpes vulpes*. *Mammalian Species* 537, 1–11.

Laughlin, W.S., 1968. Hunting: an integrating biobehavior system and its evolutionary importance, in: Lee and DeVore (eds) 1968, 304–320.

Lebreton, L., Moigne, A.-M., Filoux, A. and Perrenoud, C., 2017. A specific small game exploitation for Lower Paleolithic: The beaver (*Castor fiber*) exploitation at the Caune de l'Arago (Pyrénées-Orientales, France). *Journal of Archaeological Science: Reports* 11, 53–58.

Lebreton, L., Desclaux, E., Hanquet, C., Moigne, A.-M. and Perrenoud, C., 2016. Environmental context of the Caune de l'Arago Acheulean occupations (Tautavel, France), new insights from microvertebrates in Q–R levels. *Quaternary International* 411, 182–192.

Lebreton, V., Bertini, A., Ermolli, E.R., Stirparo, C., Orain, R., Vivarelli, M., Combourieu-Nebout, N., Peretto, C. and Arzarello, M., 2018. Tracing fire in early European prehistory: microcharcoal quantification in geological and archaeological records from Molise (southern Italy). *Journal of Archaeological Method and Theory* 26, 247–275.

Lee, R.B., 1968. What hunters do for a living, or, how to make out on scarce resources, in: Lee and DeVore (eds) 1968, 30–48.

Lee, R.B. and DeVore, I., 1968. *Man the Hunter*. Aldine, Chicago IL.

Lehman, N., Clarkson, P., Mech, L.D., Meier, T.J. and Wayne, R.K., 1992. A study of the genetic relationships within and among wolf packs using DNA fingerprinting and mitochondrial DNA. *Behavioral Ecology and Sociobiology* 30, 83–94.

Lent, P.C., 1988. *Ovibos moschatus. Mammalian Species* 302, 1–9.

Leonard, W.R. and Robertson, M.L., 1997. Comparative primate energetics and hominid evolution. *American Journal of Physical Anthropology* 102, 265–281.

Leonard, W.R., Stock, J.T. and Valeggia, C.R., 2010. Evolutionary perspectives on human diet and nutrition. *Evolutionary Anthropology: Issues, News, and Reviews* 19, 85–86.

Leonti, M., Nebel, S., Rivera, D. and Heinrich, M., 2006. Wild gathered food plants in the European Mediterranean: a comparative analysis. *Economic Botany* 60, 130–142.

Leroy, S.A.G., Arpe, K. and Mikolajewicz, U., 2011. Vegetation context and climatic limits of the Early Pleistocene hominin dispersal in Europe. *Quaternary Science Reviews* 30, 1448–1463.

Lewis, M.E., 2018. *Paleopathology of Children: identification of pathological conditions in the human skeletal remains of non-adults*. Academic Press, London.

Lewis, S.G., Ashton, N.M., Field, M.H., Hoare, P.G., Kamermans, H., Knul, M., Mücher, H.J., Parfitt, S.A., Roebroeks, W. and Sier, M.J., 2019. Human occupation of northern Europe in MIS 13: Happisburgh Site 1 (Norfolk, UK) and its European context. *Quaternary Science Reviews* 211, 34–58.

Lhomme, V., 2007. Tools, space and behaviour in the Lower Palaeolithic: discoveries at Soucy in the Paris basin. *Antiquity* 81, 536–554.

Liarsou, A., 2013. Interactions between the beaver (*Castor fiber L.*) and human societies: a long-term archaeological and historical approach. *Archaeological Review from Cambridge* 28, 171–185.

Liebenberg, L., 2008. The relevance of persistence hunting to human evolution. *Journal of Human Evolution* 55, 1156–1159.

Lieberman, D.E. and Shea, J.J., 1994. Behavioral differences between Archaic and Modern humans in the Levantine Mousterian. *American Anthropologist* 96, 300–332.

Lieberman, D.E., Bramble, D.M., Raichlen, D.A. and Shea, J.J., 2007. The evolution of endurance running and the tyranny of ethnography: a reply to Pickering and Bunn (2007). *Journal of Human Evolution* 53, 439–442.

Limondin-Lozouet, N., Nicoud, E., Antoine, P., Auguste, P., Bahain, J.-J., Dabkowski, J., Dupéron, J., Dupéron, M., Falguères, C., Ghaleb, B., Jolly-Saad, M.-C. and Mercier, N., 2010. Oldest evidence of Acheulean occupation in the Upper Seine valley (France) from an MIS 11 tufa at La Celle. *Quaternary International* 223–224, 299–311.

Limondin-Lozouet, N., Antoine, P., Bahain, J.J., Cliquet, D., Coutard, S., Dabkowski, J., Ghaleb, B., Locht, J.L., Nicoud, E. and Voinchet, P., 2015. North-west European MIS 11 malacological successions: a framework for the timing of Acheulean settlements. *Journal of Quaternary Science* 30, 702–712.

Lippert, W. and Podlech, D., 2001. *Wild Flowers of Britain & Europe*. HarperCollins, London.

Lisiecki, L.E. and Raymo, M.E., 2005. A Pliocene-Pleistocene stack of 57 globally distributed benthic $\delta^{18}O$ records. *Paleoceanography* 20, 10.1029/2004PA001071.

Lisovski, S., Ramenofsky, M. and Wingfield, J.C., 2017. Defining the degree of seasonality and its significance for future research. *Integrative and Comparative Biology* 57, 934–942.

Lister, A.M., 1984. Evolutionary and ecological origins of British deer. *Proceedings of the Royal Society of Edinburgh, Section B: Biological Sciences* 82, 205–229.

Lister, A.M., 2004. The impact of Quaternary ice ages on mammalian evolution. *Philosophical Transactions of the Royal Society B: Biological Sciences* 359, 221–241.

Loring, S., 1997. On the trail to the caribou house: some reflections on Innu caribou hunters in northern Ntessinan (Labrador). *Worldwide Archaeology Series* 6, 185–220.

Lourens, L.J., 2004. Revised tuning of Ocean Drilling Program Site 964 and KC01B (Mediterranean) and implications for the $\delta^{18}O$, tephra, calcareous nannofossil, and geomagnetic reversal chronologies of the past 1.1 Myr. *Paleoceanography* 19, PA3010.

Lowe, J.J. and Walker, M.J.C., 1997. *Reconstructing Quaternary Environments* (2nd edn). Prentice Hall, London.

Lozano, M., Bermúdez de Castro, J.M., Carbonell, E. and Arsuaga, J.L., 2008. Non-masticatory uses of anterior teeth of Sima de los Huesos individuals (Sierra de Atapuerca, Spain). *Journal of Human Evolution* 55, 713–728.

Lozano, M., Mosquera, M., Bermúdez de Castro, J.M., Arsuaga, J.L. and Carbonell, E., 2009. Right handedness of *Homo heidelbergensis* from Sima de los Huesos (Atapuerca, Spain) 500,000 years ago. *Evolution and Human Behavior* 30, 369–376.

Lugli, F., Cipriani, A., Arnaud, J., Arzarello, M., Peretto, C. and Benazzi, S., 2017. Suspected limited mobility of a Middle Pleistocene woman from Southern Italy: strontium isotopes of a human deciduous tooth. *Scientific Reports* 7, 8615.

Lummaa, V., Vuorisalo, T., Barr, R.G. and Lehtonen, L., 1998. Why cry? Adaptive significance of intensive crying in human infants. *Evolution and Human Behavior* 19, 193–202.

Lundberg, J. and McFarlane, D.A., 2007. Pleistocene depositional history in a periglacial terrane: a 500 ky record from Kents Cavern, Devon, United Kingdom. *Geosphere* 3, 199–219.

Mabey, R., 2012. *Food for Free*. Collins, London.

Macdonald, D.W. and Barrett, P., 1993. *Mammals of Britain & Europe*. HarperCollins, Glasgow.

MacDonald, K., 2007. Ecological hypotheses for human brain evolution: evidence for skill and learning processes in the ethnographic literature on hunting, in: Roebroeks, W. (ed.), *Guts and Brains: an integrative approach to the hominin record*. Leiden University Press, Leiden, 107–132.

MacDonald, K., 2018. Fire-free hominin strategies for coping with cool winter temperatures in north-western Europe from before 800,000 to circa 400,000 Years Ago. *PaleoAnthropology* 7, 7–26.

MacDonald, K., Martinón-Torres, M., Dennell, R.W. and Bermúdez de Castro, J.M., 2012. Discontinuity in the record for hominin occupation in south-western Europe: implications for occupation of the middle latitudes of Europe. *Quaternary International* 271, 84–97.

Machin, A., Hosfield, R. and Mithen, S.J., 2007. Why are some handaxes symmetrical? Testing the influence of handaxe morphology on butchery effectiveness. *Journal of Archaeological Science* 34, 883–893.

Madurell-Malapeira, J., Alba, D.M., Espigares, M.-P., Vinuesa, V., Palmqvist, P., Martínez-Navarro, B., Moya-Sola, S. and 2017. Were large carnivorans and great climatic shifts limiting factors for hominin dispersals? Evidence of the activity of *Pachycrocuta brevirostris* during the Mid-Pleistocene revolution in the Vallparadís Section (Vallès-Penedès Basin, Iberian Peninsula). *Quaternary International* 431, 42–52.

Mania, D., 1995. The earliest occupation of Europe: the Elbe-Saale region (Germany), in: Roebroeks and van Kolfschoten (eds) 1995, 85–102.

Mania, D. and Mania, U., 2003. Bilzingsleben – *Homo erectus,* his culture and his environment. The most important results of research, in: Burdukiewicz, J.M. and Ronen, A. (eds), *Lower Palaeolithic Small Tools in Europe and the Levant*. British Archaeological Report S1115, Oxford, 29–48.

Mania, D. and Mania, U., 2005. The natural and socio-cultural environment of *Homo erectus* at Bilzingsleben, Germany, in: Gamble, C.S. and Porr, M. (eds), *The Hominid Individual in Context: archaeological investigations of Lower and Middle Palaeolithic landscapes, locales and artefacts*. Routledge, London, 98–114.

Mann, N.J., 2018. A brief history of meat in the human diet and current health implications. *Meat Science* 144, 169–179.

Manzi, P., Aguzzi, A. and Pizzoferrato, L., 2001. Nutritional value of mushrooms widely consumed in Italy. *Food Chemistry* 73, 321–325.

Manzi, G., Magri, D., Milli, S., Palombo, M.R., Margari, V., Celiberti, V., Barbieri, M., Barbieri, M., Melis, R.T. and Rubini, M., 2010. The new chronology of the Ceprano calvarium (Italy). *Journal of Human Evolution* 59, 580–585.

Margari, V., Roucoux, K., Magri, D., Manzi, G. and Tzedakis, P., 2018. The MIS 13 interglacial at Ceprano, Italy, in the context of Middle Pleistocene vegetation changes in Southern Europe. *Quaternary Science Reviews* 199, 144–158.

Marlowe, F.W., 2005. Hunter-gatherers and human evolution. *Evolutionary Anthropology: issues, news, and reviews* 14(2), 54–67.

Marlowe, F.W., 2007. Hunting and gathering: the human sexual division of foraging labor. *Cross-Cultural Research* 41, 170–195.

Marquer, L., Lebreton, V., Otto, T., Valladas, H., Haesaerts, P., Messager, E., Nuzhnyi, D. and Péan, S., 2012. Charcoal scarcity in Epigravettian settlements with mammoth bone dwellings: the taphonomic evidence from Mezhyrich (Ukraine). *Journal of Archaeological Science* 39, 109–120.

Marshall, G., Dupplaw, D., Roe, D. and Gamble, C., 2002. *Lower Palaeolithic Technology, Raw Material and Population Ecology (Bifaces)* [data set]. Archaeology Data Service https://doi.org/10.5284/1000354.

Martínez, K., Garcia, J. and Carbonell, E., 2013b. Hominin multiple occupations in the Early and Middle Pleistocene sequence of Vallparadís (Barcelona, Spain). *Quaternary International* 316, 115–122.

Martínez, I., Arsuaga, J.L., Quam, R., Carretero, J., Gracia, A. and Rodríguez, L., 2008. Human hyoid bones from the Middle Pleistocene site of the Sima de los Huesos (Sierra de Atapuerca, Spain). *Journal of Human Evolution* 54, 118–124.

Martínez, I., Rosa, M., Arsuaga, J.-L., Jarabo, P., Quam, R., Lorenzo, C., Gracia, A., Carretero, J.-M., Bermúdez de Castro, J.M. and Carbonell, E., 2004. Auditory capacities in Middle Pleistocene humans from the Sierra de Atapuerca in Spain. *Proceedings of the National Academy of Sciences* 101, 9976–9981.

Martínez, I., Rosa, M., Quam, R., Jarabo, P., Lorenzo, C., Bonmatí, A., Gómez-Olivencia, A., Gracia, A. and Arsuaga, J., 2013a. Communicative capacities in Middle Pleistocene humans from the Sierra de Atapuerca in Spain. *Quaternary International* 295, 94–101.

Martínez, K., Garcia, J., Carbonell, E., Agustí, J., Bahain, J.-J., Blain, H.-A., Burjachs, F., Cáceres, I., Duval, M. and Falguères, C., 2010. A new Lower Pleistocene archeological site in Europe (Vallparadís, Barcelona, Spain). *Proceedings of the National Academy of Sciences* 107, 5762–5767.

Martinón-Torres, M., Bermúdez de Castro, J.M., Gómez-Robles, A., Prado-Simón, L. and Arsuaga, J.L., 2012. Morphological description and comparison of the dental remains from Atapuerca - Sima de los Huesos site (Spain). *Journal of Human Evolution* 62, 7–58.

Mateos, A., Goikoetxea, I., Leonard, W.R., Martín-González, J.Á., Rodríguez-Gómez, G. and Rodríguez, J., 2014. Neandertal growth: What are the costs? *Journal of Human Evolution* 77, 167–178.

Mazák, V., 1981. *Panthera tigris. Mammalian Species* 152, 1–8.

McGrew, W.C., 2010. Chimpanzee technology. *Science* 328, 579–580.

McHenry, H.M., 1996. Sexual dimorphism in fossil hominids and its socioecological implications, in: Steele, J. and Shennan, S. (eds), *The Archaeology of Human Ancestry: power, sex and tradition.* Routledge, Abingdon, 91–109.

McNabb, J., 1989. Sticks and stones: A possible experimental solution to the question of how the Clacton spear point was made. *Proceedings of the Prehistoric Society* 55, 251–257.

McNabb, J., 2007. *The British Lower Palaeolithic: Stones in Contention.* Routledge, Abingdon.

Mears, R., 2009. *Northern Wilderness.* Hodder & Stoughton, London.

Mears, R. and Hillman, G., 2007. *Wild food.* Hodder & Stoughton, London.

Mech, L.D., 1974. *Canis lupus. Mammalian Species* 37, 1–6.

Mech, L.D., Smith, D.W. and MacNulty, D.R., 2015. *Wolves on the hunt: the behavior of wolves hunting wild prey.* University of Chicago Press, Chicago IL.

Mellars, P.A., 1996. *The Neanderthal Legacy: an archaeological perspective from Western Europe.* Princeton University Press, Princeton NJ.

Mellars, P.A., 2004. Reindeer specialization in the early Upper Palaeolithic: the evidence from south west France. *Journal of Archaeological Science* 31, 613–617.

Méndez-Quintas, E., Santonja, M., Pérez-González, A., Duval, M., Demuro, M. and Arnold, L.J., 2018. First evidence of an extensive Acheulean large cutting tool accumulation in Europe from Porto Maior (Galicia, Spain). *Scientific Reports* 8, 3082.

Mercier, N., Froget, L., Miallier, D., Pilleyre, T., Sanzelle, S. and Tribolo, C., 2004. Nouvelles données chronologiques pour le site de Menez-Dregan 1 (Bretagne): l'apport de la thermoluminescence. *Quaternaire* 15, 253–261.

Messager, E., Lebreton, V., Marquer, L., Russo-Ermolli, E., Orain, R., Renault-Miskovsky, J., Lordkipanidze, D., Despriée, J., Peretto, C. and Arzarello, M., 2011. Palaeoenvironments of early hominins in temperate and Mediterranean Eurasia: new palaeobotanical data from Palaeolithic key-sites and synchronous natural sequences. *Quaternary Science Reviews* 30, 1439–1447.

Meyer, M., Arsuaga, J.-L., de Filippo, C., Nagel, S., Aximu-Petri, A., Nickel, B., Martínez, I., Gracia, A., Bermúdez de Castro, J.M., Carbonell, E., Viola, B., Kelso, J., Prüfer, K. and Pääbo, S., 2016. Nuclear DNA sequences from the Middle Pleistocene Sima de los Huesos hominins. *Nature* 531, 504–507.

Michel, V., Shen, C.-C., Woodhead, J., Hu, H.-M., Wu, C.-C., Moullé, P.-É., Khatib, S., Cauche, D., Moncel, M.-H. and Valensi, P., 2017. New dating evidence of the early presence of hominins in Southern Europe. *Scientific Reports* 7, 10074.

Milker, Y., Rachmayani, R., Weinkauf, M., Prange, M., Raitzsch, M., Schulz, M. and Kucera, M., 2013. Global and regional sea surface temperature trends during Marine Isotope Stage 11. *Climate of the Past* 9, 2231–2252.

Milks, A., 2018a. Making an impact. *Nature Ecology & Evolution* 2, 1057–1058.

Milks, A., 2018b. *Lethal Threshold: the evolutionary implications of Middle Pleistocene wooden spears.* Unpublished PhD thesis, University College London (Institute of Archaeology).

Milks, A., Parker, D. and Pope, M., 2019. External ballistics of Pleistocene hand-thrown spears: experimental performance data and implications for human evolution. *Scientific Reports* 9, 820.

Milks, A., Champion, S., Cowper, E., Pope, M. and Carr, D., 2016. Early spears as thrusting weapons: isolating force and impact velocities in human performance trials. *Journal of Archaeological Science: Reports* 10, 191–203.

Mitchell, F.J., 2005. How open were European primeval forests? Hypothesis testing using palaeoecological data. *Journal of Ecology* 93, 168–177.

Mitchell, J., 1995. Studying biface utilization at Boxgrove: Roe deer butchery with replica handaxes. *Lithics: Newsletter of the Lithic Studies Society* 16, 64–69.

Mithen, S.J., 1996. *The Prehistory of the Mind: a search for the origins of art, religion and science.* Thames & Hudson, London.

Mithen, S.J., 2005. *The Singing Neanderthals: the origins of music, language, mind and body.* Weidenfeld & Nicolson, London.

Mithen, S.J., Finlay, N., Carruthers, W., Carter, S. and Ashmore, P., 2001. Plant use in the Mesolithic: evidence from Staosnaig, Isle of Colonsay, Scotland. *Journal of Archaeological Science* 28, 223–234.

Mol, D., Post, K., Reumer, J.W., van der Plicht, J., de Vos, J., van Geel, B., van Reenen, G., Pals, J.P. and Glimmerveen, J., 2006. The Eurogeul – first report of the palaeontological, palynological and archaeological investigations of this part of the North Sea. *Quaternary International* 142, 178–185.

Moncel, M.-H., 2010. Oldest human expansions in Eurasia: Favouring and limiting factors. *Quaternary International* 223, 1–9.

Moncel, M.-H., Ashton, N.M., Lamotte, A., Tuffreau, A., Cliquet, D. and Despriée, J., 2015. The early Acheulian of north-western Europe. *Journal of Anthropological Archaeology* 40, 302–331.

Moncel, M.-H., Despriée, J., Voinchet, P., Courcimault, G., Hardy, B., Bahain, J.-J., Puaud, S., Gallet, X. and Falguères, C., 2016. The Acheulean workshop of la Noira (France, 700 ka) in the European technological context. *Quaternary International* 393, 112–136.

Moncel, M.-H., Landais, A., Lebreton, V., Combourieu-Nebout, N., Nomade, S. and Bazin, L., 2018. Linking environmental changes with human occupations between 900 and 400 ka in western Europe. *Quaternary International* 480, 78–94.

Monnier, J.-L., Ravon, A.-L., Hinguant, S., Hallégouët, B., Gaillard, C. and Laforge, M., 2016. Menez-Dregan 1 (Plouhinec, Finistère, France): un site d'habitat du Paléolithique inférieur en grotte marine. Stratigraphie, structures de combustion, industries riches en galets aménagés. *L'Anthropologie* 120, 237–262.

Morgan, T.J., Uomini, N.T., Rendell, L.E., Chouinard-Thuly, L., Street, S.E., Lewis, H.M., Cross, C.P., Evans, C., Kearney, R. and de la Torre, I., 2015. Experimental evidence for the co-evolution of hominin tool-making teaching and language. *Nature Communications* 6, 6029.

Morin, E., Meier, J., El Guennouni, K., Moigne, A.-M., Lebreton, L., Rusch, L., Valensi, P., Conolly, J. and Cochard, D., 2019. New evidence of broader diets for archaic *Homo* populations in the northwestern Mediterranean. *Science Advances* 5, eaav9106.

Moser, S., 1998. *Ancestral Images: the iconography of human origins.* Cornell University Press, New York.

Mosquera, M., Ollé, A. and Rodríguez, X.P., 2013. From Atapuerca to Europe: tracing the earliest peopling of Europe. *Quaternary International* 295, 130–137.

Mosquera, M., Ollé, A., Saladié, P., Cáceres, I., Huguet, R., Rosas, A., Villalaín, J., Carrancho, A., Bourlès, D., Braucher, R., Pineda, A. and Vallverdú, J., 2016. The Early Acheulean technology of Barranc de la Boella (Catalonia, Spain). *Quaternary International* 393, 95–111.

Mosquera, M., Saladié, P., Ollé, A., Cáceres, I., Huguet, R., Villalaín, J., Carrancho, A., Bourlès, D., Braucher, R. and Vallverdú, J., 2015. Barranc de la Boella (Catalonia, Spain): an Acheulean elephant butchering site from the European late Early Pleistocene. *Journal of Quaternary Science* 30, 651–666.

Mounier, A., Marchal, F. and Condemi, S., 2009. Is *Homo heidelbergensis* a distinct species? New insight on the Mauer mandible. *Journal of Human Evolution* 56, 219–246.

Mullenders, W.W., 1993. New palynological studies at Hoxne, in: Singer *et al.* (eds) 1993, 150–155.

Müller, W. and Pasda, C., 2011. Site formation and faunal remains of the Middle Pleistocene site Bilzingsleben. *Quartär* 58, 25–49.

Muri, H., Berger, A., Yin, Q., Karami, M.P. and Barriat, P.-Y., 2013. The climate of the MIS-13 inter-glacial according to HadCM3. *Journal of Climate* 26, 9696–9712.

Murray, C.M., Lonsdorf, E.V., Eberly, L.E. and Pusey, A.E., 2009. Reproductive energetics in free-living female chimpanzees (*Pan troglodytes schweinfurthii*). *Behavioral Ecology* 20, 1211–1216.

Musiani, M., Okarma, H. and Jędrzejewski, W., 1998. Speed and actual distances travelled by radio-collared wolves in Białowieża Primeval Forest (Poland). *Acta Theriologica* 43, 409–416.

Mussi, M., 2007. Women of the middle latitudes. The earliest peopling of Europe from a female perspective, in: Roebroeks, W. (ed), *Guts and Brains: an integrative approach to the hominin record.* Leiden University Press, Leiden, 165–183.

Muttoni, G., Scardia, G. and Kent, D.V., 2018. Early hominins in Europe: the Galerian migration hypothesis. *Quaternary Science Reviews* 180, 1–29.

Muttoni, G., Scardia, G., Kent, D.V., Morsiani, E., Tremolada, F., Cremaschi, M. and Peretto, C., 2011. First dated human occupation of Italy at ~0.85 Ma during the late Early Pleistocene climate transition. *Earth and Planetary Science Letters* 307, 241–252.

Mysterud, A., Bartoń, K., Jędrzejewska, B., Krasiński, Z., Niedziałkowska, M., Kamler, J., Yoccoz, N. and Stenseth, N., 2007. Population ecology and conservation of endangered megafauna: the case of European bison in Białowieża Primeval Forest, Poland. *Animal Conservation* 10, 77–87.

Negro, J.J., Blasco, R., Rosell, J. and Finlayson, C., 2016. Potential exploitation of avian resources by fossil hominins: an overview from ethnographic and historical data. *Quaternary International* 421, 6–11.

Nelson, E., Rolian, C., Cashmore, L. and Shultz, S., 2010. Digit ratios predict polygyny in early apes, *Ardipithecus*, Neanderthals and early modern humans but not in *Australopithecus*. *Proceedings of the Royal Society B: Biological Sciences* 278, 1556–1563.

Nelson, J.L., Zavaleta, E.S. and Chapin, F.S., 2008. Boreal fire effects on subsistence resources in Alaska and adjacent Canada. *Ecosystems* 11, 156–171.

Neubauer, S., 2014. Endocasts: possibilities and limitations for the interpretation of human brain evolution. *Brain, Behavior and Evolution* 84, 117–134.

Nicholas, G.P., 1998. Wetlands and hunter-gatherers: a global perspective. *Current Anthropology* 39, 720–731.

Nicholson, A. and Cane, S., 1991. Desert camps: analysis of Australian Aboriginal proto-historic campsites, in: Gamble and Boismier (eds) 1991, 263–354.

Niklasson, M., Zin, E., Zielonka, T., Feijen, M., Korczyk, A.F., Churski, M., Samojlik, T., Jędrzejewska, B., Gutowski, J.M. and Brzeziecki, B., 2010. A 350-year tree-ring fire record from Białowieża Primeval Forest, Poland: implications for Central European lowland fire history. *Journal of Ecology* 98, 1319–1329.

Nitychoruk, J., Bińka, K., Sienkiewicz, E., Szymanek, M., Chodyka, M., Makos, M., Ruppert, H. and Tudryn, A., 2018. A multiproxy record of the Younger Holsteinian Oscillation (YHO) in the Ossówka profile, eastern Poland. *Boreas* 47, 855–868.

Nowell, A., 2010. Working memory and the speed of life. *Current Anthropology* 51, S121–S133.

Nowell, A. and White, M., 2010. Growing up in the Middle Pleistocene: life history strategies and their relationship to Acheulian industries, in: Nowell, A. and Davidson, I. (eds), *Stone Tools and the Evolution of Human Cognition*. University Press of Colorado, Boulder CO, 67–82.

Núñez-Lahuerta, C., Cuenca-Bescós, G. and Huguet, R., 2016. First report on the birds (*Aves*) from level TE7 of Sima del Elefante (Early Pleistocene) of Atapuerca (Spain). *Quaternary International* 421, 12–22.

O'Connor, A., 2000. Brixham Cave and the antiquity of man: reassessing the archaeological and historical significance of a British cave site. *Lithics: The Newsletter of the Lithic Studies Society* 21, 20–28.

Oakley, K.P., 1952. Swanscombe man. *Proceedings of the Geologists' Association* 63, 271–300.

Oakley, K.P., Andrews, P., Keeley, L.H. and Clark, J.D., 1977. A reapprasial of the Clacton spearpoint. *Proceedings of the Prehistoric Society* 43, 13–30.

Okarma, H., Jędrzejewska, B., Jędrzejewski, W., Krasinski, Z.A. and Milkowski, L., 1995. The roles of predation, snow cover, acorn crop, and man-related factors on ungulate mortality in Białowieża Primeval Forest, Poland. *Acta Theriologica* 40, 197–217.

Okarma, H., Jędrzejewski, W., Schmidt, K., Kowalczyk, R. and Jędrzejewska, B., 1997. Predation of Eurasian lynx on roe deer and red deer in Białowieża Primeval Forest, Poland. *Acta Theriologica* 42, 203–224.

Okarma, H., Jędrzejewski, W., Schmidt, K., Sniezko, S., Bunevich, A.N. and Jędrzejewska, B., 1998. Home ranges of wolves in Białowieża Primeval Forest, Poland, compared with other Eurasian populations. *Journal of Mammalogy* 79, 842–852.

Ollé, A., Mosquera, M., Rodríguez-Álvarez, X.P., García-Medrano, P., Barsky, D., de Lombera-Hermida, A. and Carbonell, E., 2016. The Acheulean from Atapuerca: Three steps forward, one step back. *Quaternary International* 411, 316–328.

Oms, O., Anadón, P., Agustí, J. and Julià, R., 2011. Geology and chronology of the continental Pleistocene archeological and paleontological sites of the Orce area (Baza basin, Spain). *Quaternary International* 243, 33–43.

Orain, R., Lebreton, V., Russo Ermolli, E., Sémah, A.-M., Nomade, S., Shao, Q., Bahain, J.-J., Thun Hohenstein, U. and Peretto, C., 2013. Hominin responses to environmental changes during the Middle Pleistocene in central and southern Italy. *Climate of the Past* 9, 687–697.

Otte, M., 2012. The management of space during the Paleolithic. *Quaternary International* 247, 212–229.

Palombo, M.R., 2010. A scenario of human dispersal in the northwestern Mediterranean throughout the Early to Middle Pleistocene. *Quaternary International* 223–224, 179–194.

Parfitt, S.A., 1998. The interglacial mammalian fauna from Barnham, in: Ashton *et al.* (eds) 1998, 111–147.

Parfitt, S.A., 1999a. Mammalia, in: Roberts and Parfitt (eds) 1999, 197–290.

Parfitt, S.A., 1999b. Palaeontology – introduction, in: Roberts and Parfitt (eds) 1999, 157–163.

Parfitt, S.A., 2008. Pakefield Cliffs: archaeology and palaeoenvironment of the Cromer Forest-bed Formation, in: Candy, I., Lee, J.R. and Harrison, A.M. (eds), *The Quaternary of Northern East Anglia Field Guide*. Quaternary Research Association, London, 130–136.

Parfitt, S.A. and Roberts, M.B., 1999. Human modification of faunal remains, in: Roberts and Parfitt (eds) 1999, 398–419.

Parfitt, S.A., Ashton, N.M., Lewis, S.G., Abel, R.L., Coope, G.R., Field, M.H., Gale, R., Hoare, P.G., Larkin, N.R., Lewis, M.D., Karloukovski, V., Maher, B.A., Peglar, S.M., Preece, R.C., Whittaker, J.E. and Stringer, C.B., 2010. Early Pleistocene human occupation at the edge of the boreal zone in northwest Europe. *Nature* 466, 229–233.

Parfitt, S.A., Barendregt, R.W., Breda, M., Candy, I., Collins, M.J., Coope, G.R., Durbridge, P., Field, M.H., Lee, J.R., Lister, A.M., Mutch, R., Penkman, K.E.H., Preece, R.C., Rose, J., Stringer, C.B., Symmons, R., Whittaker, J.E., Wymer, J.J. and Stuart, A.J., 2005. The earliest record of human activity in northern Europe. *Nature* 438, 1008–1012.

Parsons, K.C., 1993. *Human Thermal Environments: the effects of hot, moderate, and cold environments on human health, comfort, and performance*. Taylor & Francis, London.

Pasitschniak-Arts, M., 1993. *Ursus arctos*. *Mammalian Species* 439, 1–10.

Patou-Mathis, M., 2000. Neanderthal subsistence behaviours in Europe. *International Journal of Osteoarchaeology* 10, 379–395.

Pawłowski, B., 2005. Heat loss from the head during infancy as a cost of encephalization. *Current Anthropology* 46, 136–141.

Peel, M.C., Finlayson, B.L. and McMahon, T.A., 2007. Updated world map of the Köppen-Geiger climate classification. *Hydrology and Earth System Sciences Discussions* 4, 439–473.

Peintner, U., Pöder, R. and Pümpel, T., 1998. The iceman's fungi. *Mycological Research* 102, 1153–1162.

Pellegrini, M., Donahue, R.E., Chenery, C., Evans, J., Lee-Thorp, J., Montgomery, J. and Mussi, M., 2008. Faunal migration in late-glacial central Italy: implications for human resource exploitation. *Rapid Communications in Mass Spectrometry* 22, 1714–1726.

Pereira, A., Nomade, S., Voinchet, P., Bahain, J.J., Falguères, C., Garon, H., Lefèvre, D., Raynal, J.P., Scao, V. and Piperno, M., 2015. The earliest securely dated hominin fossil in Italy and evidence of Acheulian occupation during glacial MIS 16 at Notarchirico (Venosa, Basilicata, Italy). *Journal of Quaternary Science* 30, 639–650.

Pereira, E., Barros, L., Martins, A. and Ferreira, I.C., 2012. Towards chemical and nutritional inventory of Portuguese wild edible mushrooms in different habitats. *Food Chemistry* 130, 394–403.

Peretto, C., 2006. The first peopling of southern Europe: the Italian case. *Comptes Rendus Palevol* 5, 283–290.

Peretto, C., Arnaud, J., Moggi-Cecchi, J., Manzi, G., Nomade, S., Pereira, A., Falguères, C., Bahain, J.-J., Grimaud-Hervé, D. and Berto, C., 2015. A human deciduous tooth and new [40]Ar/[39]Ar dating results from the Middle Pleistocene archaeological site of Isernia La Pineta, southern Italy. *PLOS ONE* 10, e0140091.

Pérez-Pérez, A., Bermúdez de Castro, J.M. and Arsuaga, J.L., 1999. Nonocclusal dental microwear analysis of 300,000-year-old *Homo heidelbergensis* teeth from Sima de los Huesos (Sierra de Atapuerca, Spain). *American Journal of Physical Anthropology* 108, 433–457.

Pérez, P.-J., Gracia, A., Martínez, I. and Arsuaga, J.-L., 1997. Paleopathological evidence of the cranial remains from the Sima de los Huesos Middle Pleistocene site (Sierra de Atapuerca, Spain): description and preliminary inferences. *Journal of Human Evolution* 33, 409–421.

Pérez-Pérez, A., Lozano, M., Romero, A., Martínez, L.M., Galbany, J., Pinilla, B., Estebaranz-Sánchez, F., Bermúdez de Castro, J.M., Carbonell, E. and Arsuaga, J.L., 2017. The diet of the first Europeans from Atapuerca. *Scientific Reports* 7, 43319.

Peters, R., 1978. Communication, cognitive mapping, and strategy in wolves and hominids, in: Hall, R.L. and Sharp, H.S. (eds), *Wolf and Man: evolution in parallel*. Elsevier, Cambridge, 95–107.

Peterson, R.T., Mountfort, G. and Hollom, P.A.D., 1993. *Birds of Britain and Europe* (5th edn). Collins, London.

Petraglia, M.D., 2006. The Indian Acheulian in global perspective, in: Goren-Inbar, N. (ed), *Axe Age: Acheulian tool-making from quarry to discard.* Equinox Publishing, London, 389–414.

Petraglia, M.D., Shipton, C. and Paddaya, K., 2005. Life and mind in the Acheulean, in: Gamble, C.S. and Porr, M. (eds), *The Hominid Individual in Context: archaeological investigations of Lower and Middle Palaeolithic landscapes, locales and artefacts.* Routledge, Abingdon, 197–219.

Pettitt, P.B., 1997. High resolution Neanderthals? Interpreting Middle Palaeolithic intrasite spatial data. *World Archaeology* 29, 208–224.

Pettitt, P.B., 2000. Neanderthal lifecycles: developmental and social phases in the lives of the last archaics. *World Archaeology* 31, 351–366.

Pettitt, P.B. and White, M.J., 2012. *The British Palaeolithic: human societies at the edge of the Pleistocene world.* Routledge, London.

Pettitt, P.B., Rockman, M. and Chenery, S., 2015. Creswell Crags in wider context. LA-ICP-MS trace-element analysis of Final Magdalenian lithics and mobility patterns in the British Late Upper Palaeolithic, in: Ashton, N.M. and Harris, C. (eds), *No Stone Unturned: papers in honour of Roger Jacobi.* Lithic Studies Society Occasional Paper 9, London, 101–112.

Piet, J., 2018. *Manufacturing Styles and Insulation Benefits of Neanderthal Clothing.* Unpublished undergraduate dissertation, University of Reading.

Pincock, S., 2005. The quest for pain relief, *The Scientist* 2005 (March), 30–37.

Piperno, M. and Tagliacozzo, A., 2001. The elephant butchery area at the Middle Pleistocene site of Notarchirico (Venosa, Basilicata, Italy), in: Cavarretta, G., Gioia, P., Mussi, M. and Palombo, M.R. (eds*), The World of Elephants: Proceedings of the First International Congress.* Consiglio Nazionale delle Ricerche, Rome, 230–236.

Piperno, M., Lefèvre, D., Raynal, J.-P. and Tagliacozzo, A., 1998. Notarchirico: an early Middle Pleistocene site in the Venosa basin. *Anthropologie* 36, 85–90.

Pitts, M. and Roberts, M.B., 1997. *Fairweather Eden: life in Britain half a million years ago as revealed by the excavations at Boxgrove.* Century, London.

Planer, R., 2018. Cooking, mechanical processing, and the discovery of ignition technology. *Current Anthropology* 59(2), 228.

Plard, F., Gaillard, J.M., Coulson, T., Hewison, A.M., Delorme, D., Warnant, C., Nilsen, E.B. and Bonenfant, C., 2014. Long-lived and heavier females give birth earlier in roe deer. *Ecography* 37, 241–249.

Pocock, G., Richards, C.D. and Richards, D.A., 2018. *Human Physiology* (5th edn). Oxford University Press, Oxford.

Pope, M., 2004. Behavioural implications of biface discard: assemblage variability and land-use at the Middle Pleistocene site of Boxgrove, in: Walker, E.A., Wenban-Smith, F.F. and Healy, F. (eds), *Lithics in Action: papers from the Conference Lithic Studies in the Year 2000.* Oxbow Books, Oxford, 38–47.

Pope, M. and Roberts, M.B., 2005. Observations on the relationship between Palaeolithic individuals and artefact scatters at the Middle Pleistocene site of Boxgrove, UK, in: Gamble, C.S. and Porr, M. (eds), *The Hominid Individual in Context: Archaeological investigations of Lower and Middle Palaeolithic landscapes, locales and artefacts.* Routledge, London, 81–97.

Pope, M., McNabb, J. and Gamble, C., 2018. *Crossing the Human Threshold: dynamic transformation and persistent places during the Middle Pleistocene.* Routledge, Abingdon.

Pope, M., Parfitt, S. and Roberts, M.B., in press. *GTP17. The Boxgrove Horse Butchery Site: A high-resolution record of early human behaviour.* Spoilheap Monograph, London.

Pope, M., Blundell, L., Scott, B. and Cutler, H., 2016. Behaviour and process in the formation of the North European Acheulean record: towards a unified Palaeolithic landscape approach. *Quaternary International* 411, 85–94.

Potts, R.B., 1988. *Early Hominid Activities at Olduvai*. Aldine, New York.

Potts, R.B., 2012. Environmental and behavioral evidence pertaining to the evolution of early *Homo*. *Current Anthropology* 53, S299–S317.

Potts, R.B., 1984. Home bases and early hominids. *American Scientist* 72, 338–347.

Potts, R.B., 1991. Why the Oldowan? Plio-Pleistocene toolmaking and the transport of resources. *Journal of Anthropological Research* 47, 153–176.

Power, C. and Watts, I., 1996. Female strategies and collective behaviour: the archaeology of earliest *Homo sapiens sapiens*, in: Steele, J. and Shennan, S. (eds), *The Archaeology of Human Ancestry: power, sex and tradition*. Routledge, Abingdon, 306–330.

Preece, R.C., Gowlett, J.A.J., Parfitt, S.A., Bridgland, D.R. and Lewis, S.G., 2006. Humans in the Hoxnian: habitat, context and fire use at Beeches Pit, West Stow, Suffolk, UK. *Journal of Quaternary Science* 21, 485–496.

Proctor, C.J., Berridge, P.J., Bishop, M.J., Richards, D.A. and Smart, P.L., 2005. Age of Middle Pleistocene fauna and Lower Palaeolithic industries from Kent's Cavern, Devon. *Quaternary Science Reviews* 24, 1243–1252.

Profico, A., Di Vincenzo, F., Gagliardi, L., Piperno, M. and Manzi, G., 2016. Filling the gap. Human cranial remains from Gombore II (Melka Kunture, Ethiopia; *ca.* 850 ka) and the origin of *Homo heidelbergensis*. *Journal of Anthropological Sciences* 94, 1–24.

Prossinger, H., Seidler, H., Wicke, L., Weaver, D., Recheis, W., Stringer, C. and Müller, G.B., 2003. Electronic removal of encrustations inside the Steinheim cranium reveals paranasal sinus features and deformations, and provides a revised endocranial volume estimate. *The Anatomical Record Part B: The New Anatomist* 273, 132–142.

Prüfer, K., de Filippo, C., Grote, S., Mafessoni, F., Korlević, P., Hajdinjak, M., Vernot, B., Skov, L., Hsieh, P. and Peyrégne, S., 2017. A high-coverage Neandertal genome from Vindija Cave in Croatia. *Science* 358, 655–658.

Pryor, A., Pullen, A., Beresford-Jones, D., Svoboda, J. and Gamble, C.S., 2016. Reflections on Gravettian firewood procurement near the Pavlov Hills, Czech Republic. *Journal of Anthropological Archaeology* 43, 1–12.

Pusey, A.E. and Schroepfer-Walker, K., 2013. Female competition in chimpanzees. *Philosophical Transactions of the Royal Society B: Biological Sciences* 368, 20130077.

Pushkina, D., Bocherens, H. and Ziegler, R., 2014. Unexpected palaeoecological features of the Middle and Late Pleistocene large herbivores in southwestern Germany revealed by stable isotopic abundances in tooth enamel. *Quaternary International* 339, 164–178.

Putman, R., 1988. *The Natural History of Deer*. Christopher Helm, Bromley.

Putt, S.S., Woods, A.D. and Franciscus, R.G., 2014. The role of verbal interaction during experimental bifacial stone tool manufacture. *Lithic Technology* 39, 96–112.

Rachmayani, R., Prange, M. and Schulz, M., 2016. Intra-interglacial climate variability: model simulations of Marine Isotope Stages 1, 5, 11, 13, and 15. *Climate of the Past* 12, 677–695.

Radovčić, D., Sršen, A.O., Radovčić, J. and Frayer, D.W., 2015. Evidence for Neandertal Jewelry: Modified White-Tailed Eagle Claws at Krapina. *PLOS ONE* 10, e0119802.

Ravon, A.-L., 2017. *Originalité et développement du Paléolithique inférieur à l'extrémité occidentale de l'Eurasie : le Colombanien de Menez-Dregan (Plouhinec, Finistère)*. Unpublished PhD thesis, Université Rennes (Archéologie et Préhistoire).

Ravon, A.-L., 2018. Land use in Brittany during the Middle Pleistocene, in: Pope *et al.* (eds) 2018, 128–144.

Ravon, A.-L., Gaillard, C. and Monnier, J.-L., 2016. Menez-Dregan (Plouhinec, far western Europe): The lithic industry from layer 7 and its Acheulean components. *Quaternary International* 411, 132–143.

Reby, D., Hewison, M., Izquierdo, M. and Pépin, D., 2001. Red deer (*Cervus elaphus*) hinds discriminate between the roars of their current harem-holder stag and those of neighbouring stags. *Ethology* 107, 951–959.

Reshef, H. and Barkai, R., 2015. A taste of an elephant: the probable role of elephant meat in Paleolithic diet preferences. *Quaternary International* 379, 28–34.

Rhodes, S., Walker, M., López-Jiménez, A., López-Martínez, M., Haber-Uriarte, M., Fernández-Jalvo, Y. and Chazan, M., 2016. Fire in the Early Palaeolithic: evidence from burnt small mammal bones at Cueva Negra del Estrecho del Río Quípar, Murcia, Spain. *Journal of Archaeological Science: Reports* 9, 427–436.

Richards, M.P., Pettitt, P.B., Trinkaus, E., Smith, F.H., Paunović, M. and Karavanić, I., 2000. Neanderthal diet at Vindija and Neanderthal predation: the evidence from stable isotopes. *Proceedings of the National Academy of Sciences* 97, 7663–7666.

Richerson, P.J., Boyd, R. and Henrich, J., 2010. Gene-culture coevolution in the age of genomics. *Proceedings of the National Academy of Sciences* 107, 8985–8992.

Rifkin, R.F., 2011. Assessing the efficacy of red ochre as a prehistoric hide tanning ingredient. *Journal of African Archaeology* 9, 131–158.

Rightmire, G.P., 1998. Human evolution in the Middle Pleistocene: the role of *Homo heidelbergensis*. *Evolutionary Anthropology: issues, news, and reviews* 6, 218–227.

Rightmire, G.P., 2004. Brain size and encephalization in early to Mid-Pleistocene *Homo*. *American Journal of Physical Anthropology* 124, 109–123.

Rios-Garaizar, J., López-Bultó, O., Iriarte, E., Pérez-Garrido, C., Piqué, R., Aranburu, A., Iriarte-Chiapusso, M.J., Ortega-Cordellat, I., Bourguignon, L. and Garate, D., 2018. A Middle Palaeolithic wooden digging stick from Aranbaltza III, Spain. *PLOS ONE* 13, e0195044.

Ríos, L., Kivell, T.L., Lalueza-Fox, C., Estalrrich, A., García-Tabernero, A., Huguet, R., Quintino, Y., de la Rasilla, M. and Rosas, A., 2019. Skeletal anomalies in the Neandertal family of El Sidrón (Spain) support a role of inbreeding in Neandertal extinction. *Scientific Reports* 9, 1697.

Rivals, F. and Ziegler, R., 2018. High-resolution paleoenvironmental context for human occupations during the Middle Pleistocene in Europe (MIS 11, Germany). *Quaternary Science Reviews* 188, 136–142.

Rivals, F., Schulz, E. and Kaiser, T.M., 2008. Climate-related dietary diversity of the ungulate faunas from the Middle Pleistocene succession (OIS 14–12) at the Caune de l'Arago (France). *Paleobiology* 34, 117–127.

Rivals, F., Julien, M.-A., Kuitems, M., van Kolfschoten, T., Serangeli, J., Drucker, D.G., Bocherens, H. and Conard, N.J., 2015. Investigation of equid paleodiet from Schöningen 13 II-4 through dental wear and isotopic analyses: Archaeological implications. *Journal of Human Evolution* 89, 129–137.

Roach, N.T., Venkadesan, M., Rainbow, M.J. and Lieberman, D.E., 2013. Elastic energy storage in the shoulder and the evolution of high-speed throwing in *Homo*. *Nature* 498, 483.

Roberts, M.B., 1999a. Geological summary, in: Roberts and Parfitt (eds) 1999, 149–155.

Roberts, M.B., 1999b. Quarry 2 GTP 17, in: Roberts and Parfitt (eds) 1999, 372–378.

Roberts, M.B. and Parfitt, S.A. (eds), 1999. *Boxgrove: A Middle Pleistocene Hominid Site at Eartham Quarry, Boxgrove, West Sussex*. English Heritage, London.

Roberts, M.B. and Pope, M., 2009. The archaeological and sedimentary records from Boxgrove and Slindon, in: Briant, R.M., Bates, M.R., Hosfield, R. and Wenban-Smith, F.F. (eds), *The Quaternary of the Solent Basin and West Sussex Raised Beaches*. Quaternary Research Association, London.

Roberts, M.B., Stringer, C.B. and Parfitt, S.A., 1994. A hominid tibia from Middle Pleistocene sediments at Boxgrove, UK. *Nature* 369, 311–313.

Robson, S.L. and Wood, B., 2008. Hominin life history: reconstruction and evolution. *Journal of Anatomy* 212, 394–425.

Rocca, R., Abruzzese, C. and Aureli, D., 2016. European Acheuleans: critical perspectives from the East. *Quaternary International* 411, 402–411.

Rockman, M., 2003. Knowledge and learning in the archaeology of colonization, in: Rockman, M. and Steele, J. (eds), *Colonization of Unfamiliar Landscapes: the archaeology of adaptation*. Routledge, London, 3–24.

Rodríguez-Gómez, G., Rodríguez, J., Martín-González, J.Á. and Mateos, A., 2017. Evaluating the impact of *Homo*-carnivore competition in European human settlements during the Early to Middle Pleistocene. *Quaternary Research* 88, 129–151.

Rodríguez-Gómez, G., Mateos, A., Martín-González, J.A., Blasco, R., Rosell, J. and Rodríguez, J., 2014. Discontinuity of human presence at Atapuerca during the early Middle Pleistocene: a matter of ecological competition? *PLOS ONE* 9, e101938.

Rodríguez-Gómez, G., Palmqvist, P., Rodríguez, J., Mateos, A., Martín-González, J.A., Espigares, M.P., Ros-Montoya, S. and Martínez-Navarro, B., 2016. On the ecological context of the earliest human settlements in Europe: Resource availability and competition intensity in the carnivore guild of Barranco León-D and Fuente Nueva-3 (Orce, Baza Basin, SE Spain). *Quaternary Science Reviews* 143, 69–83.

Rodríguez-Gómez, G., Rodríguez, J., Martín-González, J.Á., Goikoetxea, I. and Mateos, A., 2013. Modeling trophic resource availability for the first human settlers of Europe: The case of Atapuerca TD6. *Journal of Human Evolution* 64, 645–657.

Rodríguez-Gómez, G., Rodríguez, J., Mateos, A., Martín-González, J.Á. and Goikoetxea, I., 2012. Food web structure during the European Pleistocene. *Journal of Taphonomy* 10, 165–184.

Rodríguez-Hidalgo, A., Rivals, F., Saladié, P. and Carbonell, E., 2016. Season of bison mortality in TD10.2 bone bed at Gran Dolina site (Atapuerca): integrating tooth eruption, wear, and microwear methods. *Journal of Archaeological Science: Reports* 6, 780–789.

Rodríguez-Hidalgo, A., Saladié, P., Ollé, A. and Carbonell, E., 2015. Hominin subsistence and site function of TD10.1 bone bed level at Gran Dolina site (Atapuerca) during the late Acheulean. *Journal of Quaternary Science* 30, 679–701.

Rodríguez-Hidalgo, A., Saladié, P., Ollé, A., Arsuaga, J.L., Bermúdez de Castro, J.M. and Carbonell, E., 2017. Human predatory behavior and the social implications of communal hunting based on evidence from the TD10.2 bison bone bed at Gran Dolina (Atapuerca, Spain). *Journal of Human Evolution* 105, 89–122.

Rodríguez, J. and Mateos, A., 2018. Carrying capacity, carnivoran richness and hominin survival in Europe. *Journal of Human Evolution* 118, 72–88.

Rodríguez, J., Martín-González, J.Á., Goikoetxea, I., Rodríguez-Gómez, G. and Mateos, A., 2013. Mammalian paleobiogeography and the distribution of *Homo* in early Pleistocene Europe. *Quaternary International* 295, 48–58.

Rodríguez, J., Rodríguez-Gómez, G., Martín-González, J.Á., Goikoetxea, I. and Mateos, A., 2012. Predator–prey relationships and the role of *Homo* in Early Pleistocene food webs in Southern Europe. *Palaeogeography, Palaeoclimatology, Palaeoecology* 365, 99–114.

Rodríguez, J., Burjachs, F., Cuenca-Bescós, G., García, N., Van der Made, J., Pérez González, A., Blain, H.A., Expósito, I., López-García, J.M., García Antón, M., Allué, E., Cáceres, I., Huguet, R., Mosquera, M., Ollé, A., Rosell, J., Parés, J.M., Rodríguez, X.P., Díez, C., Rofes, J., Sala, R., Saladié, P., Vallverdú, J., Bennasar, M.L., Blasco, R., Bermúdez de Castro, J.M. and Carbonell, E., 2011. One million years of cultural evolution in a stable environment at Atapuerca (Burgos, Spain). *Quaternary Science Reviews* 30, 1396–1412.

Roe, D.A., 1981. *The Lower and Middle Palaeolithic Periods in Britain*. Routledge & Kegan Paul, London.

Roebroeks, W., 2001. Hominid behaviour and the earliest occupation of Europe: an exploration. *Journal of Human Evolution* 41, 437–461.

Roebroeks, W., 2005. Archaeology: life on the Costa del Cromer. *Nature* 438, 921.

Roebroeks, W., 2006. The human colonisation of Europe: where are we? *Journal of Quaternary Science* 21, 425–435.

Roebroeks, W. and van Kolfschoten, T., 1994. The earliest occupation of Europe: a short chronology. *Antiquity* 68, 489–503.

Roebroeks, W. and van Kolfschoten, T. (eds), 1995. *The Earliest Occupation of Europe: Proceedings of the European Science Foundation Workshop at Tautavel (France), 1993*. Leiden University Press, Leiden.

Roebroeks, W. and Villa, P., 2011. On the earliest evidence for habitual use of fire in Europe. *Proceedings of the National Academy of Sciences* 108, 5209–5214.

Roebroeks, W., Sier, M.J., Nielsen, T.K., De Loecker, D., Parés, J.M., Arps, C.E. and Mücher, H.J., 2012. Use of red ochre by early Neandertals. *Proceedings of the National Academy of Sciences* 109, 1889–1894.

Rogers, Alan R., Iltis, D. and Wooding, S., 2004. Genetic variation at the MC1R Locus and the time since loss of human body hair. *Current Anthropology* 45, 105–108.

Roksandic, M., Radović, P. and Lindal, J., 2018. Revising the hypodigm of *Homo heidelbergensis*: a view from the eastern Mediterranean. *Quaternary International* 466, 66–81.

Roksandic, M., Mihailović, D., Mercier, N., Dimitrijević, V., Morley, M.W., Rakočević, Z., Mihailović, B., Guibert, P. and Babb, J., 2011. A human mandible (BH-1) from the Pleistocene deposits of Mala Balanica cave (Sićevo Gorge, Niš, Serbia). *Journal of Human Evolution* 61, 186–196.

Rolland, N., 2004. Was the emergence of home bases and domestic fire a punctuated event? A review of the Middle Pleistocene record in Eurasia. *Asian Perspectives* 43, 248–280.

Romanowska, I., 2012. Ex oriente lux: A re-evaluation of the Lower Palaeolithic of central and eastern Europe, in: Ruebens, K., Romanowska, I. and Bynoe, R. (eds), *Unravelling the Palaeolithic: 10 years of Research at the Centre for the Archaeology of Human Origins*. British Archaeological Report S2400, Oxford, 1–12.

Ronen, A., 2006. The oldest human groups in the Levant. *Comptes Rendus Palevol* 5, 343–351.

Rosas, A. and Bermúdez de Castro, J.M., 1998. The Mauer mandible and the evolutionary significance of *Homo heidelbergensis*. *Geobios* 31, 687–697.

Rosell, J., Blasco, R., Campeny, G., Díez, J.C., Alcalde, R.A., Menéndez, L., Arsuaga, J.L., Bermúdez de Castro, J.M. and Carbonell, E., 2011. Bone as a technological raw material at the Gran Dolina site (Sierra de Atapuerca, Burgos, Spain). *Journal of Human Evolution* 61, 125–131.

Rosenberg, K. and Trevathan, W., 2002. Birth, obstetrics and human evolution. *BJOG: An International Journal of Obstetrics & Gynaecology* 109, 1199–1206.

Rots, V., Hardy, B.L., Serangeli, J. and Conard, N.J., 2015. Residue and microwear analyses of the stone artifacts from Schöningen. *Journal of Human Evolution* 89, 298–308.

Rowland, M.J., 2002. Geophagy: an assessment of implications for the development of Australian indigenous plant processing technologies. *Australian Aboriginal Studies* 1, 51–66.

Ruff, C.B., 1991. Climate and body shape in hominid evolution. *Journal of Human Evolution* 21, 81–105.

Ruff, C.B., 1993. Climatic adaptation and hominid evolution: the thermoregulatory imperative. *Evolutionary Anthropology: issues, news, and reviews* 2, 53–60.

Ruff, C.B., 1994. Morphological adaptation to climate in modern and fossil hominids. *American Journal of Physical Anthropology* 37, 65–107.

Ruff, C.B., 1995. Biomechanics of the hip and birth in early *Homo*. *American Journal of Physical Anthropology* 98, 527–574.

Russo Ermolli, E., Di Donato, V., Martín-Fernández, J.A., Orain, R., Lebreton, V. and Piovesan, G., 2015. Vegetation patterns in the Southern Apennines (Italy) during MIS 13: Deciphering pollen variability along a NW–SE transect. *Review of Palaeobotany and Palynology* 218, 167–183.

Ruxton, G.D. and Wilkinson, D.M., 2011. Thermoregulation and endurance running in extinct hominins: Wheeler's models revisited. *Journal of Human Evolution* 61, 169–175.

Ryder, M.L., 1977. Seasonal coat changes in grazing Red deer (*Cervus elaphus*). *Journal of Zoology* 181(2), 137–143.

Saccà, D., 2012. Taphonomy of *Palaeoloxodon antiquus* at Castel di Guido (Rome, Italy): proboscidean carcass exploitation in the Lower Palaeolithic. *Quaternary International* 276, 27–41.

Sakellariou, D. and Galanidou, N., 2016. Pleistocene submerged landscapes and Palaeolithic archaeology in the tectonically active Aegean region, in: Harff, J., Bailey, G. and Lüth, F. (eds), *Geology and Archaeology: submerged landscapes of the continental shelf*. Geological Society Special Publications 411, London, 145–178.

Sala, N., Pantoja-Pérez, A., Arsuaga, J.L., Pablos, A. and Martínez, I., 2016. The Sima de los Huesos Crania: analysis of the cranial breakage patterns. *Journal of Archaeological Science* 72, 25–43.

Sala, N., Arsuaga, J.L., Pantoja-Pérez, A., Pablos, A., Martínez, I., Quam, R.M., Gómez-Olivencia, A., Bermúdez de Castro, J.M. and Carbonell, E., 2015. Lethal interpersonal violence in the Middle Pleistocene. *PLOS ONE* 10, e0126589.

Saladié, P., Huguet, R., Díez, C., Rodríguez-Hidalgo, A., Cáceres, I., Vallverdú, J., Rosell, J., Bermúdez de Castro, J.M. and Carbonell, E., 2011. Carcass transport decisions in *Homo antecessor* subsistence strategies. *Journal of Human Evolution* 61, 425–446.

Saladié, P., Huguet, R., Rodríguez-Hidalgo, A., Cáceres, I., Esteban-Nadal, M., Arsuaga, J.L., Bermúdez de Castro, J.M. and Carbonell, E., 2012. Intergroup cannibalism in the European Early Pleistocene: The range expansion and imbalance of power hypotheses. *Journal of Human Evolution* 63, 682–695.

Saladié, P., Rodríguez-Hidalgo, A., Díez, C., Martín-Rodríguez, P. and Carbonell, E., 2013. Range of bone modifications by human chewing. *Journal of Archaeological Science* 40, 380–397.

Saladié, P., Rodríguez-Hidalgo, A., Huguet, R., Cáceres, I., Díez, C., Vallverdú, J., Canals, A., Soto, M., Santander, B. and Bermúdez de Castro, J.M., 2014. The role of carnivores and their relationship to hominin settlements in the TD6-2 level from Gran Dolina (Sierra de Atapuerca, Spain). *Quaternary Science Reviews* 93, 47–66.

Šálková, T., Divišová, M., Kadochová, Š., Beneš, J., Delawská, K., Kadlčková, E., Němečková, L., Pokorná, K., Voska, V. and Žemličková, A., 2011. Acorns as a food resource. An experiment with acorn preparation and taste. *Interdisciplinaria Archaeologica* II, 133–141.

Sampson, C.G., 1978. *Palaeoecology and Archaeology of an Acheulean Site at Caddington, England.* Department of Anthropology, Institute for the Study of Earth and Man, Southern Methodist University: Dallas TX.

Sampson, C.G., 2006. Acheulian quarries at hornfels outcrops in the Upper Karoo region of South Africa, in: Goren-Inbar, N. and Sharon, G. (eds), *Axe Age: Acheulian tool-making from quarry to discard.* Equinox Publishing, London, 75–107.

Sánchez Goñi, M.F., Llave, E., Oliveira, D., Naughton, F., Desprat, S., Ducassou, E., Hodell, D.A. and Hernández-Molina, F.J., 2016. Climate changes in south western Iberia and Mediterranean outflow variations during two contrasting cycles of the last 1 Myrs: MIS 31–MIS 30 and MIS 12–MIS 11. *Global and Planetary Change* 136, 18–29.

Sánchez-Quinto, F., and Lalueza-Fox, C., 2015. Almost 20 years of Neanderthal palaeogenetics: adaptation, admixture, diversity, demography and extinction. *Philosophical Transactions of the Royal Society B: Biological Sciences* 370, 20130374.

Sandel, A.A., 2013. Brief communication: Hair density and body mass in mammals and the evolution of human hairlessness. *American Journal of Physical Anthropology* 152, 145–150.

Sandgathe, D.M. and Hayden, B., 2003. Did Neanderthals eat inner bark? *Antiquity* 77, 709–718.

Sandgathe, D.M., Dibble, H.L., Goldberg, P., McPherron, S.P., Turq, A., Niven, L. and Hodgkins, J., 2011. Timing of the appearance of habitual fire use. *Proceedings of the National Academy of Sciences* 108, E298–E298.

Sankararaman, S., Mallick, S., Dannemann, M., Prüfer, K., Kelso, J., Pääbo, S., Patterson, N. and Reich, D., 2014. The genomic landscape of Neanderthal ancestry in present-day humans. *Nature* 507(7492), 354–357.

Santagata, C., 2016. Operating systems in units B and E of the Notarchirico (Basilicata, Italy) ancient Acheulean open-air site and the role of raw materials. *Quaternary International* 411, 284–300.

Santonja, M. and Villa, P., 1990. The Lower Palaeolithic of Spain and Portugal. *Journal of World Prehistory* 4, 45–94.

Santonja, M. and Villa, P., 2006. The Acheulian of western Europe, in: Goren-Inbar, N. and Sharon, G. (eds), *Axe Age: Acheulian tool-making from quarry to discard.* Equinox, London, 429–478.

Santucci, E., Marano, F., Cerilli, E., Fiore, I., Lemorini, C., Palombo, M.R., Anzidei, A.P. and Bulgarelli, G.M., 2016. *Palaeoloxodon* exploitation at the Middle Pleistocene site of La Polledrara di Cecanibbio (Rome, Italy). *Quaternary International* 406, 169–182.

Sasaki, M., Endo, H., Yamagiwa, D., Yamamoto, M., Arishima, K. and Hayashi, Y., 1999. Morphological character of the shoulder and leg skeleton in Przewalski's horse (*Equus przewalskii*). *Annals of Anatomy - Anatomischer Anzeiger* 181(4), 403–407.

Schmidt, I. and Zimmermann, A., 2019. Population dynamics and socio-spatial organization of the Aurignacian: scalable quantitative demographic data for western and central Europe. *PLOS ONE* 14, e0211562.

Schmidt, K., Jędrzejewski, W., Theuerkauf, J., Kowalczyk, R., Okarma, H. and Jędrzejewska, B., 2008. Reproductive behaviour of wild-living wolves in Białowieża Primeval Forest (Poland). *Journal of Ethology* 26, 69–78.

Schoch, W.H., Bigga, G., Böhner, U., Richter, P. and Terberger, T., 2015. New insights on the wooden weapons from the Paleolithic site of Schöningen. *Journal of Human Evolution* 89, 214–225.

Schreve, D.C., 1996. The mammalian fauna from the Waechter excavations, Barnfield Pit, Swanscombe, in: Conway *et al.* (eds) 1996, 149–162.

Schreve, D.C., 2001. Differentiation of the British late Middle Pleistocene interglacials: the evidence from mammalian biostratigraphy. *Quaternary Science Reviews* 20, 1693–1705.

Schulp, C.J.E., Thuiller, W., Verburg, P.H., 2014. Wild food in Europe: a synthesis of knowledge and data of terrestrial wild food as an ecosystem service. *Ecological Economics* 105, 292–305.

Schwartz, G.T., 2012. Growth, development, and life history throughout the evolution of *Homo*. *Current Anthropology* 53, S395–S408.

Sedwick, C., 2008. What killed the woolly mammoth? *PLOS Biology* 6, e99.

Selva, N., Jędrzejewska, B., Jędrzejewski, W. and Wajrak, A., 2003. Scavenging on European bison carcasses in Białowieża Primeval Forest (eastern Poland). *Ecoscience* 10, 303–311.

Serangeli, J. and Conard, N.J., 2015. The behavioral and cultural stratigraphic contexts of the lithic assemblages from Schöningen. *Journal of Human Evolution* 89, 287–297.

Serangeli, J., Böhner, U., van Kolfschoten, T. and Conard, N.J., 2015a. Overview and new results from large-scale excavations in Schöningen. *Journal of Human Evolution* 89, 27–45.

Serangeli, J., van Kolfschoten, T., Starkovich, B.M. and Verheijen, I., 2015b. The European saber-toothed cat (*Homotherium latidens*) found in the 'Spear Horizon' at Schöningen (Germany). *Journal of Human Evolution* 89, 172–180.

Serangeli, J., Rodríguez-Álvarez, B., Tucci, M., Verheijen, I., Bigga, G., Böhner, U., Urban, B., van Kolfschoten, T. and Conard, N.J., 2018. The project Schöningen from an ecological and cultural perspective. *Quaternary Science Reviews* 198, 140–155.

Shahack-Gross, R., Berna, F., Karkanas, P., Lemorini, C., Gopher, A. and Barkai, R., 2014. Evidence for the repeated use of a central hearth at Middle Pleistocene (300 ky ago) Qesem Cave, Israel. *Journal of Archaeological Science* 44, 12–21.

Shaw, A.D. and Scott, R., 2018. La Cotte de St Brelade: placemaking, assemblage and persistence in the Normano-Breton Gulf, in: Pope *et al.* (eds) 2018, 145–163.

Shaw, A.D. and White, M.J., 2003. Another look at the Cuxton handaxe assemblage. *Proceedings of the Prehistoric Society* 69, 305–313.

Shaw, A.D., Bates, M., Conneller, C., Gamble, C.S., Julien, M.-A., McNabb, J., Pope, M. and Scott, B., 2016. The archaeology of persistent places: the Palaeolithic case of La Cotte de St Brelade, Jersey. *Antiquity* 90, 1437–1453.

Shaw, C.N., Hofmann, C.L., Petraglia, M.D., Stock, J.T. and Gottschall, J.S., 2012. Neandertal humeri may reflect adaptation to scraping tasks, but not spear thrusting. *PLOS ONE* 7, e40349.

Shea, J.J., 2006. Child's play: reflections on the invisibility of children in the Paleolithic record. *Evolutionary Anthropology: Issues, News, and Reviews* 15, 212–216.

Shea, J.J., 2015. Making and using stone tools: advice for learners and teachers and insights for archaeologists. *Lithic Technology* 40, 231–248.

Shotton, F.W., Keen, D.H., Coope, G.R., Currant, A.P., Gibbard, P.L., Aalto, M., Peglar, S.M. and Robinson, J.E., 1993. The Middle Pleistocene deposits of Waverley Wood Pit, Warwickshire, England. *Journal of Quaternary Science* 8, 293–325.

Sidorovich, V.E., Jędrzejewska, B. and Jędrzejewski, W., 1996. Winter distribution and abundance of mustelids and beavers in the river valleys of Białowieża Primeval Forest. *Acta Theriologica* 41, 155–170.

Sillitoe, P. and Hardy, K., 2003. Living lithics: ethnoarchaeology in highland Papua New Guinea. *Antiquity* 77, 555–566.

Singer, R., Gladfelter, B.G. and Wymer, J.J., 1993. *The Lower Palaeolithic Site at Hoxne, England.* Chicago University Press, Chicago IL.

Singer, R., Wymer, J.J. and Gladfelter, B.G., 1973. Excavation of the Clactonian industry at the Golf Course, Clacton-on-Sea, Essex. *Proceedings of the Prehistoric Society* 39, 6–74.

Sirakov, N., Guadelli, J.-L., Ivanova, S., Sirakova, S., Boudadi-Maligne, M., Dimitrova, I., Fernandez, Ph., Ferrier, C., Guadelli, A., Iordanova, N., Kovatcheva, M., Krumov, I., Leblanc, J.-Cl., Miteva, V., Popov, V. Spassov, R., Taneva. S. and Tsanova, T., 2010. An ancient continuous human presence in the Balkans and the beginnings of human settlement in western Eurasia: a Lower Pleistocene example of the Lower Palaeolithic levels in Kozarnika cave (north-western Bulgaria). *Quaternary International* 223, 94–106.

Skinner, M.M., de Vries, D., Gunz, P., Kupczik, K., Klassen, R.P., Hublin, J.-J. and Roksandic, M., 2016. A dental perspective on the taxonomic affinity of the Balanica mandible (BH-1). *Journal of Human Evolution* 93, 63–81.

Smith, G.M., 2003. Damage inflicted on animal bone by wooden projectiles: experimental results and archaeological implications. *Journal of Taphonomy* 1, 105–114.

Smith, G.M., 2013. Taphonomic resolution and hominin subsistence behaviour in the Lower Palaeolithic: differing data scales and interpretive frameworks at Boxgrove and Swanscombe (UK). *Journal of Archaeological Science* 40, 3754–3767.

Soffer, O., 2004. Recovering perishable technologies through use wear on tools: preliminary evidence for Upper Paleolithic weaving and net making. *Current Anthropology* 45, 407–413.

Soffer, O., Adovasio, J.M., Hyland, D.C., Gvozdover, M.D., Habu, J., Kozlowski, J.K., McDermott, L.R., Mussi, M., Owen, L.R. and Svoboda, J., 2000. The 'Venus' figurines: textiles, basketry, gender, and status in the Upper Paleolithic. *Current Anthropology* 41, 511–537.

Solodenko, N., Zupancich, A., Cesaro, S.N., Marder, O., Lemorini, C. and Barkai, R., 2015. Fat residue and use-wear found on Acheulian biface and scraper associated with butchered elephant remains at the site of Revadim, Israel. *PLOS ONE* 10, e0118572.

Sorensen, A.C., 2017. On the relationship between climate and Neandertal fire use during the Last Glacial in south-west France. *Quaternary International* 436, 114–128.

Sorensen, A.C., Claud, E. and Soressi, M., 2018. Neandertal fire-making technology inferred from microwear analysis. *Scientific Reports* 8, 10065.

Sorensen, A.C., Roebroeks, W. and Van Gijn, A., 2014. Fire production in the deep past? The expedient strike-a-light model. *Journal of Archaeological Science* 42, 476–486.

Speth, J.D., 1990. Seasonality, resource stress, and food sharing in so-called 'egalitarian' foraging societies. *Journal of Anthropological Archaeology* 9, 148–188.

Speth, J.D., 1991a. Nuritional constraints and Late Glacial adaptive transformations: the importance of non-protein energy sources, in: Barton, N., Roberts, A.J. and Roe, D.A. (eds), *The Late Glacial in North-west Europe: human adaptaton and environmental change at the end of the Pleistocene.* Council for British Archaeology Research Report 77, York, 169–178.

Speth, J.D., 1991b. Protein selection and avoidance strategies of contemporary and ancestral foragers: unresolved issues. *Philosophical Transactions of the Royal Society B: Biological Sciences* 334, 265–270.

Speth, J.D., 2015. When did humans learn to boil. *PaleoAnthropology* 2015, 54–67.

Speth, J.D., 2017. Putrid meat and fish in the Eurasian Middle and Upper Paleolithic: Are we missing a key part of Neanderthal and modern human diet? *PaleoAnthropology* 2017, 44–72.

Speth, J.D. and Spielmann, K.A., 1983. Energy source, protein metabolism, and hunter-gatherer subsistence strategies. *Journal of Anthropological Archaeology* 2, 1–31.

Spiess, A.E., 1979. *Reindeer and Caribou Hunters; an archaeological study*. Academic Press, New York.

Spikins, P., 2012. Goodwill hunting? Debates over the 'meaning' of Lower Palaeolithic handaxe form revisited. *World Archaeology* 44, 378–392.

Spikins, P., Hitchens, G., Needham, A. and Rutherford, H., 2014. The cradle of thought: growth, learning, play and attachment in Neanderthal children. *Oxford Journal of Archaeology* 33, 111–134.

Spikins, P., Needham, A., Wright, B., Dytham, C., Gatta, M. and Hitchens, G., 2019. Living to fight another day: the ecological and evolutionary significance of Neanderthal healthcare. *Quaternary Science Reviews* 217, 98–118.

Stahlschmidt, M.C., Miller, C.E., Ligouis, B., Goldberg, P., Berna, F., Urban, B. and Conard, N.J., 2015a. The depositional environments of Schöningen 13 II-4 and their archaeological implications. *Journal of Human Evolution* 89, 71–91.

Stahlschmidt, M.C., Miller, C.E., Ligouis, B., Hambach, U., Goldberg, P., Berna, F., Richter, D., Urban, B., Serangeli, J. and Conard, N.J., 2015b. On the evidence for human use and control of fire at Schöningen. *Journal of Human Evolution* 89, 181–201.

Stapert, D., 2007. Neanderthal children and their flints. *PalArch's Journal of Archaeology of Northwest Europe* 1, 16–39.

Starkovich, B.M. and Conard, N.J., 2015. Bone taphonomy of the Schöningen 'Spear Horizon South' and its implications for site formation and hominin meat provisioning. *Journal of Human Evolution* 89, 154–171.

Steegmann Jr, A.T., Cerny, F.J. and Holliday, T.W., 2002. Neandertal cold adaptation: physiological and energetic factors. *American Journal of Human Biology* 14, 566–583.

Steele, J., 1996. On predicting hominid group sizes, in: Steele, J. and Shennan, S.J. (eds), *The Archaeology of Human Ancestry: power, sex and tradition*. Routledge, London, 230–252.

Stepanchuk, V. and Moigne, A.-M., 2016. MIS 11-locality of Medzhibozh, Ukraine: archaeological and paleozoological evidence. *Quaternary International* 409, 241–254.

Steudel-Numbers, K.L. and Wall-Scheffler, C.M., 2009. Optimal running speed and the evolution of hominin hunting strategies. *Journal of Human Evolution* 56, 355–360.

Stevens, R.E., Lister, A.M. and Hedges, R.E., 2006. Predicting diet, trophic level and palaeoecology from bone stable isotope analysis: a comparative study of five red deer populations. *Oecologia* 149, 12–21.

Stewart, J.R., 2005. The ecology and adaptation of Neanderthals during the non-analogue environment of Oxygen Isotope Stage 3. *Quaternary International* 137, 35–46.

Stewart, J.R., van Kolfschoten, T., Markova, A. and Musil, R., 2003. The mammalian faunas of Europe during Oxygen Isotope Stage 3, in: Van Andel and Davies (eds) 2003, 103–130.

Stewart, J.R., García-Rodríguez, O., Knul, M.V., Sewell, L., Montgomery, H., Thomas, M.G. and Diekmann, Y., 2019. Palaeoecological and genetic evidence for Neanderthal power locomotion as an adaptation to a woodland environment. *Quaternary Science Reviews* 217, 310–315.

Stiner, M.C. and Kuhn, S.L., 2006. Changes in the 'connectedness' and resilience of Paleolithic societies in Mediterranean ecosystems. *Human Ecology* 34, 693–712.

Stiner, M.C., Gopher, A. and Barkai, R., 2011. Hearth-side socioeconomics, hunting and paleoecology during the late Lower Paleolithic at Qesem Cave, Israel. *Journal of Human Evolution* 60, 213–233.

Stopp, M.P., 1993. Taphonomic analysis of the faunal assemblage, in: Singer *et al.* (eds) 1993, 139–149.

Stopp, M.P., 2002. Ethnohistoric analogues for storage as an adaptive strategy in northeastern subarctic prehistory. *Journal of Anthropological Archaeology* 21, 301–328.

Stout, D. and Chaminade, T., 2012. Stone tools, language and the brain in human evolution. *Philosophical Transactions of the Royal Society B: Biological Sciences* 367, 75–87.

Stout, D., Apel, J., Commander, J. and Roberts, M.B., 2014. Late Acheulean technology and cognition at Boxgrove, UK. *Journal of Archaeological Science* 41, 576–590.

Stout, D., Toth, N., Schick, K. and Chaminade, T., 2008. Neural correlates of Early Stone Age tool-making: technology, language and cognition in human evolution. *Philosophical Transactions of the Royal Society B: Biological Sciences* 363, 1939–1949.

Straus, L.G., 1987. Hunting in Late Upper Paleolithic western Europe, in: Nitecki, M. and Nitecki, D. (eds), *The Evolution of Human Hunting*. Plenum Press, New York, 147–176.

Stringer, C.B., 2012. The status of *Homo heidelbergensis* (Schoetensack 1908). *Evolutionary Anthropology: issues, news, and reviews* 21, 101–107.

Stringer, C.B., 2006. *Homo Britannicus: the incredible story of human life in Britain*. Penguin Books, London.

Stringer, C.B. and Gamble, C.S., 1993. *In Search of the Neanderthals: solving the puzzle of human origins*. Thames and Hudson, London.

Stringer, C.B. and Hublin, J.-J., 1999. New age estimates for the Swanscombe hominid, and their significance for human evolution. *Journal of Human Evolution* 37, 873–877.

Stringer, C.B., Trinkaus, E., Roberts, M.B., Parfitt, S.A. and Macphail, R.I., 1998. The Middle Pleistocene human tibia from Boxgrove. *Journal of Human Evolution* 34, 509–547.

Stuart, A.J., 1992. The High Lodge mammalian fauna, in: Ashton *et al.* (eds) 1992, 120–123.

Stuart, A.J., Wolff, R.G., Lister, A.M., Singer, R. and Egginton, J.M., 1993. Fossil vertebrates, in: Singer *et al.* (eds) 1993, 163–206.

Sturdy, D. and Webley, D., 1988. Palaeolithic geography: or where are the deer? *World Archaeology* 19, 262–280.

Szemethy, L., Heltai, M., Mátrai, K. and Peto, Z., 1999. Home ranges and habitat selection of red deer (*Cervus elaphus*) on a lowland area. *Gibier Faune Sauvage* 15, 607–616.

Szymanek, M., 2017. Palaeotemperature estimation in the Holsteinian Interglacial (MIS 11) based on oxygen isotopes of aquatic gastropods from eastern Poland. *Acta Geologica Polonica* 67, 585–605.

Szymanek, M. and Julien, M.-A., 2018. Early and Middle Pleistocene climate-environment conditions in Central Europe and the hominin settlement record. *Quaternary Science Reviews* 198, 56–75.

Szymanek, M., Bińka, K., and Nitychoruk, J., 2016. Stable ^{18}O and ^{13}C isotope records of *Viviparus diluvianus* (Kunth, 1865) shells from Holsteinian (MIS 11) lakes of eastern Poland as palaeoenvironmental and palaeoclimatic proxies. *Boreas* 45, 109–121.

Tardío, J., Pascual, H. and Morales, R., 2005. Wild food plants traditionally used in the province of Madrid, Central Spain. *Economic Botany* 59, 122–136.

Taylor, T., 1997. *The Prehistory of Sex: four million years of human sexual culture*. Fourth Estate, London.

Tejero, J.-M., Christensen, M. and Bodu, P., 2012. Red deer antler technology and early modern humans in Southeast Europe: an experimental study. *Journal of Archaeological Science* 39, 332–346.

Theberge, J.B. and Falls, J.B., 1967. Howling as a means of communication in timber wolves. *American Zoologist* 7(2), 331–338.

Théry-Parisot, I. and Meignen, L., 2000. Économie des combustibles (bois et lignite) dans l'abri moustérien des Canalettes [L'expérimentation à la simulation des besoins énergétiques]. *Gallia Préhistoire* 42, 45–55.

Theuerkauf, J., Rouys, S. and Jedrzejewski, W., 2003. Selection of den, rendezvous, and resting sites by wolves in the Białowieża Forest, Poland. *Canadian Journal of Zoology* 81, 163–167.

Thieme, H., 1997. Lower Palaeolithic hunting spears from Germany. *Nature* 385, 807–810.

Thieme, H., 2005. The Lower Palaeolithic art of hunting: the case of Schöningen 13 II-4, Lower Saxony, Germany, in: Gamble, C.S. and Porr, M. (eds), *The Hominid Individual in Context: archaeological investigations of Lower and Middle Palaeolithic landscapes, locales and artefacts*. Routledge, London, 115–132.

Thoma, A., 1972. Cranial capacity, taxonomical and phylogenetical status of Vértesszőlős man. *Journal of Human Evolution* 1, 511–512.

Thomas, H. and Nisbet, T.R., 2007. An assessment of the impact of floodplain woodland on flood flows. *Water and Environment Journal* 21, 114–126.

Thompson, J.L. and Nelson, A.J., 2011. Middle childhood and modern human origins. *Human Nature* 22, 249–280.

Thompson, S., Vehkaoja, M. and Nummi, P., 2016. Beaver-created deadwood dynamics in the boreal forest. *Forest Ecology and Management* 360, 1–8.

Titton, S., Barsky, D., Bargallo, A., Vergès, J.M., Guardiola, M., Solano, J.G., Jimenez Arenas, J.M., Toro-Moyano, I. and Sala-Ramos, R., 2018. Active percussion tools from the Oldowan site of Barranco León (Orce, Andalusia, Spain): the fundamental role of pounding activities in hominin lifeways. *Journal of Archaeological Science* 96, 131–147.

Toro-Moyano, I., Martínez-Navarro, B., Agustí, J., Souday, C., Bermúdez de Castro, J.M., Martinón-Torres, M., Fajardo, B., Duval, M., Falguères, C., Oms, O., Parés, J.M., Anadón, P., Julià, R., García-Aguilar, J.M., Moigne, A.-M., Espigares, M.P., Ros-Montoya, S. and Palmqvist, P., 2013. The oldest human fossil in Europe, from Orce (Spain). *Journal of Human Evolution* 65, 1–9.

Toups, M.A., Kitchen, A., Light, J.E and Reed, D.L., 2010. Origin of clothing lice indicates early clothing use by anatomically modern humans in Africa. *Molecular Biology and Evolution* 28, 29–32.

Tourloukis, V. and Harvati, K., 2018. The Palaeolithic record of Greece: A synthesis of the evidence and a research agenda for the future. *Quaternary International* 466, 48–65.

Trevathan, W. and Rosenberg, K., 2000. The shoulders follow the head: postcranial constraints on human childbirth. *Journal of Human Evolution* 39, 583–586.

Treves, A. and Naughton-Treves, L., 1999. Risk and opportunity for humans coexisting with large carnivores. *Journal of Human Evolution* 36, 275–282.

Trinkaus, E., 2012. Neandertals, early modern humans, and rodeo riders. *Journal of Archaeological Science* 39, 3691–3693.

Trinkaus, E., 2018. An abundance of developmental anomalies and abnormalities in Pleistocene people. *Proceedings of the National Academy of Sciences* 115, 11941–11946.

Trinkaus, E. and Buzhilova, A., 2012. The death and burial of Sunghir 1. *International Journal of Osteoarchaeology* 22, 655–666.

Trinkaus, E., Ruff, C.B., Churchill, S.E. and Vandermeersch, B., 1998. Locomotion and body proportions of the Saint-Césaire 1 Châtelperronian Neandertal. *Proceedings of the National Academy of Sciences* 95, 5836–5840.

Trinkaus, E., Stringer, C.B., Ruff, C.B., Hennessy, R.J., Roberts, M.B. and Parfitt, S.A., 1999. Diaphyseal cross-sectional geometry of the Boxgrove 1 Middle Pleistocene human tibia. *Journal of Human Evolution* 37, 1–25.

Tuffreau, A. and Antoine, P., 1995. The earliest occupation of Europe: continental northwestern Europe, in: Roebroeks and van Kolfschoten (eds) 1995, 147–163.

Tuffreau, A., Lamotte, A. and Marcy, J.-L., 1997. Land-use and site function in Acheulean complexes of the Somme Valley. *World Archaeology* 29, 225–241.

Turner, A., 1992. Large carnivores and earliest European hominids: changing determinants of resource availability during the Lower and Middle Pleistocene. *Journal of Human Evolution* 22, 109–126.

Turner, A. and Antón, M., 1996. The giant hyaena, *Pachycrocuta brevirostris* (Mammalia, Carnivora, Hyaenidae). *Geobios* 29, 455–468.

Turner, E., 1999. Lithic artefacts and animal bones in floodplain deposits at Miesenheim I (Central Rhineland, Germany), in: Gaudzinski, S. and Turner, E. (eds), *The Role of Early Humans in the Accumulation of European Lower and Middle Palaeolithic Bone Assemblages*. Habelt, Bonn, 103–119.

Twomey, T., 2013. The cognitive implications of controlled fire use by early humans. *Cambridge Archaeological Journal* 23, 113–128.

Tyldesley, J.A. and Bahn, P., 1983. Use of plants in the European Palaeolithic: a review of the evidence. *Quaternary Science Reviews* 2, 53–81.

Tzedakis, P.C., 1993. Long-term tree populations in northwest Greece through multiple Quaternary climatic cycles. *Nature* 364, 437–440.

Tzedakis, P.C., Hooghiemstra, H. and Pälike, H., 2006. The last 1.35 million years at Tenaghi Philippon: revised chronostratigraphy and long-term vegetation trends. *Quaternary Science Reviews* 25, 3416–3430.

UCJRC, 2011. *Forest Fires in Europe 2010*. European Union (European Commission Joint Research Centre: Institute for Environment & Sustainability), Luxembourg.

Uomini, N.T. and Meyer, G.F., 2013. Shared brain lateralization patterns in language and Acheulean stone tool production: a functional transcranial Doppler ultrasound study. *PLOS ONE* 8, e72693.

Urban, B. and Bigga, G., 2015. Environmental reconstruction and biostratigraphy of late Middle Pleistocene lakeshore deposits at Schöningen. *Journal of Human Evolution* 89, 57–70.

Vaesen, K., Scherjon, F., Hemerik, L. and Verpoorte, A., 2019. Inbreeding, Allee effects and stochasticity might be sufficient to account for Neanderthal extinction. *PLOS ONE* 14(11), p.e0225117.

Vallverdú, J., Saladié, P., Rosas, A., Huguet, R., Cáceres, I., Mosquera, M., Garcia-Tabernero, A., Estalrrich, A., Lozano-Fernández, I., Pineda-Alcalá, A., Carrancho, Á., Villalaín, J.J., Bourlès, D., Braucher, R., Lebatard, A., Vilalta, J., Esteban-Nadal, M., Bennàsar, M.L., Bastir, M., López-Polín, L., Ollé, A., Vergés, J.M., Ros-Montoya, S., Martínez-Navarro, B., García, A., Martinell, J., Expósito, I., Burjachs, F., Agustí, J. and Carbonell, E., 2014. Age and Date for Early Arrival of the Acheulian in Europe (Barranc de la Boella, la Canonja, Spain). *PLOS ONE* 9, e103634.

Valoch, K., 1995. The earliest occupation of Europe: Eastern Central and Southeastern Europe, in: Roebroeks and van Kolfschoten (eds) 1995, 67–84.

Valoch, K., 2013. The Early Palaeolithic site Stránská Skála I near Brno (Czechoslovakia). *Anthropologie* 51, 67–86.

Van Andel, T.H. and Davies, W. (eds), 2003. *Neanderthals and Modern Humans in the European Landscape During the Last Glaciation*. McDonald Institute for Archaeological Research, Cambridge.

Van Andel, T.H. and Tzedakis, P.C., 1996. Palaeolithic landscapes of Europe and environs, 150,000–25,000 years ago: an overview. *Quaternary Science Reviews* 15, 481–500.

van Asperen, E.N., 2010. Ecomorphological adaptations to climate and substrate in late Middle Pleistocene caballoid horses. *Palaeogeography, Palaeoclimatology, Palaeoecology* 297, 584–596.

van Asperen, E.N., 2013. Position of the Steinheim interglacial sequence within the marine oxygen isotope record based on mammal biostratigraphy. *Quaternary International* 292, 33–42.

van der Made, J., 2011. Biogeography and climatic change as a context to human dispersal out of Africa and within Eurasia. *Quaternary Science Reviews* 30, 1353–1367.

van Kolfschoten, T. and Laban, C., 1995. Pleistocene terrestrial mammal faunas from the north Sea area. *Mededelingen Rijks Geologische Dienst* 52, 135–151.

van Kolfschoten, T., Buhrs, E. and Verheijen, I., 2015a. The larger mammal fauna from the Lower Paleolithic Schöningen Spear site and its contribution to hominin subsistence. *Journal of Human Evolution* 89, 138–153.

van Kolfschoten, T., Parfitt, S.A., Serangeli, J. and Bello, S.M., 2015b. Lower Paleolithic bone tools from the 'Spear Horizon' at Schöningen (Germany). *Journal of Human Evolution* 89, 226–263.

Van Valkenburgh, B., Hayward, M.W., Ripple, W.J., Meloro, C. and Roth, V.L., 2016. The impact of large terrestrial carnivores on Pleistocene ecosystems. *Proceedings of the National Academy of Sciences* 113, 862–867.

Vandenberghe, J., 1995. Timescales, climate and river development. *Quaternary Science Reviews* 14, 631–638.

Vandevelde, S., Brochier, J., Desachy, B., Petit, C. and Slimak, L., 2018. Sooted concretions: a new micro-chronological tool for high temporal resolution archaeology. *Quaternary International* 474, 103–118.

Vaňková, D., Bartoš, L. and Málek, J., 1997. The role of vocalization in the communication between red deer hinds and calves. *Ethology* 103, 795–808.

Venditti, F., Cristiani, E., Nunziante-Cesaro, S., Agam, A., Lemorini, C. and Barkai, R., 2019. Animal residues found on tiny Lower Paleolithic tools reveal their use in butchery. *Scientific Reports* 9, 13031.

Vialet, A., Modesto-Mata, M., Martinón-Torres, M., de Pinillos, M.M. and Bermúdez de Castro, J.M., 2018. A reassessment of the Montmaurin-La Niche mandible (Haute Garonne, France) in the context of European Pleistocene human evolution. *PLOS ONE* 13, e0189714.

Villa, P., 1982. Conjoinable pieces and site formation processes. *American Antiquity* 47, 276–290.

Villa, P., 1990. Torralba and Aridos: elephant exploitation in Middle Pleistocene Spain. *Journal of Human Evolution* 19, 299–309.

Villa, P., 2001. Early Italy and the colonization of western Europe. *Quaternary International* 75, 113–130.

Villa, P. and Lenoir, M., 2009. Hunting and hunting weapons of the Lower and Middle Paleolithic of Europe, in: Hublin, J.-J. and Richards, M. (eds), *The Evolution of Hominin Diets: integrating approaches to the study of Palaeolithic subsistence*. Springer, New York, 59–85.

Villa, P., Soto, E., Santonja, M., Pérez-González, A., Mora, R., Parcerisas, J. and Sesé, C., 2005. New data from Ambrona: closing the hunting versus scavenging debate. *Quaternary International* 126, 223–250.

Vita-Finzi, C., Higgs, E.S., Sturdy, D., Harriss, J., Legge, A. and Tippett, H., 1970. Prehistoric economy in the Mount Carmel area of Palestine: site catchment analysis. *Proceedings of the Prehistoric Society* 36, 1–37.

Vlček, E., 1978. A new discovery of *Homo erectus* in Central Europe. *Journal of Human Evolution* 7, 239–251.

Voinchet, P., Moreno, D., Bahain, J.J., Tissoux, H., Tombret, O., Falguères, C., Moncel, M.H., Schreve, D.C., Candy, I. and Antoine, P., 2015. New chronological data (ESR and ESR/U-series) for the earliest Acheulian sites of north-western Europe. *Journal of Quaternary Science* 30, 610–622.

Voormolen, B., 2008. *Ancient Hunters, Modern Butchers: Schöningen 13 II-4, a kill-butchery site dating from the northwest European Lower Palaeolithic*. Unpublished PhD thesis, University of Leiden.

Vreeman, R.C. and Carroll, A.E., 2008. Festive medical myths. *British Medical Journal* 337, a2769.

Wagner, G.A., Maul, L.C., Löscher, M. and Schreiber, H.D., 2011. Mauer – the type site of *Homo heidelbergensis*: palaeoenvironment and age. *Quaternary Science Reviews* 30, 1464–1473.

Wales, N., 2012. Modeling Neanderthal clothing using ethnographic analogues. *Journal of Human Evolution* 63, 781–795.

Walker, M.J., Anesin, D., Angelucci, D.E., Avilés-Fernández, A., Berna, F., Buitrago-López, A., Fernández-Jalvo, Y., Haber-Uriarte, M., López-Jiménez, A. and López-Martínez, M., 2016. Combustion at the late Early Pleistocene site of Cueva Negra del Estrecho del Río Quípar (Murcia, Spain). *Antiquity* 90, 571–589.

Wall-Scheffler, C.M. and Myers, M.J., 2013. Reproductive costs for everyone: how female loads impact human mobility strategies. *Journal of Human Evolution* 64, 448–456.

Wall-Scheffler, C.M., Geiger, K. and Steudel-Numbers, K.L., 2007. Infant carrying: the role of increased locomotory costs in early tool development. *American Journal of Physical Anthropology* 133, 841–846.

Wani, B.A., Bodha, R. and Wani, A., 2010. Nutritional and medicinal importance of mushrooms. *Journal of Medicinal Plants Research* 4, 2598–2604.

Wanner, H., Brönnimann, S., Casty, C., Gyalistras, D., Luterbacher, J., Schmutz, C., Stephenson, D.B. and Xoplaki, E., 2001. North Atlantic Oscillation – concepts and studies. *Surveys in Geophysics* 22, 321–381.

Warren, S.H., 1911. On a Palaeolithic (?) wooden spear. *Quarterly Journal of the Geological Society of London* 67, 119.

Weaver, T.D. and Hublin, J.-J., 2009. Neandertal birth canal shape and the evolution of human childbirth. *Proceedings of the National Academy of Sciences* 106(20), 8151–8156.

Wells, J.C.K., 2012. The capital economy in hominin evolution: how adipose tissue and social relationships confer phenotypic flexibility and resilience in stochastic environments. *Current Anthropology* 53, S466–S478.

Wenban-Smith, F.F., 1989. The use of canonical variates for determination of biface manufacturing technology at Boxgrove Lower Palaeolithic site and the behavioural implications of this technology. *Journal of Archaeological Science* 16, 17–26.

Wenban-Smith, F.F., 1998. Clactonian and Acheulean industries in Britain: their chronology and significance reconsidered, in: Ashton, N.M., Healy, F., Pettitt, P.B. (eds), *Stone Age Archaeology: essays in honour of John Wymer*. Oxbow Books, Oxford, 90–97.

Wenban-Smith, F.F., 1999. Knapping technology, in: Roberts and Parfitt (eds) 1999, 384–394.

Wenban-Smith, F.F., 2004. Bringing behaviour into focus: archaic landscapes and lithic technology, in: Walker, E.A., Wenban-Smith, F.F. and Healy, F. (eds), *Lithics in Action*. Oxbow Books, Oxford, 48–56.

Wenban-Smith, F.F., 2013. *The Ebbsfleet Elephant: excavations at Southfleet Road, Swanscombe in advance of High Speed I, 2003-4*. Oxford Archaeology, Oxford.

Wenban-Smith, F.F., Allen, P., Bates, M.R., Parfitt, S.A., Preece, R.C., Stewart, J.R., Turner, C. and Whittaker, J.E., 2006. The Clactonian elephant butchery site at Southfleet Road, Ebbsfleet, UK. *Journal of Quaternary Science* 21, 471–483.

Westaway, R., 2011. A re-evaluation of the timing of the earliest reported human occupation of Britain: the age of the sediments at Happisburgh, eastern England. *Proceedings of the Geologists' Association* 122(3), 383–396.

Weston, L.R., 2008. Palaeoliths from the Thames headwaters in Gloucestershire and North Wiltshire. *Lithics: The Journal of the Lithic Studies Society* 29, 36–54.

Weyrich, L.S., Duchene, S., Soubrier, J., Arriola, L., Llamas, B., Breen, J., Morris, A.G., Alt, K.W., Caramelli, D. and Dresely, V., 2017. Neanderthal behaviour, diet, and disease inferred from ancient DNA in dental calculus. *Nature* 544, 357–361.

Wheeler, P.E., 1996. The environmental context of functional body hair loss in hominids (a reply to Amaral, 1996). *Journal of Human Evolution* 30, 367–371.

White, M.J., 1998. On the significance of Acheulean biface variability in southern Britain. *Proceedings of the Prehistoric Society* 64, 15–44.

White, M.J., 2000. The Clactonian question: on the interpretation of core-and-flake assemblages in the British Lower Palaeolithic. *Journal of World Prehistory* 14, 1–63.

White, M.J., 2006. Things to do in Doggerland when you're dead: surviving OIS3 at the northwestern-most fringe of Middle Palaeolithic Europe. *World Archaeology* 38, 547–575.

White, M.J., 2015. 'Dancing to the rhythms of the biotidal zone': settlement history and culture history in Middle Pleistocene Britain, in: Coward, F., Hosfield, R., Pope, M. and Wenban-Smith, F.F. (eds), *Settlement, Society and Cognition in Human Evolution: Landscapes in Mind*. Cambridge University Press, Cambridge, 154–173.

White, M.J. and Foulds, F., 2018. Symmetry is its own reward: on the character and significance of Acheulean handaxe symmetry in the Middle Pleistocene. *Antiquity* 92, 304–319.

White, M.J. and Plunkett, S., 2004. *Miss Layard Excavates: a Palaeolithic site at Foxhall Road, Ipswich, 1903-1905*. Western Academic & Specialist Press, Liverpool.

White, M.J. and Schreve, D.C., 2000. Island Britain – peninsula Britain: palaeogeography, colonisation and the Lower Palaeolithic settlement of the British Isles. *Proceedings of the Prehistoric Society* 66, 1–28.

White, M.J., Pettitt, P.B. and Schreve, D.C., 2016. Shoot first, ask questions later: interpretative narratives of Neanderthal hunting. *Quaternary Science Reviews* 140, 1–20.

White, M.J., Scott, R. and Ashton, N.M., 2006. The Early Middle Palaeolithic in Britain: archaeology, settlement history and human behaviour. *Journal of Quaternary Science* 21, 525–541.

White, M.J., White, T.S., Howard, A.J. and Bridgland, D.R., 2014. Archaeology: the Lower and Middle Palaeolithic record from the Trent catchment, in: Bridgland, D.R., Howard, A.J., White, M.J. and White, T.S. (eds), *Quaternary of the Trent*. Oxbow Books, Oxford, 243–294.

Wikenros, C., Sand, H., Wabakken, P., Liberg, O. and Pedersen, H.C., 2009. Wolf predation on moose and roe deer: chase distances and outcome of encounters. *Acta Theriologica* 54, 207–218.

Wilson, L., 1988. Petrography of the Lower Palaeolithic Tool Assemblage of the Caune de l'Arago (France). *World Archaeology* 19, 376–387.

Wilson, L., 2007. Terrain difficulty as a factor in raw material procurement in the Middle Palaeolithic of France. *Journal of Field Archaeology* 32, 315–324.

Wobst, H.M., 1974. Boundary conditions for Paleolithic social systems: a simulation approach. *American Antiquity* 39, 147–178.

Wood, B., and Lonergan, N., 2008. The hominin fossil record: taxa, grades and clades. *Journal of Anatomy* 212(4), 354–376.

Woodburn, J., 1968. An introduction to Hadza ecology, in: Lee and DeVore (eds) 1968, 49–55.

Woodward, J., 2009. Quaternary geography and the human past, in: Castree, N., Demeritt, D., Liverman, D. and Rhoads, B. (eds), *A Companion to Environmental Geography*. Wiley-Blackwell, Chichester, 198–222.

Woodward, J., 2014. *The Ice Age: a very short introduction*. Oxford University Press, Oxford.

Wrangham, R., 2009. *Catching fire: how cooking made us human*. Basic Books, New York.

Wymer, J.J., 1968. *Lower Palaeolithic Archaeology in Britain, as Represented by the Thames Valley*. John Baker, London.

Wymer, J.J., 1999. *The Lower Palaeolithic Occupation of Britain*. Wessex Archaeology & English Heritage, Salisbury.

Wynn, T. and Gowlett, J.A.J., 2018. The handaxe reconsidered. *Evolutionary Anthropology: issues, news, and reviews* 27, 21–29.

Yankielun, N.E., 2007. *How to Build an Igloo: and other snow shelters*. WW Norton & Company, New York.

Yellen, J.E., 1977. *Archaeological Approaches to the Present: models for reconstructing the past*. Academic Press, New York.

Yokoyama, Y., Falgueres, C. and Quaegebeur, J.-P., 1985. ESR dating of quartz from Quaternary sediments: first attempt. *Nuclear Tracks and Radiation Measurements* 10, 921–928.

Yravedra, J., Domínguez-Rodrigo, M., Santonja, M., Pérez-González, A., Panera, J., Rubio-Jara, S. and Baquedano, E., 2010. Cut marks on the Middle Pleistocene elephant carcass of Áridos 2 (Madrid, Spain). *Journal of Archaeological Science* 37, 2469–2476.

Zimov, S.A., Zimov, N.S., Tikhonov, A.N. and Chapin, F.S., 2012. Mammoth steppe: a high-productivity phenomenon. *Quaternary Science Reviews* 57, 26–45.

Zink, K.D. and Lieberman, D.E., 2016. Impact of meat and Lower Palaeolithic food processing techniques on chewing in humans. *Nature* 531, 500–503.

Zollikofer, C.P.E. and De León, M.S.P., 2013. Pandora's growing box: Inferring the evolution and development of hominin brains from endocasts. *Evolutionary Anthropology: issues, news, and reviews* 22, 20–33.

Zutovski, K. and Barkai, R., 2016. The use of elephant bones for making Acheulian handaxes: A fresh look at old bones. *Quaternary International* 406(Part B), 227–238.

Zveryaev, I.I., 2004. Seasonality in precipitation variability over Europe. *Journal of Geophysical Research: Atmospheres* 109, https://doi.org/10.1029/2003JD003668.